Applied Proba
Con
Econom
Information and Communication
Modeling and Identification
Numerical Techniques
Optimization

Edited by   A.

Advisory Board   E. Dyn
G. Kall
K. Krick
G. I. Marc
R. Radner

T. Hida

Brownian Motion

Translated by the Author and
T. P. Speed

With 13 Illustrations

Springer-Verlag
New York Heidelberg Berlin

QA274.75 .H5213

T. Hida
Department of Mathematics
Faculty of Science
Nagoya University
Chikasu-Ku, Nagoya 464
Japan

T. P. Speed
Department of Mathematics
University of Western Australia
Nedlands, W.A. 6009
Australia

Editor

A. V. Balakrishnan
Systems Science Department
University of California
Los Angeles, California 90024
USA

AMS Subject Classification (1980): 60j65

Library of Congress Cataloging in Publication Data

Hida, Takeyuki, 1927–
  Brownian motion.
  (Applications of Mathematics; Vol. 11)
  Bibliography: p.
  Includes index.
  1. Brownian motion processes. I. Title.
QA274.75.H5213   519.2′82   79-16742

Originally published in Japanese by Iwanami Shoten, Publishers, Tokyo, 1975.

All rights reserved.

No part of this book may be translated or reproduced in any
form without written permission from the copyright holder.

© 1980 by Takeyuki Hida.

Printed in the United States of America.

9 8 7 6 5 4 3 2 1

ISBN 0-387-90439-5 Springer-Verlag New York
ISBN 3-540-90439-5 Springer-Verlag Berlin Heidelberg

Preface to the English Edition

Following the publication of the Japanese edition of this book, several interesting developments took place in the area. The author wanted to describe some of these, as well as to offer suggestions concerning future problems which he hoped would stimulate readers working in this field. For these reasons, Chapter 8 was added.

Apart from the additional chapter and a few minor changes made by the author, this translation closely follows the text of the original Japanese edition.

We would like to thank Professor J. L. Doob for his helpful comments on the English edition.

T. Hida
T. P. Speed

Preface

The physical phenomenon described by Robert Brown was the complex and erratic motion of grains of pollen suspended in a liquid. In the many years which have passed since this description, Brownian motion has become an object of study in pure as well as applied mathematics. Even now many of its important properties are being discovered, and doubtless new and useful aspects remain to be discovered. We are getting a more and more intimate understanding of Brownian motion.

The mathematical investigation of Brownian motion involves:

1. a probabilistic aspect, viewing it as the most basic stochastic process;
2. a discussion of the analysis on a function space on which a most interesting measure, *Wiener measure*, is introduced using Brownian motion;
3. the development of tools to describe random events arising in the natural environment, for example, the function of biological organs; and
4. a presentation of the background to a wide range of applications in which Brownian motion is involved in mathematical models of random phenomena.

It is hoped that this exposition can also serve as an introduction to these topics.

As far as (1) is concerned, there are many outstanding books which discuss Brownian motion, either as a Gaussian process or as a Markov process, so that there is no need for us to go into much detail concerning these viewpoints. Thus we only discuss them briefly. Topics related to (2) are the most important for this book, and comprise the major part of it. Our aim is to discuss the analysis arising from Brownian motion, rather than Brownian motion itself regarded as a stochastic process. Having established this analysis, we turn to several applications in which non-linear *functionals of*

Brownian motion (often called *Brownian functionals*) are involved. We can hardly wait for a systematic approach to (3) and (4) to be established, aware as we are of recent rapid and successful developments. In anticipation of their fruitful future, we present several topics from these fields, explaining the ideas underlying our approach as the occasion demands.

It seems appropriate to begin with a brief history of the theory. Our plan is not to write a comprehensive history of the various developments, but rather to sketch a history of the study of Brownian motion from our specific viewpoint. We locate the origin of the theory, and examine how Brownian motion passed into Mathematics.

The story began in the 1820's. In the months of June, July and August 1827 Robert Brown F.R.S. made microscopic observations on the minute particles contained in the pollen of plants, using a simple microscope with one lens of focal length about 1 mm. He observed the highly irregular motion of these particles which we now call "Brownian motion", and he reported all this in R. Brown (1828). After making further observations involving different materials, he believed that he had discovered active molecules in organic and inorganic bodies. Following this, many scientists attempted to interpret this strange phenomenon. It was established that finer particles move more rapidly, that the motion is stimulated by heat, and that the movement becomes more active with a decrease in viscosity of the liquid medium. It was not until late in the last century that the true cause of the movement became known. Indeed such irregular motion comes from the extremely large number of collisions of the suspended pollen grains with molecules of the liquid.

Following these observations and experiments, but apparently independent of them, a theoretical and quantitative approach to Brownian motion

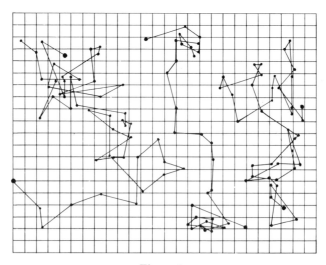

Figure 1

was given for the first time by A. Einstein. This was in 1905, the same year in which Einstein published his famous special theory of relativity.

It is interesting to recall the mathematical framework for Brownian motion set up by Einstein; for simplicity we consider only the projection of the motion onto a line. The density of the pollen grains per unit length at an instant t will be denoted by $u(x, t)$, $x \in \mathbf{R}$, and it will be supposed that the movement occurs uniformly in both time and space, so that the proportion of the pollen grains moved from x to $x + y$ in a time interval of length τ may be written $\varphi(\tau, y)$. For the time interval t to $t + \tau$ ($\tau > 0$) we thus obtain

$$u(x, t + \tau) \, dx = dx \int_{-\infty}^{\infty} u(x - y, t) \varphi(\tau, y) \, dy, \tag{0.1}$$

where the functions u and φ can be assumed smooth. Further, the function φ can be supposed symmetric in space about the origin, with variance proportional to τ:

$$\int_{-\infty}^{\infty} y^2 \varphi(\tau, y) \, dy = D\tau, \quad D \text{ constant.}$$

The Taylor expansion of (0.1) for small τ gives

$$u(x, t) + \tau u_t(x, t) + o(\tau)$$

$$= \int_{-\infty}^{\infty} \left\{ u(x, t) - y u_x(x, t) + \frac{1}{2} y^2 u_{xx}(x, t) - \cdots \right\} \varphi(\tau, y) \, dy,$$

which, under the assumptions above, leads to the heat equation

$$u_t = \frac{1}{2} D u_{xx}. \tag{0.2}$$

If the initial state of a grain is at some point y say, so that $u(x, 0) = \delta(x - y)$, then from (0.2) we have

$$u(x, t) = (2\pi D t)^{-1/2} \exp\left[-\frac{(x - y)^2}{2Dt} \right]. \tag{0.3}$$

The $u(x, t)$ thus obtained turns out to be the transition probability function of Brownian motion viewed as a Markov process (see §2.4).

Let us point out that formulae (0.2) and (0.3) were obtained in a purely theoretical manner. Similarly the constant D is proved to be

$$D = \frac{RT}{Nf}, \tag{0.4}$$

where R is a universal constant depending on the suspending material, T the absolute temperature, N the Avogadro number and f the coefficient of friction. It is worth noting that in 1926 Jean Perrin was able to use the formula (0.4) in conjunction with a series of experiments to obtain a reasonably accurate determination of the Avogadro number. In this we find a beautiful interplay between theory and experiment.

Although we will not give any details, we should not forget that around the year 1900 L. Bachelier tried to establish the framework for a mathematical theory of Brownian motion.

Next we turn to the celebrated work of P. Lévy. As soon as one hears the term Brownian motion in a mathematical context, Lévy's 1948 book (second edition in 1965) comes to mind. However our aim is to start with Lévy's much earlier work in functional analysis, referring to the book P. Lévy (1951) in which he has organised his work along these lines dating back to 1910. Around that time he started analysing functionals on the Hilbert space $L^2([0, 1])$, and the need to compute a mean value or integral of a functional $\varphi(x)$, $x \in L^2([0, 1])$ arose. Unfortunately, as is well known, there exists no measure in this situation analogous to Lebesgue measure on a finite-dimensional Euclidean space, so that Lévy introduced the concept of mean (valeur moyenne) of such a functional. This is defined as follows: given a functional φ, we wish to compute the mean on the ball $S(R)$ with radius R and centre 0 (the origin). To this end, let $\varphi_n(x^{(n)})$, $x^{(n)} \in S_n$, denote the restriction of φ to the n-dimensional section S_n of $S(R)$, and let m_n denote the mean value of φ_n over S_n. If the sequence $\{m_n\}$ converges to a limit μ as $n \to \infty$, then it would be natural to define the mean of φ over $S(R)$ to be μ. More precisely, let us approximate $x(t)$ in the sense of L^2 by a sequence of step functions $\{x^{(n)}(t)\}$, where $x^{(n)}(t)$ takes constant values x_k on the interval $[k/n, (k + 1)/n]$, $0 \le k \le n - 1$. Then the original assumption that the L^2-norm of $x(t)$ is less than R carries over to the requirement $\sum_{k=0}^{n-1} x_k^2 \le nR^2$ on the nth approximation. If we view the step function $x^{(n)}(t)$ as an n-dimensional vector, this inequality defines the n-dimensional ball S_n with radius $n^{1/2}R$. Thus m_n is the average or mean of $\varphi_n(x^{(n)})$ relative to the uniform probability measure over S_n, and the limit of the sequence $\{m_n\}$, if it exists, is understood to be the mean of $\varphi(x)$ over $S(R)$.

The following simple example, due to P. Lévy (1951), illustrates the above procedure for obtaining a mean, and at the same time shows how the Gaussian distribution arises in classical functional analysis. Take an arbitrary point $\tau \in [0, 1]$ and fix it, and take a function f on $\mathbf{R}$. Setting $\varphi(x) = f(x(\tau))$, $x \in L^2([0, 1])$, we can see that $\varphi_n(x^{(n)})$ is a function of only one component of $x^{(n)}$, and the mean m_n is therefore the expectation of this with respect to the uniform probability measure on S_n. It then follows (see Example 3 in §1.2) that for large n, the probability that one coordinate of a point on the sphere lies between aR and bR is approximately

$$(2\pi)^{-1/2} \int_a^b \exp\left(-\frac{1}{2} y^2\right) dy. \tag{0.5}$$

In this way a Gaussian distribution arises, and the mean m of the functional becomes

$$m = (2\pi)^{-1/2} \int_{-\infty}^{\infty} f(Ry) \exp\left(-\frac{1}{2} y^2\right) dy. \tag{0.6}$$

Such an intuitive approach remains possible for the more general class of essentially finite-dimensional functionals, and we are led to recognise the general mean as the integral with respect to the measure of white noise to be introduced in Chapter 3. [An interpretation of this fact may be found in T. Hida and H. Nomoto (1964)]. P. Lévy (1951) also discussed the Laplacian operator, as well as harmonic functionals on the Hilbert space $L^2([0, 1])$, and it is interesting to note that the germ of the notion of the infinite-dimensional rotation group (see Chapter 5 below) can also be found in Lévy's book.

After establishing the theory of sums of independent random variables, Lévy proceeded to study continuous sums of independent infinitesimal random variables, and was able to obtain the canonical decomposition of an additive process. Brownian motion, written $\{B(t): t \geq 0\}$, or more simply as $\{B(t)\}$, is just an additive process whose distribution is Gaussian. More fully, a Brownian motion is defined to be a stochastic process satisfying the following two conditions (see Definition 2.1):

a. $B(0) = 0$;
b. $\{B(t): t \geq 0\}$ is a Gaussian process, and for any t, h with $t + h > 0$, the difference $B(t + h) - B(t)$ has expectation 0 and variance $|h|$.

It follows from this definition that $\{B(t)\}$ is an additive process.

P. Lévy used the method of interpolation (described in §2.3 i) below) to obtain an analytical expression for Brownian motion, and this method became a powerful tool for gaining insight into its interesting complexity. The series of great works on Brownian motion by Lévy are unrivalled, beginning with papers in the 1930's and including his book in 1948. As part of this work will be illustrated in Chapter 2 below, the reader will be able to get some impression of his importance in the theory of probability. Following Lévy, we will discuss sample path properties and the fine structure of Brownian motion as a Markov process, and then briefly explain why we should give linear representations of general Gaussian processes with Brownian motion as a base. Moreover we shall take a quick look at Brownian motion with a multi-dimensional parameter, introduced by P. Lévy to display its intrinsically interesting probabilistic structure.

The investigations of Brownian motion as a stochastic process and the work of Lévy on functional analysis may appear unrelated, but they are in fact two aspects of the same thing, as can be seen in the course of analysing Brownian functionals. This can be roughly explained as follows. Any functional of a Brownian motion $\{B(t): t \geq 0\}$ may equally well be regarded as a functional of $\{\dot{B}(t): t \geq 0\}$ (where $\dot{B}(t) = dB(t)/dt$); the latter turns out to be easier to deal with. Such a functional, which is just a random variable, has an expectation which coincides with the mean of P. Lévy just explained above [after (0.6)]. Similarly we find that other aspects of the discussion of functionals of $\{\dot{B}(t): t \geq 0\}$ always have their counterpart in Lévy's functional analysis, and a systematic study involving these interpretations is carried out in Chapters 3 and 4.

Now let us return to the days in which N. Wiener's paper "Differential space" (now called "Wiener space") was published. This paper was a landmark in the study of Brownian motion, and Wiener acknowledges in the preface that he was greatly inspired by the works of R. Gâteaux and P. Lévy. Indeed it seems to have been a private communication with Lévy on integration over an infinite-dimensional vector space which led Wiener to write that famous paper. Since almost all Brownian sample paths are continuous, we may assert that, roughly speaking, the probability distribution of $\{B(t)\}$ should be defined over a space of continuous functions. The measure space thus obtained is nothing but Wiener space, and since that time it has been developed by Wiener (amongst others), and has also made a significant contribution to natural science more generally.

The great work carried out along these lines by Wiener culminated in the well-known and popular notions of "Cybernetics". In his famous book of the same title, Wiener indicates via the subtitle "Control and communication in the animal and the machine" the importance he attached to interdisciplinary investigations. He actually discovered many interesting problems in other fields in this way, and we would like to understand his ideas and do likewise. Indeed if we examined the mathematical aspects of his work, we would recognise how his investigation of brain waves led him to an application of his results on non-linear functionals of Brownian motion, and even to a discussion of their analysis. A similar approach is detectable in his work on engineering and communication theory, where the disturbances due to noise and hence the prediction theory of stochastic processes arise. Again his discussion of the flow of Brownian motion made a great contribution to ergodic theory. All in all a new branch of mathematics—analysis on an infinite-dimensional function space—was originated by him, and further investigations have illuminated modern analysis.

A book published in 1958 contains the notes of lectures in which he discussed developments of the theory proposed in *Cybernetics*, including the analysis of non-linear functionals on Wiener space and their applications. We were very impressed with these books, for in them we can see the beautiful process of a field of probability theory being built up from actual problems, with Brownian motion playing a key role. It should be emphasised that the theory thus obtained can be naturally applied to the original problem, and as a next step one might expect a new problem. This process of feedback would then be repeated again and again.

Another thing to be noted here is the important role played by Fourier transforms in Wiener's approach. Surprisingly enough, the Fourier transform itself does have a close connection with Brownian motion, although this is implicit rather than explicit, and it will be noticed every now and then in the pages which follow.

Summing up what we have said so far, we can say that the mathematical study of Brownian motion originated with Einstein, was highly developed by Lévy and Wiener, and is now being continued by many scientists. Since its inception, the theory of Brownian motion has always had an intimate connection with sciences other than mathematics, and the present author believes that these relationships will continue into the future.

The topics of this book were chosen primarily to expound the work of Lévy and Wiener, but after finishing the manuscript, the author now realises how difficult it was to carry out this task, and is frustrated by his inability in this regard. Nevertheless, with the encouragement of the beautiful works of K. Itô (1951b, 1953a, and others), this manuscript has been completed. We also note the work of H. Yoshizawa (1969) and Y. Umemura (1965), who introduced the concept of the infinite-dimensional rotation group, and demonstrated its important role in the study of white noise. The author is grateful to these works.

Chapters 6 and 7 are devoted to the complexification of white noise and the rotation group. If these two are regarded as the basic concepts of what we might call infinite-dimensional harmonic analysis, it seems natural to pass to their complexifications.

As can be seen from the foregoing, each chapter except Chapter 1 may be said to be along the lines of our original aim. In some chapters the motivation and general ideas are explained before the main part of the discussion, and it is hoped that these explanations will aid the understanding of the material, as well as demonstrating some connections between the chapters. Certain material which does not lie in the main stream of our development, and some basic formulae, have been collected in the Appendix.

I greatly appreciate the many comments received at both the manuscript and proof stage from Professors H. Nomoto, M. Hitsuda and S. Takenaka, and am particularly indebted to Mr. N. Urabe at the Iwanami Publishing Company who suggested that I write this book. He has helped me during the writing, and even at the proof stage, and without his help the book would never have appeared. Having now completed the task, I would like to share the congratulations with him.

I would like to dedicate this book to my former teachers Professor K. Yosida and Professor K. Itô, who have encouraged me in the present work.

May, 1974 Takeyuki Hida
Nagoya

Contents

1	**Background**	1
	1.1 Probability Spaces, Random Variables, and Expectations	1
	1.2 Examples	3
	1.3 Probability Distributions	7
	1.4 Conditional Expectations	19
	1.5 Limit Theorems	24
	1.6 Gaussian Systems	31
	1.7 Characterisations of Gaussian Distributions	36
2	**Brownian Motion**	44
	2.1 Brownian Motion. Wiener Measure	44
	2.2 Sample Path Properties	51
	2.3 Constructions of Brownian Motion	63
	2.4 Markov Properties of Brownian Motion	75
	2.5 Applications of the Hille–Yosida Theorem	86
	2.6 Processes Related to Brownian Motion	99
3	**Generalised Stochastic Processes and Their Distributions**	114
	3.1 Characteristic Functionals	114
	3.2 The Bochner-Minlos Theorem	116
	3.3 Examples of Generalised Stochastic Processes and Their Distributions	122
	3.4 White Noise	127
4	**Functionals of Brownian Motion**	132
	4.1 Basic Functionals	132
	4.2 The Wiener–Itô Decomposition of (L^2)	134

4.3	Representations of Multiple Wiener Integrals	137
4.4	Stochastic Processes	142
4.5	Stochastic Integrals	151
4.6	Examples of Applications	165
4.7	The Fourier–Wiener Transform.	179

5 The Rotation Group — 185

5.1	Transformations of White Noise (I): Rotations	185
5.2	Subgroups of the Rotation Group	188
5.3	The Projective Transformation Group	192
5.4	Projective Invariance of Brownian Motion	196
5.5	Spectral Type of One-Parameter Subgroups	198
5.6	Derivation of Properties of White Noise Using the Rotation Group	207
5.7	Transformations of White Noise (II): Translations	212
5.8	The Canonical Commutation Relations of Quantum Mechanics	223

6 Complex White Noise — 232

6.1	Complex Gaussian Systems	233
6.2	Complexification of White Noise	237
6.3	The Complex Multiple Wiener Integral	240
6.4	Special Functionals in (L_c^2)	247

7 The Unitary Group and Its Applications — 252

7.1	The Infinite-Dimensional Unitary Group	252
7.2	The Unitary Group $U(\mathscr{S}_c)$	254
7.3	Subgroups of $U(\mathscr{S}_c)$	256
7.4	Generators of the Subgroups	264
7.5	The Symmetry Group of the Heat Equation	266
7.6	Applications to the Schrödinger Equation	274

8 Causal Calculus in Terms of Brownian Motion — 280

8.1	Summary of Known Results	281
8.2	Coordinate Systems in $(\mathscr{S}^*, \mu)$	283
8.3	Generalised Brownian Functionals	286
8.4	Generalised Random Measures	287
8.5	Causal Calculus	289

Appendix — 293

A.1	Martingales	293
A.2	Brownian Motion with a Multidimensional Parameter	298
A.3	Examples of Nuclear Spaces	301
A.4	Wiener's Non-Linear Circuit Theory	308
A.5	Formulae for Hermite Polynomials	310

Bibliography — 315

Index — 321

Background 1

In this chapter we present some of the basic concepts from probability theory necessary for the main part of this book. No attempt has been made at either generality or completeness. Those concepts which provide motivation, or which are basic to our approach, are illustrated to some extent, whilst others will only be touched upon briefly. For example, certain specific properties of an infinite-dimensional probability measure (§1.3, (iii)) are discussed in some detail, as are some characterisations of Gaussian systems of random variables. Many theorems and propositions whose proofs can be found readily in standard texts will be stated without proof, or with only an outline of the proof. For further details of these, as well as related topics, the reader is referred to such books as K. Itô (1953c), W. Feller (1968, 1971), and J. L. Doob (1953).

1.1 Probability Spaces, Random Variables, and Expectations

The theory of probability is based upon the notion of a probability space or probability triple. Firstly, we have a non-empty set Ω, and in many actual cases it is possible to regard each element $\omega \in \Omega$ as a parameter indexing realizations of the random phenomenon in question. Next we take a family **B** of subsets of Ω satisfying the following three conditions:

1. $\Omega \in \mathbf{B}$;
2. If $B_n \in \mathbf{B}$, $n = 1, 2, \ldots$, then $\bigcup_n B_n \in \mathbf{B}$;
3. If $B \in \mathbf{B}$, then $B^c \in \mathbf{B}$, where $B^c = \Omega \backslash B$.

In other words **B** forms a σ-field or σ-algebra of subsets of Ω. Finally we have a countably additive set function P defined on **B** satisfying the following conditions

1. $0 \le P(B) \le 1$ for every $B \in \mathbf{B}$;
2. If $B_n \in \mathbf{B}$, $n = 1, 2, \ldots$, are such that $B_i \cap B_j = \emptyset$ when $i \ne j$, then

$$P\left(\bigcup_n B_n\right) = \sum_n P(B_n);$$

3. $P(\Omega) = 1$.

A triple $(\Omega, \mathbf{B}, P)$ is called a *probability space* if each component satisfies the conditions stated above. Elements $B \in \mathbf{B}$ are called *events*, and $P(B)$ is called the *probability* of the event B. The event B^c is said to be the event *complementary* to the event B. If the events in the class $\{B_\alpha : \alpha \in A\}$ are pairwise disjoint, i.e. if $B_\alpha \cap B_{\alpha'} = \emptyset$ whenever $\alpha \ne \alpha'$, then they are often termed *mutually exclusive*.

The choice of the σ-field **B** depends upon the nature of the random phenomenon to be described, the simplest being $\mathbf{B} = \{\emptyset, \Omega\}$, which corresponds to the deterministic case, and is of no interest to us in this work. In general the larger **B** is as a class of sets, the more events there are that can be considered, and so the more minutely can the random phenomenon under discussion be described. For any set Ω the class $\mathbf{B} = 2^\Omega$ consisting of all subsets of Ω is clearly the richest such class, but unfortunately we cannot always define a suitable P on it.

It is clear that the P of any probability space $(\Omega, \mathbf{B}, P)$ is simply a measure on the measurable space $(\Omega, \mathbf{B})$ which satisfies the further condition that the total measure is unity, i.e. $P(\Omega) = 1$. Because of this probability theory frequently uses measure-theoretic terminology; for example "measurable set" and "almost everywhere" are sometimes used as alternatives to "event" and "almost surely", respectively. The reason why the probabilistic terminology is preferred is that frequently its use gives us an intuitive feel for the topic under discussion.

There is a concept which is very important in probability theory but which does not figure prominently in measure theory, and this is the independence of events. In a probability space $(\Omega, \mathbf{B}, P)$ we say that two events B_1 and B_2 are *independent* if

$$P(B_1 \cap B_2) = P(B_1)P(B_2). \tag{1.1}$$

This notion can be generalised to finitely many events, indeed to an arbitrary class of events as follows: a class $\{B_\alpha : \alpha \in A\}$ of events is *independent* if for any finite set $\{\alpha_1, \alpha_2, \ldots \alpha_n\} \subset A$ of indices we have

$$P\left(\bigcap_{k=1}^n B_{\alpha_k}\right) = \prod_{k=1}^n P(B_{\alpha_k}). \tag{1.2}$$

More generally, the family $\{\mathbf{B}_\alpha, \alpha \in A\}$ of σ-fields is said to be independent if for every choice of $B_\alpha \in \mathbf{B}_\alpha$ ($\alpha \in A$) the class $\{B_\alpha: \alpha \in A\}$ is independent.

A real or complex-valued measurable function $X(\omega)$ defined on a probability space $(\Omega, \mathbf{B}, P)$ is called a *random variable*. Recall that $\omega \in \Omega$ is a parameter denoting a random element, and thus $X(\omega)$ is regarded as the numerical value to be associated with the random element ω. The notion of random variable is easily extended to the cases of vector-, function- or even generalised function-valued random variables, and in the function-valued case we often write

$$X(t, \omega), \quad t \in T, \omega \in \Omega, \tag{1.3}$$

where T is a finite or infinite time interval. The details concerning generalised function-valued random variables will be given later (Chapter 3).

If a complex-valued random variable $X(\omega)$ is integrable with respect to P, then we say that the expectation of X exists, and

$$E(X) = \int_\Omega X(\omega)\, dP(\omega) \tag{1.4}$$

is called the *expectation* or *mean* of X. Further, if $|X|^n$ is integrable, then $E(X^n)$ is called the nth order moment of X. In particular, when a real-valued random variable X has a second moment, then

$$V(X) = E([X - E(X)]^2) = E(X^2) - E(X)^2 \tag{1.5}$$

is called the *variance* of X.

Let a system $\{X_\alpha: \alpha \in A\}$ of random variables on the probability space $(\Omega, \mathbf{B}, P)$ be given, and for each $\alpha \in A$ denote by $\mathbf{B}(X_\alpha)$ the smallest sub-σ-field of $\mathbf{B}$ with respect to which X_α is measurable. If the family $\{\mathbf{B}(X_\alpha): \alpha \in A\}$ is independent in the sense defined above, then the system $\{X_\alpha: \alpha \in A\}$ is said to be *independent*.

The systems $\{X_\alpha: \alpha \in A_1\}$ and $\{X_\beta: \beta \in A_2\}$ of random variables defined on the same probability space are said to be independent if every $B \in \mathbf{B}(\{X_\alpha\})$ and $B' \in \mathbf{B}(\{X_\beta\})$ are independent. In particular if $\{X_\alpha\}$ consists of the single random variable X, and if $\{X\}$ and $\{X_\beta: \beta \in A_2\}$ are independent, then we often say that X is independent of $\mathbf{B}(\{X_\beta\})$. The independence of three or more, as well as infinitely many such systems, can be defined by analogy with the case of two systems.

1.2 Examples

In constructing or determining a probability space we first clarify the type of random phenomena to be analysed and the probabilistic structures to be investigated, and then we fix Ω, $\mathbf{B}$ and P to fit in with these aims. We illustrate this approach with several examples, which also play a role in motivating our main topics.

EXAMPLE 1. A simple counting model using four letters [after W. Feller (1968)].

The probability space describing the random ordering of the four letters a, b, c and d is constructed by the following procedure. When we want to discuss the most detailed way of ordering these letters, Ω must be taken as the set of all $4! = 24$ permutations say $\omega_1, \omega_2, \ldots, \omega_{24}$, of the letters, and **B** the class of all subsets of Ω. Obviously **B** satisfies the conditions (1), (2) and (3) required. An ordering being random suggests that one can expect every permutation ω_i to appear as frequently as every other, and as such a requirement must be described in terms of P, we see that P should be defined in such a way that

$$P(B) = \frac{\#(B)}{24}, \quad \#(B) = \text{the number of elements in } B. \quad (1.6)$$

This P obviously satisfies the requirements (1), (2) and (3), and so we have a probability space $(\Omega, \mathbf{B}, P)$ describing the random ordering of four letters.

In terms of this probability space we can derive the following sample results. Letting A denote the event "a comes first" (i.e. the set of ω_i which begin with a), we find that $P(A) = 3!/4! = 1/4$. Again if B_1 denotes the event "a precedes b" and B_2 the event "c precedes d", then we find that $P(B_1) = P(B_2) = 1/2$ and $P(B_1 \cap B_2) = 1/4$. Thus (1.1) holds for these events and so they are independent.

EXAMPLE 2 (Wiener's probability space). The set Ω consists of the unit interval $[0, 1]$. **B** is the class of all Lebesgue-measurable subsets of Ω, and P is Lebesgue measure. This triple is a probability space, and succinctly describes the random choice of a point from the interval $[0, 1]$. Because of its surprisingly rich measure-theoretic structure it is one of the most important and useful probability spaces; since it appears frequently and was used by N. Wiener, we may name it *Wiener's probability space*.

Using the binary expansion of $\omega \in \Omega$ we define a sequence $\{X_n(\omega)\}$ as follows: if ω admits two different binary expansions, put $X_n(\omega) = 0$ for all n; otherwise $X_n(\omega)$ is 1 or -1 according as the n-th digit in the binary expansion of ω is 1 or 0, $n = 1, 2, \ldots$. The $\{X_n(\omega)\}$ are called the *Rademacher functions* and each X_n is **B**-measurable, so that it is a random variable on $(\Omega, \mathbf{B}, P)$. More importantly, $\{X_n(\omega), n = 1, 2, \ldots\}$ is a sequence of independent random variables.

The random phenomenon known as coin-tossing, that is, the carrying out of successive and independent tosses of an unbiased coin, can be described mathematically in terms of the $\{X_n\}$. For if ω denotes the random element describing the realisation of a sequence of such tosses, and $X_n(\omega) = 1$ (or -1) corresponding to the n-th toss being a head (or tail), then the event that the n-th toss is a head is given by $\{\omega: X_n(\omega) = 1\}$ and has probability 1/2. Another example is given by the event E_k that the first head occurs at the k-th toss; this is given by $\{\omega: X_1(\omega) = X_2(\omega) = \cdots = X_{k-1}(\omega) = -1,$

1.2 Examples

$X_k(\omega) = 1\}$ and has probability $P(E_k) = 2^{-k}$. The sets $\{E_k, k = 1, 2, \ldots\}$ are mutually exclusive and the union $\bigcup_k E_k$ denotes the event that a head ultimately occurs. As we would expect

$$P\left(\bigcup_k E_k\right) = \sum_k P(E_k) = 1.$$

The symmetric *random walk* can also be formed in terms of the $\{X_n\}$. Set

$$S_n(\omega) = \sum_{k=1}^n X_k(\omega), \qquad n = 0, 1, \ldots, \tag{1.7}$$

where $S_0(\omega)$ is taken as 0. Then on $(\Omega, \mathbf{B}, P)$ we see that $\{S_n(\omega)\}$ describes the usual symmetric random walk.

EXAMPLE 3 (A model of the monatomic ideal gas [after M. Kac (1959)]). We consider an isolated monatomic ideal gas consisting of N molecules, and seek to describe its velocity distribution. Each particle is supposed to have the same mass m, and a velocity denoted by v_k, $1 \leq k \leq N$. The energy of the gas is solely kinetic and thus is the sum of the kinetic energies of the individual particles. As this sum has to be constant ($= E$), it can be expressed in the form

$$\frac{1}{2}m\sum_{k=1}^N \|v_k\|^2 = E, \tag{1.8}$$

where $\|v\|$ is the norm of a 3-dimensional vector v, and we denote the velocity vector by $v_k = (v_{k,x}, v_{k,y}, v_{k,z})$. Equation (1.8) means that $\{v_k\}$ is represented by a point on the surface of the sphere with radius $(2E/m)^{1/2}$ in $3N$-dimensional space. Let us set $R = (2E/m)^{1/2}$ and denote this sphere by $S_{3N}(R)$.

As our interest is concentrated upon the velocity distribution, it is quite natural to set $\Omega = S_{3N}(R)$. When discussing velocity components such as $v_{k,x}$ we see that all subsets of the form

$$B = \{\omega = (v_{1,x}, v_{1,y}, v_{1,z}, \ldots, v_{N,z}) : a < v_{k,x} < b\}$$

should be considered, and we therefore take $\mathbf{B}$ to be the σ-field of all Borel subsets of Ω. Finally we choose P to be the uniform measure on Ω, because all the molecules are essentially the same and are moving around without any specific orientation.

It is natural to suppose that the energy E is proportional to the number N of molecules

$$E = \kappa N.$$

Then the set B above has the probability

$$P(B) = \frac{\int_a^b (1 - mx^2/2\kappa N)^{(3N-3)/2} \, dx}{\int_{-R}^R (1 - mx^2/2\kappa N)^{(3N-3)/2} \, dx},$$

which is obtained by computing the surface area of the appropriate spherical region. If the number N of molecules is sufficiently large, then we have an asymptotic expression for $P(B)$

$$P(B) \sim \left(\frac{3m}{2\pi \cdot 2\kappa}\right)^{1/2} \int_a^b \exp\left(-\frac{1}{2}\frac{3m}{2\kappa}x^2\right) dx.$$

Setting $\kappa = 3cT/2$ (c a universal constant; T the absolute temperature), we see that the above formula agrees with the familiar Maxwell formula [see M. Kac (1959) Chapter 1].

EXAMPLE 4 (Density as a set function on the natural numbers [after M. Kac (1959)]). We are going to discuss a mathematical model of the experiment consisting of choosing a natural number from the set **N** of natural numbers, all choices being equally likely, and our interest focusses on constructing a suitable probability space describing this experiment. We would expect the proposed probability space to lead us to a probability of 1/2 for the event that an even number is chosen, and more generally a probability of $1/p$ for the event that the number chosen is a multiple of the prime number p.

The basic set Ω should surely be taken to be **N** in this case, and the requirement that all choices be equally likely leads us to try to define P by a formula like (1.6). However it is here that we meet the difficulty that for certain $B \subset \Omega$ of interest $\#(B)$ as well as $\#(\Omega)$ is infinite. Thus we modify the definition to

$$P(B) = \lim_{N \to \infty} \frac{\#(B_N)}{N}, \qquad (1.9)$$

where $B_N = \{n \in B : n \leq N\}$. Such a limit will not always exist and so, for convenience, we define **B** to be the class of all subsets $B \subset \mathbf{N}$ for which the limit on the right of (1.9) does exist, and then $P(B)$ is defined by (1.9). It follows, for example, that the set B^p consisting of all multiples of the prime number p is a member of **B**, and that $P(B^p) = 1/p$. Our aim seems to be achieved.

Unfortunately the conditions (2) for **B** and (2) for P in §1.1 both fail when **B** and P are defined as above. Here is a simple counterexample to (2) for P. The set $\{n\}$ consisting of the single natural number n certainly belongs to **B** and it can easily be seen that $P(\{n\}) = 0$. But $\Omega = \bigcup_n \{n\}$ and $P(\Omega) = 1$, whence

$$1 = P(\Omega) \neq \sum_n P(\{n\}).$$

In other words the triple $(\Omega, \mathbf{B}, P)$ is not a probability space in the sense of the previous section. However this P is finitely additive in the following sense: if $B_1, B_2, \ldots, B_n$ are mutually exclusive elements of **B**, then $\bigcup_1^n B_k$ belongs to **B** and

$$P(B) = \sum_1^n P(B_k). \qquad (1.10)$$

In view of this property $(\Omega, \mathbf{B}, P)$ might be called a probability space in the *weak sense*.

Let us now consider some properties of the integers obtained by using our probability space in the weak sense. Denote by $p_1, p_2, \ldots p_k, \ldots$ the increasing sequence of prime numbers. Then any $n \in \mathbf{N}$ can be expressed in the form

$$n = p_1^{\alpha_1(n)} p_2^{\alpha_2(n)} \cdots p_k^{\alpha_k(n)} \cdots,$$

and the uniqueness of the factorisation into primes uniquely defines $\alpha_k(n)$, $k = 1, 2, \ldots$, as functions of n. Indeed these α_k might be regarded as random variables on $(\Omega, \mathbf{B}, P)$, for in a sense α_k is **B**-measurable. From the definitions we can prove that

$$P(\alpha_k = l) = p_k^{-l}(1 - p_k^{-1}),$$

where the left side is an abbreviation of $P(\{n: \alpha_k(n) = l\})$. In the same notation we can prove that

$$P(\alpha_1 = l_1, \alpha_2 = l_2, \ldots, \alpha_k = l_k) = \prod_{j=1}^{k} p_j^{-l_j}(1 - p_j^{-1}).$$

As the right side is the product of the terms $P(\alpha_j = l_j)$ we may regard α_1, $\alpha_2, \ldots$ as a sequence of independent random variables on $(\Omega, \mathbf{B}, P)$.

What we have done in this example is to show that even when a genuine probability space cannot be defined because of the stringency of the requirements on **B** and P, there is still some merit in discussing a suitably weakened notion of probability space. Of course special care must be taken when passing to limits, for in general neither P nor **B** are countably additive. In Chapter 3 we will discuss more important examples similar to this one.

1.3 Probability Distributions

(i) Finite-Dimensional Distributions

If $X(\omega)$, $\omega \in \Omega$, is a real-valued random variable defined on a probability space $(\Omega, \mathbf{A}, P)$, then we can assign a probability to the event that the value of X falls into a given interval. More generally, we may regard X as a mapping from Ω into $\mathbf{R}$ and define a probability measure Φ on the measurable space $(\mathbf{R}, \mathbf{B})$, where **B** denotes the σ-field of Borel subsets of $\mathbf{R}$, in such a way that for every $B \in \mathbf{B}$ $\Phi(B)$ is the P-measure of the set $X^{-1}(B) = \{\omega: X(\omega) \in B\}$, i.e.

$$\Phi(B) = P(X^{-1}(B)). \tag{1.11}$$

The set $X^{-1}(B)$ is often written $(X \in B)$ and accordingly $P(X^{-1}(B))$ is written $P(X \in B)$. The measure Φ so obtained is called the (probability) *distribution* of X.

Next let us write

$$F(x) = \Phi((-\infty, x]). \tag{1.12}$$

The function $F(x)$ is called the *distribution function* of X and enjoys the following properties:

1. $F(x)$ is right-continuous;
2. $F(x)$ is monotone non-decreasing; $\hspace{4cm}$ (1.13)
3. $\lim_{x \to \infty} F(x) = 1$, $\lim_{x \to -\infty} F(x) = 0$.

Conversely, given any function $F(x)$ satisfying these three properties, we can define an interval function Φ by writing $\Phi((a, b]) = F(b) - F(a)$, $a < b$, and this can then be extended uniquely to a probability measure Φ on $(\mathbf{R}, \mathbf{B})$ such that (1.12) holds. Thus F and Φ correspond uniquely, and we can pass from one to the other as convenience dictates.

We now introduce the Fourier-Stieltjes transform $\varphi(z)$, $z \in \mathbf{R}$, of Φ (and of F):

$$\varphi(z) = \int_{\mathbf{R}} e^{izx} \, dF(x) = \int_{\mathbf{R}} e^{izx} \Phi(dx), \quad z \in \mathbf{R}. \tag{1.14}$$

The function $\varphi(z)$ is called the *characteristic function* of X, of the distribution Φ, or of the distribution function F, and is readily seen to be simply the expectation of $\exp[izX]$ with respect to the measure P. From its definition we immediately see that the following properties hold:

1. φ is positive definite: for any finite sets $\{z_1, \ldots, z_n\} \subset \mathbf{R}$ and $\{\alpha_1, \ldots, \alpha_n\} \subset \mathbf{C}$ we have

$$\sum_{j,k} \alpha_j \bar{\alpha}_k \varphi(z_j - z_k) \geq 0; \tag{1.15}$$

2. φ is uniformly continuous;
3. $\varphi(0) = 1$.

The most important fact concerning characteristic functions is that the converse to the above result is true, that is:

Theorem 1.1 (S. Bochner). *If φ is any function satisfying the three conditions of (1.15), then there is a unique probability measure Φ on $(\mathbf{R}, \mathbf{B})$ such that*

$$\varphi(z) = \int_{\mathbf{R}} e^{izx} \Phi(dx). \tag{1.16}$$

For a proof see, for example, S. Bochner (1932) or K. Yosida (1951).

The following theorem is P. Lévy's inversion formula which gives a method of obtaining the distribution Φ from a given characteristic function $\varphi(z)$.

1.3 Probability Distributions

Theorem 1.2 (P. Lévy). *Let $\varphi(z)$ be the characteristic function of a distribution Φ on $\mathbf{R}$. Then for any $a < b$ the following identity holds,*

$$\int_{-\infty}^{\infty} \bar{\chi}_{[a,\,b]}(x)\, d\Phi(x) = \lim_{c\to\infty} \frac{1}{2\pi} \int_{-c}^{c} \frac{e^{-ibz} - e^{-iaz}}{-iz} \varphi(z)\, dz, \qquad (1.17)$$

where $\bar{\chi}_{[a,\,b]}$ is given by

$$\bar{\chi}_{[a,\,b]}(x) = \begin{cases} 1 & a < x < b, \\ \tfrac{1}{2} & x = a,\; x = b, \\ 0 & \text{otherwise.} \end{cases}$$

By using this formula we can explicitly obtain Φ on $(\mathbf{R}, \mathbf{B})$. Thus the three quantities Φ, F and φ determined by X correspond in a one-to-one manner

$$\Phi \leftrightarrow F \leftrightarrow \varphi.$$

Consequently we see that the convergence of distributions or of distribution functions may be replaced by that of characteristic functions, but before we come to this, let us be clear about the meaning of convergence of distributions. Since distributions are measures on $(\mathbf{R}, \mathbf{B})$, *weak convergence* will be used; that is, $\Phi_n \to \Phi$ means that for all $f \in \mathscr{C}_0$,

$$\lim_{n\to\infty} \int_{-\infty}^{\infty} f(x)\Phi_n(dx) = \int_{-\infty}^{\infty} f(x)\Phi(dx),$$

where $\mathscr{C}_0$ is the Banach space consisting of all continuous functions on $\mathbf{R}$ vanishing at $+\infty$ and $-\infty$ equipped with the topology of uniform convergence.

Let φ_n, φ be the characteristic functions of Φ_n, Φ respectively. We list some of the main results concerning their convergence.

Theorem 1.3 (P. Lévy, V. Glivenko).

1. *If $\Phi_n \to \Phi$, then $\varphi_n(z)$ converges to $\varphi(z)$ uniformly on each compact subset of $\mathbf{R}$.*
2. *If $\varphi_n(z)$ converges to $\varphi(z)$, then $\Phi_n \to \Phi$.*

In the following theorem it is not assumed in advance that $\varphi(z)$ is a characteristic function.

Theorem 1.4 (P. Lévy). *Suppose that the sequence $\varphi_n(z)$ of characteristic functions of distributions Φ_n converges to a function $\varphi(z)$, this convergence being uniform on some neighbourhood of $z = 0$. Then $\varphi(z)$ is also a characteristic function, and in addition $\Phi_n \to \Phi$, where Φ is the distribution associated with φ.*

What we have discussed so far can easily be extended to the case where $X(\omega)$ is a multidimensional (vector-valued) random variable. As far as the definition (1.11) of Φ, everything is the same, but after this care is required on several points. Some of these are the following: the definition (1.12) of F is replaced by

$$F(x) = F(x_1, x_2, \ldots, x_n) = \Phi\left(\prod_1^n (-\infty, x_k]\right),$$

$x = (x_1, x_2, \ldots, x_n) \in \mathbf{R}^n$. The properties (1), (2) in (1.13) should now be understood to be in each variable x_k, and (3) is replaced by

$$\lim_{x_1, \ldots, x_n \to \infty} F(x_1, \ldots, x_n) = 1, \quad \text{and for all } k, \quad \lim_{x_k \to -\infty} F(x_1, \ldots, x_n) = 0.$$

In addition, $F(x_1 + h_1, x_2 + h_2, \ldots, x_n + h_n) - F(x_1, x_2 + h_2, \ldots, x_n + h_n) - \cdots - F(x_1 + h_1, x_2 + h_2, \ldots, x_n) + F(x_1, x_2, x_3 + h_3, \ldots, x_n + h_n) + \cdots + (-1)^n \times F(x_1, x_2, \ldots, x_n) \geq 0$ for $h_1, h_2, \ldots, h_n \geq 0$. The characteristic function $\varphi(z)$ is given by the relation

$$\varphi(z) = \int_{\mathbf{R}^n} e^{i(z, x)} \Phi(dx), \qquad z \in \mathbf{R}^n, \tag{1.18}$$

where $(\cdot, \cdot)$ is the inner product on $\mathbf{R}^n$. Here we may regard the variable z as running over the dual space $(\mathbf{R}^n)^*$ of the space $\mathbf{R}^n$ over which the variable x ranges. With suitable modifications the above theorems hold in this case as well.

Since our distributions are probability measures on Euclidean spaces, basic tools from analysis such as the Fourier-Stieltjes transform of a Borel measure, enable us to give clear descriptions of their properties. When we turn to the case where $X(\omega)$ is an infinite-dimensional random variable, suitable tools are not available, and we have to be careful when analogues of finite-dimensional results are discussed. This will be illustrated in the next two subsections.

(ii) Stochastic Processes and Their Distributions

We begin with several definitions. A system $\{X(t, \omega): t \in T\}$ of random variables, where t is to be thought of as a time parameter, is a mathematical model of some random phenomenon fluctuating in time.

Definition 1.1. A system $\{X(t, \omega): t \in T\}$ is called a *stochastic process* when the parameter set T is ordered.

We usually write $\{X(t): t \in T\}$ or just $\{X(t)\}$ and call it simply a process. If the parameter set T is a finite or infinite interval subset of the real numbers $\mathbf{R}$, the process $\{X(t)\}$ is said to be a *continuous parameter* process, whilst if T is $\mathbf{Z}$, the set of all integers, or the set $\mathbf{N}$ of all natural numbers, $\{X(t)\}$ is called a *discrete parameter* process. Other possible sets for T include n-dimen-

1.3 Probability Distributions

sional Euclidean space ($n > 1$) or an open connected subset of such a space, or even a Riemannian space, but this book deals mainly with continuous parameter stochastic processes where T is $\mathbf{R}$ or an interval subset of $\mathbf{R}$ such as $[0, \infty)$ or $[0, 1]$. For fixed ω we have a function $X(t, \omega)$ of t which is called a *sample function* or *sample path*.

Viewing $\{X(t, \omega): t \in T\}$ as an infinite-dimensional random vector taking values in $\mathbf{R}^T$, its distribution can be defined, as in the finite-dimensional case, as follows. A subset of $\mathbf{R}^T$ of the form

$$A = \{x \in \mathbf{R}^T: (x(t_1), x(t_2), \ldots, x(t_n)) \in B_n\} \quad (1.19)$$

where B_n is a Borel subset of $\mathbf{R}^n$ is called a *cylinder set*. For $t_1, t_2, \ldots, t_n$ fixed, the class of all cylinder sets obtained as B_n varies over the Borel subsets of $\mathbf{R}^n$ forms a σ-field which we write $\mathfrak{B}^{(t_1, t_2, \ldots, t_n)}$. For any set A of the form (1.19) we write

$$\Phi_{t_1, t_2, \ldots, t_n}(A) = P((X(t_1), X(t_2), \ldots, X(t_n)) \in B_n), \quad (1.20)$$

giving us a set function $\Phi_{t_1, t_2, \ldots, t_n}$ on the measurable space $(\mathbf{R}^T, \mathfrak{B}^{(t_1, t_2, \ldots, t_n)})$. By varying the choice of the finite subset $\{t_1, t_2, \ldots, t_n\} \subset T$ we get a class $\Phi = \{\Phi_{t_1, t_2, \ldots, t_n}\}$ of such measures, and this class satisfies the following consistency condition: if the cylinder set A of (1.19) has another expression, say

$$A = \{x \in \mathbf{R}^T: (x(s_1), x(s_2), \ldots, x(s_m)) \in B_m\},$$

then we have the equality

$$\Phi_{t_1, t_2, \ldots, t_n}(A) = \Phi_{s_1, s_2, \ldots, s_m}(A).$$

Denoting by $\mathfrak{A}^T$ the field of subsets of $\mathbf{R}^T$ consisting of all cylinder sets, the above discussion shows that we are given a finitely additive measure $\tilde{\Phi}$ on $(\mathbf{R}^T, \mathfrak{A}^T)$ such that the restriction of $\tilde{\Phi}$ to $\mathfrak{B}^{(t_1, t_2, \ldots, t_n)}$ coincides with $\Phi_{t_1, t_2, \ldots, t_n}$. Writing $\mathfrak{B}^T$ for the smallest σ-field containing $\mathfrak{A}^T$, we call an element of $\mathfrak{B}^T$ a Borel subset of $\mathbf{R}^T$. The following theorem is known as the *Kolmogorov extension theorem*.

Theorem 1.5. *The set function $\tilde{\Phi}$ on $(\mathbf{R}^T, \mathfrak{A}^T)$ defined above is uniquely extendable to a probability measure Φ on $(\mathbf{R}^T, \mathfrak{B}^T)$.*

The proof will not be given here, but analogous results are discussed in detail in §2.1 and §3.2 below.

Definition 1.2. The measure Φ obtained in Theorem 1.5 is called the *distribution* of $\{X(t): t \in T\}$.

We have now given the definition of the distribution of $\{X(t)\}$. Conversely, for any probability measure Φ on $(\mathbf{R}^T, \mathfrak{B}^T)$ there exists a stochastic process with distribution Φ, and this can be easily described in the following manner: set $\Omega = \mathbf{R}^T$, elements of Ω being denoted by $x = (x(t): t \in T)$, $\mathbf{B} = \mathcal{B}^T$, and $P = \Phi$. Then the relation $X(t, x) = x(t)$, $t \in T$, $x \in \Omega$, clearly defines a stochastic process and the distribution of $\{X(t): t \in T\}$ is readily seen to be Φ.

It is now appropriate to make an important remark. The space $\mathbf{R}^T$, on which the distribution Φ of $\{X(t)\}$ is defined, is really quite a large space in general, and the subset which actually supports Φ is often only a small part of it. For example, if some continuity of sample functions is assumed, and there are several possible kinds of continuity, then it is possible to use a rather clever idea to obtain a reasonable subset of $\mathbf{R}^T$ supporting Φ.

An important aspect of the structure of any given stochastic process $\{X(t): t \in T\}$ involves the manner in which the random variables $X(t)$, $t \in T$, depend on one another, and we now turn to a discussion of this topic.

Definition 1.3. Suppose that the parameter set T is either $\mathbf{R}$ or $\mathbf{Z}$.

1. If the distribution Φ_h of $\{X(t + h): t \in T\}$ is the same as the distribution Φ of $\{X(t): t \in T\}$ for any $h \in T$, then $\{X(t)\}$ is called a *strictly stationary process*, or simply a *stationary process*.
2. If each $X(t)$ has a second-order moment, and if

$$E(X(t)) = m \quad \text{(constant)},$$
$$E((X(t + h) - m)(X(t) - m)) = \gamma(h) \quad \text{(independent of } t\text{)},$$

then $\{X(t): t \in T\}$ is called a *weakly stationary process*, and $\gamma(h)$ is called the *covariance function* of $\{X(t)\}$.

A strictly stationary process with finite second-order moments is obviously a weakly stationary process. We usually assume that a weakly stationary process is mean-square continuous, i.e. that

$$\lim_{h \to 0} E(|X(t + h) - X(t)|^2) = 0. \tag{1.21}$$

With this assumption $\{X(t)\}$ turns out to be a continuous screw line in the Hilbert space $L^2(\Omega, \mathbf{B}, P)$, so that the theory of Hilbert spaces may be applied. Let $M_t(X)$ be the closed linear subspace of $L^2(\Omega, \mathbf{B}, P)$ spanned by $X(s)$, $s \leq t$. If $M_t(X)$ is a constant subspace, that is, if it does not vary with t, then $\{X(t)\}$ is said to be *deterministic*. On the other hand, if

$$\bigcap_t M_t(X) = \{1\}, \tag{1.22}$$

where $\{1\}$ denotes the one-dimensional subspace spanned by the constants, then it is said to be *purely non-deterministic*. The Wold decomposition consists of a decomposition of a general weakly stationary process into two parts, one deterministic, and one purely non-deterministic. There is of course a wide intermediate class, namely, those processes which are not deterministic, and these are called *non-deterministic*.

Another powerful tool available in the investigation of stationary processes is spectral analysis, and when this theory is applied, it is convenient to suppose that $X(t)$ is complex-valued. Accordingly the covariance function in this case is taken to be $E((X(t+h) - m)\overline{(X(t) - m)})$.

When we classify stochastic processes on the basis of the dependence between the random variables $X(t), t \in T$, we meet another important class of processes called *Markov processes*. These are discussed in detail in a later section (§2.4).

As we will see frequently in the sequel, the importance of Gaussian processes cannot be over-emphasised. A stochastic process $\{X(t)\}$ is termed *Gaussian* if any member of the class Φ introduced above is a Gaussian distribution. This class of processes will also be discussed in detail in later sections, from §1.6 onwards.

There is a generalisation of the notion of continuous parameter stochastic process to what is called a *random distribution* or *generalised random process*. This is understood to be a family $\{X(\xi, \omega): \xi \in E\}$ of random variables on a probability space $(\Omega, \mathbf{B}, P)$ with parameter set a certain function space E (usually assumed to be nuclear; see Appendix §A.3) such that for almost all ω, $X(\xi, \omega)$ is a continuous linear functional in ξ; that is, $X(\cdot, \omega)$ is a generalised function. In order to define the distribution of $\{X(\xi)\}$ we have to overcome several different kinds of difficulties. White noise, which is one of the main topics in this book, is the paradigm of a generalised random process, and we will discuss it briefly but systematically in Chapter 3.

(iii) Infinite Dimensional Distributions

There are two points which should be kept in mind whenever we introduce probability measures (or distributions) into an infinite-dimensional space. The first concerns Kakutani's theorem on infinite product measures, such a product measure arising as the distribution of an infinite sequence of independent random variables, and the second concerns the support (carrier) of any measure introduced into an infinite-dimensional vector space. Neither result could be guessed easily solely on the basis of an analogy with finite-dimensional measures. This subsection is devoted to Kakutani's theorem and its applications, and the question of the support of such measures will be discussed in later sections.

Let us take two probability measures P, P' on a measurable space $(\Omega, \mathbf{B})$ and choose a probability measure Q with respect to which both P and P' are absolutely continuous, i.e. $P \ll Q$ and $P' \ll Q$. For example Q could be taken as $\frac{1}{2}(P + P')$. Set

$$\psi = \left(\frac{dP}{dQ}\right)^{1/2}, \quad \psi' = \left(\frac{dP'}{dQ}\right)^{1/2},$$

where dP/dQ denotes the Radon-Nikodym density of P with respect to Q. Since ψ and ψ' are square-integrable with respect to Q the following is well-defined:

$$\int_\Omega \psi(\omega)\psi'(\omega)\, dQ(\omega) = \rho(P, P'). \tag{1.23}$$

The function ρ does not depend upon the Q chosen, and to see this we argue as follows. Suppose that $P \ll Q'$ and $P' \ll Q'$ for another measure Q' on $(\Omega, \mathbf{B})$, and assume that $Q \ll Q'$. Then the value of ρ obtained using Q' is

$$\int_\Omega \left(\frac{dP}{dQ'}\right)^{1/2} \left(\frac{dP'}{dQ'}\right)^{1/2} dQ',$$

which can be expanded as

$$\int_\Omega \left(\frac{dP}{dQ}\right)^{1/2} \left(\frac{dQ}{dQ'}\right)^{1/2} \left(\frac{dP'}{dQ}\right)^{1/2} \left(\frac{dQ}{dQ'}\right)^{1/2} dQ' = \int_\Omega \left(\frac{dP}{dQ}\right)^{1/2} \left(\frac{dP'}{dQ}\right)^{1/2} dQ.$$

Thus this is the same as that given by (1.23) using Q. For the general case where $Q \ll Q'$ is not necessarily true, we choose a third measure Q'' such that $Q \ll Q''$ and $Q' \ll Q''$, and then use the above result. We then see that the ρ obtained using Q'' has to be the same as that obtained using Q, and similarly for Q', thus proving that Q and Q' define the same value of ρ.

Proposition 1.1. *The following properties are satisfied by the function $\rho(P, P')$ defined above*:

1. $\rho(P, P') = \rho(P', P)$;
2. $0 \leq \rho(P, P') \leq 1$;
3. $\rho(P, P') = 1$ *is equivalent to the equality* $P = P'$,
4. $\rho(P, P') = 0$ *is equivalent to the fact that P and P' are singular.*

PROOF
1. is obvious.
2. The inequality $0 \leq \rho(P, P')$ is also obvious. To prove the other inequality we apply the Schwartz inequality to (1.23) obtaining

$$\rho(P, P')^2 \leq \int_\Omega \psi(\omega)^2\, dQ(\omega) \int_\Omega \psi'(\omega)^2\, dQ(\omega) = 1.$$

Incidentally we have equality in this relation if and only if $\psi(\omega)/\psi'(\omega) = $ constant $(= 1)$ a.e. Q, which proves (3).

4. The equality $\rho(P, P') = 0$ is equivalent to

$$\psi(\omega)\psi'(\omega) = 0 \quad \text{a.e.} \quad Q$$

which means that P and P' are singular. $\square$

Remark. Intuitively speaking, the closer $\rho(P, P')$ is to 1, the more similar are P and P', and if a metric along these lines is desired for some class of probabilities, then

$$d(P, P') = \left\{ \int_\Omega |\psi(\omega) - \psi'(\omega)|^2 \, dQ(\omega) \right\}^{1/2}$$

is appropriate. It is a metric and is connected to ρ by the relation

$$d(P, P') = \{2[1 - \rho(P, P')]\}^{1/2}. \tag{1.24}$$

In terms of the function ρ we can give a criterion for the equivalence or singularity of two infinite product measures. Let $(\Omega_n, \mathbf{B}_n)$, $n = 1, 2, \ldots$ be a sequence of measurable spaces with product space $(\Omega, \mathbf{B})$ where

$$\Omega = \prod_n \Omega_n, \quad \mathbf{B} = \prod_n \mathbf{B}_n.$$

Suppose that P_n and P'_n are probability measures on $(\Omega_n, \mathbf{B}_n)$ and that they satisfy:

Assumption. P_n and P'_n are equivalent for every n, $n = 1, 2, \ldots$.

We now obtain probability measures P and P' by forming the direct product of the P_n, $n \geq 1$, and the P'_n, $n \geq 1$, respectively:

$$P = \prod_n P_n, \quad P' = \prod_n P'_n.$$

Theorem 1.6 [S. Kakutani (1948)].

1. *P and P' are either equivalent or singular. A necessary and sufficient condition for them to be equivalent [resp. singular] is*

$$\prod_n \rho(P_n, P'_n) > 0 \quad [\text{resp. } \prod_n \rho(P_n, P'_n) = 0]. \tag{1.25}$$

2. *We have the relation*

$$\rho(P, P') = \prod_n \rho(P_n, P'_n).$$

PROOF. We distinguish two cases.
 a. The case when $\prod_n \rho(P_n, P'_n) > 0$.
 a-1. We first note that when P and P' are equivalent we may take the Q in the definition of ρ to be P. By assumption P_n and P'_n are equivalent for every n, so that we may form

$$\varphi_n(\omega_n) = \frac{dP'_n}{dP_n}(\omega_n), \quad \omega_n \in \Omega_n.$$

Denoting the coordinate projection from Ω onto Ω_n by p_n, we may write $\varphi_n(\omega_n) = \varphi_n(p_n \omega)$ and so view φ_n as a function on Ω, and we retain the same notation. Since P is the product measure, the sequence $\{\varphi_n(\omega): n \geq 1\}$ is a system of independent random variables on $(\Omega, \mathbf{B}, P)$ and we set

$$\psi_k(\omega) = \prod_1^k \{\varphi_j(\omega)\}^{1/2}.$$

Recalling that $\psi_k^2 = d(\prod_1^k P'_j)/d(\prod_1^k P_j)$, we see that for every $k \geq 1$

$$\psi_k \in L^2(\Omega, \mathbf{B}, P), \quad \text{and} \quad \|\psi_k\| = 1.$$

Furthermore, for $k < l$ the fact that the $\{\varphi_j\}$ are independent gives

$$\|\psi_k - \psi_l\|^2 = 2\left|1 - \prod_{k+1}^l \int_{\Omega_j} \sqrt{\varphi_j(\omega_j)}\, dP_j(\omega_j)\right|$$

$$= 2\left|1 - \prod_{k+1}^l \rho(P_j, P'_j)\right|. \tag{1.26}$$

Now we can use our assumption $\prod_n \rho(P_n, P'_n) > 0$, for it follows from this that the right side of (1.26) approaches zero as $l \to \infty$, i.e. that $\{\psi_k\}$ is a Cauchy sequence in $L^2(\Omega, \mathbf{B}, P)$. But this guarantees the existence of a mean-square limit $\bar{\psi}$ such that

$$\lim_{k \to \infty} \psi_k = \bar{\psi}. \tag{1.27}$$

a-2. We now prove that P and P' are equivalent, and that $dP'/dP = \bar{\psi}^2$. Let $B \subset \Omega$ be a cylinder set of the form $B_k \times \Omega_{k+1} \times \Omega_{k+2} \times \ldots$, where B_k is a $\prod_1^k \mathbf{B}_j$-measurable set. Recalling that $\varphi_j(\omega_j)$ can be written $\varphi_j(p_j \omega)$ $(\equiv \varphi_j(\omega))$ we have

$$P'(B) = \int_{B_k} d\left(\prod_1^k P'_j\right) = \int_{B_k} \left[\prod_1^k \varphi_j(\omega_j)\right] d\left(\prod_1^k P_j\right) = \int_B \left[\prod_1^k \varphi_j(\omega)\right] dP(\omega).$$

For $l > k$ we may regard B as $B_k \times \Omega_{k+1} \times \cdots \times \Omega_l \times \Omega_{l+1} \times \ldots$, whence

$$P'(B) = \int_{B_k} \psi_k(\omega)^2 \, dP(\omega) = \int_{B_k \times \Omega_{k+1} \times \cdots \times \Omega_l} \psi_l(\omega)^2 \, dP(\omega).$$

Since $l > k$ is arbitrary, we may let $l \to \infty$ and use (1.27) to obtain

$$P'(B) = \int_B \bar{\psi}(\omega)^2 \, dP(\omega). \tag{1.28}$$

This equation holds for any cylinder set, and it can be extended to hold for any $\mathbf{B}$-measurable set, implying that

$$P' \ll P \quad \text{and} \quad dP'/dP = \bar{\psi}^2.$$

Since ρ is symmetric the relation $P \ll P'$ also follows and we have proved that P and P' are equivalent.

a-3. The equality $\rho(P, P') = \prod_1^\infty \rho(P_n, P'_n)$ follows from the following computations.

1.3 Probability Distributions

$$\rho(P, P') = \int_\Omega \bar{\psi}(\omega)\, dP(\omega) = \lim_{n\to\infty} \int_\Omega \psi_n(\omega) \cdot 1\, dP(\omega)$$

(since the strong convergence of the ψ_n implies the weak convergence)

$$= \lim_{n\to\infty} \int_{\Omega_1 \times \Omega_2 \times \cdots \times \Omega_n} \prod_1^n \sqrt{\varphi_j(\omega_j)}\, d\left(\prod_1^n P_j\right)$$

$$= \lim_{n\to\infty} \prod_1^n \rho(P_j, P'_j).$$

b. The case $\prod_1^\infty \rho(P_n, P'_n) = 0$. This assumption implies that for any $\varepsilon > 0$ there is a natural number $k = k(\varepsilon)$ such that

$$\prod_1^k \rho(P_j, P'_j) < \varepsilon.$$

If we set $B_k = \{(\omega_1, \omega_2, \ldots, \omega_k): \prod_1^k \varphi_j(\omega_j) \geq 1\}$ we see that

$$\left(\prod_1^k P_j\right)(B_k) = \int_{B_k} 1\, d\left(\prod_1^k P_j\right) \leq \int_{B_k}\left[\prod_1^k \sqrt{\varphi_j(\omega_j)}\right] d\left(\prod_1^k P_j\right)$$

$$\leq \prod_1^k \rho(P_j, P'_j) < \varepsilon.$$

Furthermore we have

$$\left(\prod_1^k P'_j\right)\left(\prod_1^k \Omega_j \setminus B_k\right)$$

$$= \int_{\prod_1^k \Omega_j \setminus B_k} \left[\prod_1^k \varphi_j(\omega_j)\right] d\left(\prod_1^k P_j\right)$$

$$\leq \int_{\prod_1^k \Omega_j \setminus B_k} \left[\prod_1^k \sqrt{\varphi_j(\omega_j)}\right] d\left(\prod_1^k P_j\right) \quad \text{since } \prod_1^k \varphi_j(\omega_j) < 1 \quad \text{off } B_k$$

$$\leq \prod_1^k \rho(P_j, P'_j) < \varepsilon.$$

Thus we have shown that for any $\varepsilon > 0$ there exists a cylinder set $B_\varepsilon = B_k \times \Omega_{k+1} \times \cdots$ such that $P(B_\varepsilon) < \varepsilon$ and $P'(B_\varepsilon) > 1 - \varepsilon$, that is P and P' are singular.

The necessity of the condition (2.25) is obvious and therefore the theorem is fully proved. □

Remark. As we have just seen two infinite product measures are either equivalent or singular, no intermediate stage being permitted. This dichotomy is a unique feature of infinite product measures.

EXAMPLE. We apply theorem to the distribution of a sequence of independent random variables (i.e. a discrete parameter stochastic process) $\{X_n, n \geq 1\}$ when each X_n is Gaussian.

1. The first topic is a comparison theorem between the distributions of $\{X_n, n \geq 1\}$ and $\{c_n X_n, n \geq 1\}$, $\{c_n\}$ a sequence of constants.

Take $\Omega_n = \mathbf{R}$ for every n, and suppose that the distribution P_n of X_n is Gaussian with mean 0 and variance $\sigma_n^2 > 0$. Accordingly the distribution P'_n of $c_n X_n$ is Gaussian and has mean 0 and variance $c_n^2 \sigma_n^2$. Since

$$\frac{dP'_n}{dP_n} = |c_n|^{-1} \exp\left[-\frac{1}{2} x^2 (c_n^{-2} \sigma_n^{-2} - \sigma_n^{-2})\right],$$

we have

$$\rho(P_n, P'_n) = (2\pi)^{-1/2} \sigma_n^{-1} \int_{-\infty}^{\infty} \left|\frac{dP'_n}{dP_n}(x)\right|^{1/2} \exp\left[-\frac{x^2}{2\sigma_n^2}\right] dx$$

$$= (2\pi |c_n|)^{-1/2} \sigma_n^{-1} \int_{-\infty}^{\infty} \exp\left[-\frac{x^2(c_n^{-2}+1)}{4\sigma_n^2}\right] dx$$

$$= \left|\frac{2|c_n|}{1+c_n^2}\right|^{1/2}.$$

Since $P = \prod_n P_n$, $P' = \prod_n P'_n$, we have the following equivalent assertions

$$P \sim P' \text{ (equivalence)} \Leftrightarrow \prod_n \left|\frac{2|c_n|}{1+c_n^2}\right|^{1/2} > 0$$

$$\Leftrightarrow \sum_n \left[\frac{(1-c_n^2)^2}{(1+c_n^2)^2}\right] < \infty$$

$$\Leftrightarrow \sum_n (1-|c_n|)^2 < \infty.$$

If the last series diverges, then of course P and P' are singular.

2. Next we consider translations, comparing the distribution of $\{X_n + m_n : n \geq 1\}$, $\{m_n\}$ a sequence of constants, with that of $\{X_n : n \geq 1\}$.

Let P be a Gaussian distribution with mean 0 and variance $\sigma_n^2 > 0$, whilst P' is also Gaussian, with mean m_n and variance σ_n^2, to be thought of as the distribution of X_n and $X_n + m_n$ respectively. It is easy to derive the result

$$\frac{dP'_n}{dP_n} = \exp\left[\frac{2m_n x - m_n^2}{2\sigma_n^2}\right],$$

so that

$$\rho(P_n, P'_n) = (2\pi)^{-1/2} \sigma_n^{-1} \int_{-\infty}^{\infty} \exp\left[\frac{2m_n x - m_n^2}{4\sigma_n^2} - \frac{x^2}{2\sigma_n^2}\right] dx$$

$$= \exp\left[-\frac{m_n^2}{8\sigma_n^2}\right].$$

With this established we can prove that

$$P \sim P' \Leftrightarrow \sum_n \frac{m_n^2}{\sigma_n^2} < \infty.$$

Remark. In the case $\sigma_n = 1$ for all n, P is the distribution of a discrete parameter white noise (see §3.4). Our theorem says a permissible translation of the space vector, expressed in terms of the infinite-dimensional vector (m_n), leaves the absolute continuity undisturbed if and only if

$$(m_n) \in l^2.$$

1.4 Conditional Expectations

Let $(\Omega, \mathbf{B}, P)$ be a probability space, and suppose that $\mathbf{C}$ is a sub-σ-field of $\mathbf{B}$. We denote by $P_\mathbf{C}$ the restriction of P to $\mathbf{C}$ and suppose that a non-negative $\mathbf{B}$-measurable function $X(\omega) \geq 0$ is given on $(\Omega, \mathbf{B}, P)$.

Proposition 1.2. *Let $\mathbf{C}$ and $X(\omega)$ be as above. Then there exists a unique function $Y(\omega)$ satisfying the following two conditions:*

1. *$Y(\omega)$ is $\mathbf{C}$-measurable and $0 \leq Y(\omega) \leq \infty$, a.e. $P_\mathbf{C}$;*
2. *For every $C \in \mathbf{C}$ we have*

$$\int_C Y(\omega) \, dP(\omega) = \int_C X(\omega) \, dP(\omega).$$

In particular if $X(\omega)$ has a finite expectation, then $0 \leq Y(\omega) < \infty$, a.e. $P_\mathbf{C}$.

PROOF. We define a set function Q on the measurable space $(\Omega, \mathbf{C})$ by

$$Q(C) = \int_C X(\omega) \, dP(\omega), \qquad C \in \mathbf{C}.$$

Then $(\Omega, \mathbf{C}, Q)$ becomes a measure space, although not necessarily a finite one. It is clear from its definition that $Q \ll P_\mathbf{C}$ and so by the Radon-Nikodym theorem there exists a $\mathbf{C}$-measurable density function $Y(\omega) = dQ/dP_\mathbf{C}$ such that

$$Q(C) = \int_C Y(\omega) \, dP_\mathbf{C} = \int_C X(\omega) \, dP(\omega). \tag{1.29}$$

This function $Y(\omega)$ is the one we seek. Its uniqueness is shown in the following way. Take another function $Y'(\omega)$ satisfying (1) and (2) above, and an arbitrary element $C \in \mathbf{C}$. Then we have

$$\int_C Y(\omega) \, dP_\mathbf{C}(\omega) = \int_C Y'(\omega) \, dP_\mathbf{C}(\omega) \left[= \int_C X(\omega) \, dP(\omega) \right]$$

which implies that $Y(\omega) = Y'(\omega)$, a.e. $P_\mathbf{C}$.

The final assertion is an immediate consequence of the fact that Q is a finite measure if $X(\omega)$ has a finite expectation. $\square$

Remark. The uniqueness of $Y(\omega)$ is in the sense that it is determined uniquely up to sets of zero $P_\mathbf{C}$-measure.

Definition 1.4. The function $Y(\omega)$ given by Proposition 1.2 is called the *conditional expectation of $X(\omega)$ relative to (or given) $\mathbf{C}$*, and is denoted by $E(X|\mathbf{C})(\omega)$ or simply $E(X|\mathbf{C})$. In the case $\mathbf{C} = \mathbf{B}(Z)$, the conditional expectation of X given $\mathbf{C}$ is often written $E(X|Z)$.

For a general not necessarily positive random variable $X(\omega)$ we have to make the additional assumption that $X(\omega)$ has a finite expectation. In this case we can decompose $X(\omega)$ into its positive part $X^+(\omega)$ and negative part $X^-(\omega)$,

$$X(\omega) = X^+(\omega) - X^-(\omega), \qquad X^+(\omega), \ X^-(\omega) \geq 0,$$

both of which must have finite expectations. We can then obtain the conditional expectations $E(X^+|\mathbf{C})$ and $E(X^-|\mathbf{C})$ of X^+ and X^- given $\mathbf{C}$, respectively, and define the conditional expectation of X itself given $\mathbf{C}$ by

$$E(X|\mathbf{C}) = E(X^+|\mathbf{C}) - E(X^-|\mathbf{C}).$$

Proposition 1.3. *The conditional expectation operator has the following properties, all random variables having finite expectations:*
1. *For all constants a and b*

$$E(aX_1 + bX_2|\mathbf{C}) = aE(X_1|\mathbf{C}) + bE(X_2|\mathbf{C}). \tag{1.30}$$

2. *If $X_n(\omega) \geq 0$, $X_n(\omega) \leq X_{n+1}(\omega)$, $n = 1, 2, \ldots$, and if the limit*

$$\lim_{n \to \infty} X_n(\omega) = X(\omega)$$

exists a.e. P, then

$$E(X_n|\mathbf{C}) \leq E(X_{n+1}|\mathbf{C}) \qquad n = 1, 2, \ldots,$$

and a.e. P we have

$$\lim_{n \to \infty} E(X_n|\mathbf{C}) = E(X|\mathbf{C}). \tag{1.31}$$

3. *For $\mathbf{C}_2 \supseteq \mathbf{C}_1$ we have the relation*

$$E(E(X|\mathbf{C}_2)|\mathbf{C}_1) = E(X|\mathbf{C}_1), \tag{1.32}$$

and in particular, for every sub-σ-field $\mathbf{C}$

$$E(E(X|\mathbf{C})) = E(X).$$

4. *If $X(\omega)$ is independent of $\mathbf{C}$ then a.e. $P_\mathbf{C}$.*

$$E(X|\mathbf{C}) = E(X). \tag{1.33}$$

1.4 Conditional Expectations

5. *For every bounded **C**-measurable function $Z(\omega)$ we have*

$$E(ZX|\mathbf{C}) = ZE(X|\mathbf{C}). \tag{1.34}$$

OUTLINE OF PROOF. (1) follows immediately from the definition and (2) comes from the definition and Fatou's lemma.

3. If $C \in \mathbf{C}_1$ then $C \in \mathbf{C}_2$ also, and thus

$$\int_C E(E(X|\mathbf{C}_2)|\mathbf{C}_1) \, dP = \int_C E(X|\mathbf{C}_2) \, dP = \int_C X \, dP.$$

As this last integral coincides with the integral of $E(X|\mathbf{C}_1)$ over the set C, and this is true for all $C \in \mathbf{C}_1$, (1.32) follows.

4. Suppose that $X(\omega) = \chi_A(\omega)$, the indicator function of a set A independent of $\mathbf{C}$. Then for all $C \in \mathbf{C}$ we have

$$\int_C E(X|\mathbf{C}) \, dP(\omega) = \int_C \chi_A(\omega) \, dP(\omega) = P(A \cap C)$$

$$= P(A)P(C) = \int_C E(X) \, dP(\omega),$$

proving a special case of (1.33). To get (1.33) for a general such X, we approximate it by a sequence of simple functions and use (1), (2) and the result just established.

The proof of (5) also begins by considering the case $Z(\omega) = \chi_B(\omega)$. For every $C \in \mathbf{C}$ we prove that

$$\int_C E(ZX|\mathbf{C}) \, dP(\omega) = \int_C \chi_B(\omega) X(\omega) \, dP(\omega) = \int_{B \cap C} X(\omega) \, dP(\omega)$$

$$= \int_{B \cap C} E(X|\mathbf{C}) \, dP(\omega) = \int_C Z E(X|\mathbf{C}) \, dP(\omega),$$

which implies (1.34) in this special case. We get the result for a general bounded **C**-measurable function $Z(\omega)$ as in the proof of (4). □

Definition 1.5. The conditional expectation $E(X|\mathbf{C})$ in the case $X(\omega) = \chi_B(\omega)$, $B \in \mathbf{B}$, is called the *conditional probability of B given $\mathbf{C}$* and written $P(B|\mathbf{C})$.

In addition to being a **C**-measurable function, the conditional probability $P(B|\mathbf{C})$ can be viewed as a set function on **B**, and as such has properties similar to those of a measure. Apart from countable additivity, which is not true in general, these properties can easily be derived from those of conditional expectations in general stated in Proposition 1.3, and so the details are omitted.

EXAMPLE 1. We return to Example 2 of §1.2. Let $\mathbf{B}_n$ be the smallest σ-field with respect to which the random variables $S_1(\omega), S_2(\omega), \ldots, S_n(\omega)$ are all measurable. It can easily be seen to be generated by the subintervals of $[0, 1]$ of the form $C_k = [k2^{-n}, (k+1)2^{-n})$, $0 \leq k < 2^n$, and so coincides with the class of events defined by the random variables $X_1(\omega), X_2(\omega), \ldots, X_n(\omega)$. We next compute the conditional expectation $E(S_{n+1}|\mathbf{B}_n)$. Now S_n is constant on the interval C_k whilst, as shown in Fig. 2, S_{n+1} is equal to $S_n - 1$ on the left half of C_k. By definition

$$\int_{C_k} E(S_{n+1}|\mathbf{B}_n)\,dP(\omega) = \int_{C_k} S_{n+1}(\omega)\,dP(\omega)$$

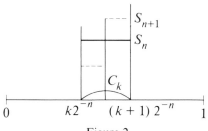

Figure 2

From an observation above, the right side integral turns out to be the value of S_n on C_k multiplied by the length of C_k. On the other hand $E(S_{n+1}|\mathbf{B}_n)$ is constant on C_k as it is $\mathbf{B}_n$-measurable, and so we must have

$$E(S_{n+1}|\mathbf{B}_n) = S_n, \quad \text{a.e.} \tag{1.35}$$

on C_k. Since C_k is an arbitrary element of $\mathbf{B}_n$ and both sides of (1.35) are $\mathbf{B}_n$-measurable functions, this relation must hold on the whole space $[0, 1]$.

Remark. Let $\{X_n: n \geq 1\}$ be an arbitrary stochastic process which satisfies $E(|X_n|) < \infty$ for every $n \geq 1$, and denote by $\mathbf{B}_n$ the smallest σ-field with respect to which $X_1, X_2, \ldots, X_n$ are measurable. If the relation

$$E(X_{n+1}|\mathbf{B}_n) = X_n$$

holds for every n, then $\{X_n\}$ is said to be a *martingale* adapted to $\{\mathbf{B}_n\}$; for further details see §A.1. An example of a martingale is the process $\{S_n: n \geq 0\}$ defined above. As $\{S_n\}$ can be regarded as a mathematical model of the tossing an unbiased memoryless coin, so we can think of a martingale as a model for some arbitrary "fair" game.

EXAMPLE 2. Suppose that $X(\omega)$ and $Y(\omega)$ are discrete random variables on a probability space $(\Omega, \mathbf{B}, P)$ with values x_j and y_k $j, k = 1, 2, \ldots$, respectively, and set

$$A_j = \{\omega: X(\omega) = x_j\}, \quad P(A_j) = p_j,$$
$$B_{j,k} = \{\omega: X(\omega) = x_j, Y(\omega) = y_k\}, \quad P(B_{j,k}) = q_{j,k}.$$

1.4 Conditional Expectations

The σ-field generated by the $A_j, j \geq 1$, will be denoted by **C** and, assuming the existing of the expectation $E(Y)$, we shall obtain an explicit formula for $E(Y|\mathbf{C})$. From the definitions we have

$$\int_{A_j} E(Y|\mathbf{C}) \, dP(\omega) = \int_{A_j} Y \, dP(\omega) = \sum_k y_k q_{j,k}.$$

But $E(Y|\mathbf{C})$ is **C**-measurable and so constant on A_j, and therefore, provided that $p_j = P(A_j) > 0$, we obtain

$$E(Y|\mathbf{C}) = \frac{\sum_k y_k q_{j,k}}{p_j}, \qquad \omega \in A_j.$$

If $p_j = 0$ then also $q_{j,k} = 0$ for all k and so in such a case we may define $E(Y|\mathbf{C})$ to be 0 on A_j.

Turning now to the next topic, we begin by recalling that given any probability space $(\Omega, \mathbf{B}, P)$, we can form a real Hilbert space $L^2(\Omega, \mathbf{B}, P)$. This may be simply viewed as the collection of all real-valued random variables on Ω with finite variance, and in this space the inner product of a random variable X with the constant function 1 is the expectation of X, whilst the square of the norm of X is nothing but its second order moment.

Let **C** be a sub-σ-field of the σ-field **B** which contains all the P-null sets of **B**.

Proposition 1.4. $E(\cdot | \mathbf{C})$ *is the orthogonal projection operator from* $L^2(\Omega, \mathbf{B}, P)$ *onto the subspace* $L^2(\Omega, \mathbf{C}, P)$.

PROOF. Let X be an arbitrary element of $L^2(\Omega, \mathbf{B}, P)$. For every $Y \in L^2(\Omega, \mathbf{C}, P)$ we have

$$E\{[X - E(X|\mathbf{C})]Y\} = E(XY) - E\{YE(X|\mathbf{C})\}$$
$$= E(XY) - E\{E(XY|\mathbf{C})\} = 0,$$

that is, $X - E(X|\mathbf{C})$ is orthogonal to $L^2(\Omega, \mathbf{C}, P)$. On the other hand $E(X|\mathbf{C})$ can be seen to belong to $L^2(\Omega, \mathbf{C}, P)$, and these facts, together with (1.30), (1.32) and (1.34) allow us to prove that the map $X \to E(X|\mathbf{C})$ is a linear, idempotent and Hermitian operator. □

The following corollary is a straightforward consequence of the fact that a conditional expectation is an orthogonal projection. We suppose that all sub-σ-fields contain the P-null sets of **B**.

Corollary. *Let* $\mathbf{B}_n, n \geq 1$ *be an increasing family of sub-σ-fields of* **B** *with limit* $\mathbf{B}_\infty$, *i.e.*

$$\mathbf{B}_n \to \mathbf{B}_\infty \quad \text{as} \quad n \to \infty. \tag{1.36}$$

Then for every random variable $X \in L^2(\Omega, \mathbf{B}, P)$ *we have (in mean square)*

$$\lim_{n \to \infty} E(X|\mathbf{B}_n) = E(X|\mathbf{B}_\infty).$$

This result is a consequence of the fact that the sequence $L^2(\Omega, \mathbf{B}_n, P)$ of associated Hilbert subspaces is also increasing, and has the limit $L^2(\Omega, \mathbf{B}_\infty, P)$.

1.5 Limit Theorems

We turn now to several topics relating to the limit theorems of probability. As usual there will be a probability space $(\Omega, \mathbf{B}, P)$ on which all the random variables $X(\omega)$, $X_n(\omega) \ldots$, etc. will be defined, but when no confusion is likely we shall omit the parameter ω in the notation for such random variables.

(i) Law of Large Numbers

The following three lemmas are well known.

Lemma 1.1 (*the Borel-Cantelli lemma*). *Let $\{A_n\}$ be a sequence of events.*
  a. *If $\sum_n P(A_n) < \infty$, then*
$$P\left(\limsup_{n\to\infty} A_n\right) = 0, \qquad \left[P\left(\liminf_{n\to\infty} A_n^c\right) = 1\right].$$
  b. *If $\sum_n P(A_n) = \infty$ and, in addition, the sequence $\{A_n\}$ is independent, then*
$$P\left(\limsup_{n\to\infty} A_n\right) = 1, \qquad \left[P\left(\liminf_{n\to\infty} A_n^c\right) = 0\right].$$

Lemma 1.2 (Tchebychev's inequality). *Suppose that X is a random variable with finite variance $V(X)$. Then we have the following inequality:*
$$P(|X - E(X)| \geq t) \leq \frac{V(X)}{t^2}, \qquad t > 0. \tag{1.37}$$

Lemma 1.3 (Kolmogorov's inequality). *Suppose that $\{X_n\}$ is an independent sequence of random variables with finite variance, and $S_n = \sum_1^n X_k$. Then we have the inequality*
$$P\left(\max_{1 \leq k \leq n} |S_k - E(S_k)| \geq t\right) \leq \frac{V(S_n)}{t^2}, \qquad t > 0. \tag{1.38}$$

The *weak law of large numbers* stated in the next proposition is an immediate consequence of Lemma 1.2.

Proposition 1.5. *Let $\{X_k\}$ be an independent sequence of random variables with $E(X_k) = m_k$ and $V(X_k) \leq V$, $k \geq 1$. Then for any $\varepsilon > 0$ we have*
$$\lim_{n\to\infty} P\left(\left|n^{-1} \sum_1^n (X_k - m_k)\right| > \varepsilon\right) = 0. \tag{1.39}$$

1.5 Limit Theorems

We now have the *strong law of large numbers* in the following.

Theorem 1.7. *Let $\{X_k\}$ be an independent sequence of random variables with $E(X_k) = m_k$ and $V(X_k) \leq V$, $k \geq 1$. Then*

$$\lim_{n \to \infty} n^{-1} \sum_{1}^{n} (X_k - m_k) = 0 \quad \text{a.e.} \tag{1.40}$$

PROOF. Replacing X_k by $X_k - m_k$ we may assume that $E(X_k) = 0$ for all k. If we put

$$M_n(\omega) = \max_{1 \leq k \leq n} \left| \sum_{j=1}^{k} X_j(\omega) \right|,$$

then Lemma 1.3 implies that

$$P(M_{8^k} \geq 4^k) \leq \frac{8^k V}{4^{2k}} = 2^{-k} V.$$

We now set $A_k = \{8^{-k} M_{8^k} \geq 2^{-k}\}$ and note that since $\sum_k P(A_k) \leq \sum_k 2^{-k} V < \infty$, the Borel-Cantelli lemma implies that

$$P\left(\limsup_{n \to \infty} A_n \right) = 0.$$

But this means that for almost all ω there exists a natural number $N = N(\omega)$ such that for all $k > N$, $8^{-k} M_{8^k}(\omega) < 2^{-k}$. Thus

$$\lim_{k \to \infty} 8^{-k} M_{8^k}(\omega) = 0, \quad \text{a.e.}$$

For a general n we choose $k = k(n)$ such that $8^{k-1} \leq n < 8^k$ and observe that for such a value of k

$$0 \leq n^{-1} |S_n(\omega)| \leq 8^{-(k-1)} M_{8^k}(\omega),$$

which converges to 0 a.e. as $n \to \infty$. $\square$

Corollary. *Let $\{X_k\}$ be an independent sequence of identically distributed random variables with finite variance and common mean $m = E(X_k)$. Then we have*

$$\lim_{n \to \infty} n^{-1} S_n = m \quad \text{a.e.} \tag{1.41}$$

(ii) The Central Limit Theorem

As its name indicates, this theorem holds a central position in probability theory, and through it we see the importance of Gaussian distributions, for they arise as limits. There are many and varied generalisations of this theorem, but we restrict our attention here to a typical and easy case in which the random variables involved are one-dimensional.

Let $\{X_k\}$ be an independent sequence of random variables. Not only will we suppose that each X_k has a finite expectation m_k and variance V_k, but also a finite third-order moment $c_k = E\{|X_k - m_k|^3\}$ about the mean. The partial sums of these moment sequences are denoted by A_n, B_n and C_n, and we also write $S_n = \sum_1^n X_k$.

Theorem 1.8. *If $B_n \to \infty$, $B_n^{-1}(\max_{1 \leq k \leq n} V_k) \to 0$ and $C_n^2/B_n^3 \to 0$ as $n \to \infty$, then the distribution of $Y_n = B_n^{-1/2}(S_n - A_n)$ converges to the standard Gaussian distribution as $n \to \infty$.*

PROOF. We shall prove that the characteristic function $\varphi_n(z)$ of Y_n converges pointwise (and this is sufficient) to the characteristic function $\exp(-\frac{1}{2}z^2)$ of the standard Gaussian distribution. To this end we introduce the notation $Y_{n,k} = B_n^{-1/2}(X_k - m_k)$, and observe that $E(Y_{n,k}) = 0$, $E(Y_{n,k}^2) = B_n^{-1}V_k$. Furthermore,

$$E\{\exp[izY_{n,k}]\} = 1 - \tfrac{1}{2}z^2 B_n^{-1}V_k + \tfrac{1}{6}z^3(B_n^{-3/2}c_k)\theta_{n,k},$$

where $|\theta_{n,k}| \leq 1$. Denoting the right side of this equation by $1 + \alpha_{n,k}(z)$ we see from the assumptions that

$$\max_{1 \leq k \leq n} |\alpha_{n,k}(z)| \leq \tfrac{1}{2}z^2 B_n^{-1}\left(\max_{1 \leq k \leq n} V_k\right) + \tfrac{1}{6}|z|^3 B_n^{-3/2}C_n \to 0,$$

as $n \to \infty$, whilst

$$\sum_{k=1}^n |\alpha_{n,k}(z)|^2 \leq \left\{\max_{1 \leq k \leq n} |\alpha_{n,k}(z)|\right\} \sum_{k=1}^n |\alpha_{n,k}(z)|$$

$$\leq \left\{\max_{1 \leq k \leq n} |\alpha_{n,k}(z)|\right\} [\tfrac{1}{2}z^2 + \tfrac{1}{6}(B_n^{-3/2}C_n)|z|^3] \to 0,$$

and

$$\sum_1^n \alpha_{n,k}(z) = -\tfrac{1}{2}z^2 + \tfrac{1}{6}\theta_n(B_n^{-3/2}C_n)z^3 \to -\tfrac{1}{2}z^2,$$

as $n \to \infty$, since $|\theta_n| \leq 1$. However $Y_n = \sum_{k=1}^n Y_{n,k}$ is a sum of independent random variables, and so we have

$$\varphi_n(z) = \prod_k E\{\exp[izY_{n,k}]\} = \prod_k (1 + \alpha_{n,k}(z)).$$

Using the results above concerning $\alpha_{n,k}$ we finally obtain the result

$$\lim_{n \to \infty} \varphi_n(z) = \exp[-\tfrac{1}{2}z^2]. \qquad \square$$

Remark. The assumption $B_n^{-1}(\max_{1 \leq k \leq n} V_k) \to 0$ as $n \to \infty$ can in fact be derived from $B_n^{-3}C_n^2 \to 0$ as $n \to \infty$.

We now list a few sufficient conditions for the central limit theorem to hold.

  a. If the sequence $\{X_k\}$ consists of identically distributed random variables it suffices for them to have a third order moment.

  b. Let V_k be the variance of X_k and set $Y_k = X_k - E(X_k)$. Another sufficient condition is: for every $t > 0$

$$\left(\sum_1^n V_k\right)^{-1} \sum_1^n \int_{(|Y_k|^2 > t \sum_1^n V_k)} Y_k^2(\omega)\, dP(\omega) \to 0 \quad \text{as} \quad n \to \infty.$$

  c. Suppose that the $\{X_k\}$ are uniformly bounded, i.e. there is a constant K such that for all $k \geq 1$,

$$|X_k(\omega)| \leq K, \quad \text{a.e.}$$

In this case $B_n \to \infty$ is a sufficient condition.

A special case of the last two conditions is the famous De Moivre-Laplace theorem for the binomial distribution, see Feller (1968) Chapter VII.

As we have seen the central limit theorem guarantees that the distribution of the normalised sum of a suitable sequence of random variables converges to a Gaussian distribution. This fact puts the Gaussian distribution in a premier position amongst probability distributions, and at the same time gives us a conviction that Gaussian random variables and Gaussian distributions are indeed important.

(iii) Convergence of a Sequence of Random Variables

We shall discuss several kinds of convergence for sequences $\{X_n(\omega)\}$ of random variables and for the series $\sum_n X_n(\omega)$ derived from them. In this subsection the independence of $\{X_n\}$ is no longer assumed.

  a. X_n is said to converge *in law* to X if the distribution of X_n converges (in the sense of §1.3) to the distribution of X.

  b. $X_n(\omega)$ converges *almost surely* or *with probability one* to $X(\omega)$ if for almost all ω, $X_n(\omega)$ converges to $X(\omega)$. We saw this in the strong law of large numbers.

  c. X_n is said to converge *in probability* to X if for any $\varepsilon > 0$

$$\lim_{n \to \infty} P(|X_n - X| > \varepsilon) = 0. \tag{1.42}$$

  d. Assume that X_n and X have a finite moment of order p, $p \geq 1$. We say that X_n converges to X *in the sense of* L^p if

$$\lim_{n \to 0} \int |X_n(\omega) - X(\omega)|^p\, dP(\omega) = 0. \tag{1.43}$$

In the particular case of $p = 2$, X_n is said to converge *in mean square* to X.

The modes of convergence (a), (b), (c) and (d) also apply to a series $\sum_n X_n$

as well as a sequence $\{X_n\}$, and we continue to use the terms introduced above for this case.

As we are really just considering the special case in measure theory in which the total measure is unity, the following properties of convergence of sequences or series follow immediately from known results.

  a. Convergence almost everywhere $\Rightarrow$ convergence in probability
  $\Rightarrow$ convergence in law;

  b. L^q-convergence $\Rightarrow$ L^p-convergence if $q > p \geq 1$;
  $\Rightarrow$ the existence of an almost surely convergent subsequence;

  c. L^p-convergence $\Rightarrow$ convergence in probability.

The following assertion follows straightforwardly from the Borel-Cantelli lemma.

Proposition 1.6. *If there are sequences $\varepsilon_n > 0$ and $\eta_n > 0$ such that both $\sum_n \varepsilon_n$ and $\sum_n \eta_n$ are convergent, and for all $n \geq 1$,*

$$P(|X_n| > \eta_n) < \varepsilon_n,$$

then $\sum_n X_n$ converges almost surely.

We find that in general neither of the converses to the assertions in (a) above hold, but we can prove a somewhat weaker assertion.

Proposition 1.7. *If $\{X_n\}$ converges in probability to X, then there exists a subsequence of $\{X_n\}$ which converges almost surely to X.*

OUTLINE OF PROOF. From the assumptions we can choose $\{n_k\}$ and $\varepsilon_k > 0$ such that $\sum_k \varepsilon_k < \infty$ and

$$P(|X_{n(k)} - X| > \varepsilon_k) < \varepsilon_k.$$

The almost sure convergence of $\{X_{n(k)}\}$ now follows from Proposition 1.6. □

Proposition 1.8. *Suppose that for any $\varepsilon > 0$ we have*

$$P\left(\max_{n \leq k \leq m} |X_n - X_k| > \varepsilon\right) \to 0 \quad \text{as} \quad m, n \to \infty.$$

Then $\{X_n\}$ converges almost surely.

OUTLINE OF PROOF. For any $\varepsilon > 0$ we can choose N such that for all $m, n \geq N$ with $m > n$

$$P\left(\max_{n \leq k,\, l \leq m} |X_k - X_l| > 2\varepsilon\right) < \varepsilon.$$

Letting $m \to \infty$ we obtain

$$P\left(\sup_{k,l \geq n} |X_k - X_l| > 2\varepsilon\right) \leq \varepsilon,$$

and allowing $n \to \infty$

$$P\left(\lim_{n \to \infty} \sup_{k,l \geq n} |X_k - X_l| > 2\varepsilon\right) \leq \varepsilon.$$

But this last inequality does not depend on N and so we may let $\varepsilon \to 0$ to obtain

$$P\left(\lim_{n \to \infty} \sup_{k,l \geq n} |X_k - X_l| > 0\right) = 0,$$

and so $\lim_k X_k(\omega)$ exists for almost all $\omega \in \Omega$. $\square$

Proposition 1.9. *A necessary and sufficient condition for $\{X_n\}$ to converge in probability is: for every $\varepsilon > 0$*

$$P(|X_m - X_n| > \varepsilon) \to 0 \quad \text{as} \quad m, n \to \infty.$$

OUTLINE OF PROOF. We show only the sufficiency of the condition. By assumption we can find a sequence $\{n(k)\}$ such that

$$P(|X_{n(k+1)} - X_{n(k)}| > 2^{-k}) < 2^{-k},$$

and so by Proposition 1.6 the series $\sum_k (X_{n(k+1)} - X_{n(k)})$ converges almost surely. But then $\lim_{k \to \infty} X_{n(k)}(\omega) = X(\omega)$ exists for almost every ω. Again using the hypothesis we can find an N for every $\varepsilon > 0$ such that for all $m, n \geq N$, $P(|X_m - X_n| > \varepsilon) < \varepsilon$. From these facts

$$P(|X_n - X| > \varepsilon) = P\left(\lim_{k \to \infty} |X_n - X_{n(k)}| > \varepsilon\right)$$

$$\leq \liminf_{k \to \infty} P(|X_n - X_{n(k)}| > \varepsilon) \leq \varepsilon,$$

and so X_n converges in probability to X as was to be proven. $\square$

(iv) Sums of Independent Random Variables

When the sequence $\{X_n\}$ is independent, the convergence of the series $\sum_n X_n$ has some striking aspects to it, one remarkable one being due to P. Lévy. To explain his theorem we need two preliminary lemmas. Write $\mathbf{B}^n = \mathbf{B}(X_k; k \geq n)$.

Lemma 1.4 (Kolmogorov's 0–1 law). *If an event A belongs to $\mathbf{B}^n$ for every $n \geq 1$, then $P(A)$ is either 0 or 1.*

Events satisfying the condition in this lemma are called *tail events*.

EXAMPLE. Suppose that $\{X_n\}$ is an independent sequence. Then the set of $\omega \in \Omega$ for which $\sum_n X_n(\omega)$ converges, and the set for which $\lim_n X_n(\omega) = a$ ($a \in \mathbf{R}$), are both tail events, and thus have probability 0 or 1.

Let us continue to suppose that $\{X_n\}$ is an independent sequence, and put $S_n = \sum_1^n X_k$.

Lemma 1.5 (Ottaviani's inequality). *If for every $k \leq n$ we have*

$$P(|S_n - S_k| > c) \leq \frac{1}{2},$$

where $c > 0$, then

$$P\left(\max_{1 \leq k \leq n} |S_k| > 2c\right) \leq 2P(|S_n| > c). \tag{1.44}$$

Theorem 1.9. *Suppose that $\{X_n\}$ is an independent sequence of random variables. Then almost sure convergence, convergence in probability and convergence in law of the series $\sum_n X_n$ are all equivalent.*

OUTLINE OF PROOF. Recalling the assertions, (a), (b) and (c) prior to Proposition 1.6, we see that it suffices to prove only

a. If $\sum_n X_n$ converges in law, then it converges in probability.
b. If $\sum_n X_n$ converges in probability, then it converges almost surely.

PROOF OF (a). Let $\varphi_n(z)$ be the characteristic function of X_n and set

$$\Phi_n(z) = \prod_{k=1}^n \varphi_k(z) \quad \text{and} \quad \Phi_{n,m}(z) = \prod_{k=n+1}^m \varphi_k(z), \quad m > n.$$

Then by assumption we have the existence of the limit

$$\lim_{n \to \infty} \Phi_n(z) = \Phi(z) \quad \text{(pointwise convergence)} \tag{1.45}$$

and the fact that $\Phi(z)$ is again a characteristic function. Now for $m > n$ we have $\Phi_n(z)\Phi_{n,m}(z) = \Phi_m(z)$, and since $\Phi(0) = 1$ implies that $\Phi(z) \neq 0$ in some neighbourhood of $z = 0$, the result

$$\Phi_{n,m}(z) \to 1 \quad \text{as} \quad n, m \to \infty, \tag{1.46}$$

follows from (1.45). To prove (1.46) for z away from 0 we use the inequality

$$|\Phi_{n,m}(z) - 1| \leq N\{2|\Phi_{n,m}(N^{-1}z) - 1|\}^{1/2}$$

valid for large N.

Suppose now that $\sum_n X_n$ was not convergent in probability. Then we could find $\varepsilon > 0$, $\{n(k)\}$ and $\{m(k)\}$ with $n(k) < m(k) < n(k+1)$ for all k, such that for every k

$$P(|S_{m(k)} - S_{n(k)}| > \varepsilon) \geq \varepsilon. \tag{1.47}$$

Obviously (1.47) contradicts (1.46) and so we must have $\sum_n X_n$ converging in probability.

PROOF OF (b). It follows from the hypothesis that $P(|S_m - S_n| > \varepsilon) \to 0$ as $m, n \to \infty$, and hence for sufficiently large m, n

$$P(|S_m - S_n| > \varepsilon) \le \frac{1}{2}.$$

Applying Lemma 1.5 to the random variables $X_{n+1}, X_{n+2}, \ldots, X_m$ gives

$$P\left(\max_{n \le k \le m} |S_k - S_n| > 2\varepsilon\right) \le 2P(|S_m - S_n| > \varepsilon) \to 0$$

as $m, n \to \infty$, which proves that $\sum_n X_n$ converges almost surely. □

Corollary. *Let $\{X_n\}$ be an independent sequence such that for all $n \ge 1$, $V(X_n) < \infty$. If both $\sum_n E(X_n)$ and $\sum_n V(X_n)$ converge, then $\sum_n X_n$ converges almost surely.*

EXAMPLE. Let $\{X_n\}$ be the independent sequence with $P(X_n = 1) = P(X_n = -1) = \frac{1}{2}$ for all n. Then the above corollary implies that $\sum_n n^{-1} X_n$ converges almost surely.

1.6 Gaussian Systems

A real-valued random variable defined on probability space is said to be *Gaussian* if its distribution function $F(x)$ is of the form

$$F(x) = (2\pi\sigma^2)^{-1/2} \int_{-\infty}^{x} \exp\left[-\frac{(y-m)^2}{2\sigma^2}\right] dy \tag{1.48}$$

where σ^2 and m are constants. More precisely such a random variable $X(\omega)$ follows a Gaussian distribution with mean m and variance σ^2. Equivalently, X has a characteristic function of the form

$$E\{\exp[izX]\} = \exp\left[imz - \frac{1}{2}z^2\sigma^2\right], \quad z \in \mathbf{R}. \tag{1.49}$$

Definition 1.6. A system $\mathbf{X} = \{X_\lambda(\omega): \lambda \in \Lambda\}$ of real-valued random variables is said to be *Gaussian*, and $\mathbf{X}$ is a *Gaussian system*, if any finite real linear combination of elements X_λ of $\mathbf{X}$ has a Gaussian distribution.

Any subsystem of a Gaussian system is again a Gaussian system. In particular the joint distribution of $(X_1, X_2, \ldots, X_n)$ any finite subsystem $\{X_j: 1 \le j \le n\}$ of a Gaussian system $\mathbf{X}$ is a multi-dimensional Gaussian distribution. This distribution extends over the whole of $\mathbf{R}^n$ or over a hyperplane (at most $(n-1)$-dimensional). We can obtain details of this distribution in the following way. Since any linear combination $\sum_{j=1}^{n} z_j X_j$ has a Gaussian distribution, the characteristic function of $(X_1, X_2, \ldots, X_n)$ must be of the form

$$E\left\{\exp\left(i\sum_{j=1}^{n} z_j X_j\right)\right\} = \exp\left[im(z) - \frac{1}{2}\sigma(z)^2\right], \quad (1.50)$$

where $z = (z_1, z_2, \ldots, z_n)$, and

$$m(z) = \sum_{j=1}^{n} z_j E(X_j), \quad \sigma(z)^2 = E\left(\sum_{j=1}^{n} z_j X_j\right)^2 - m(z)^2.$$

Let m_j be the expectation of X_j and let $V = (V_{jk})$ be the (positive definite) covariance matrix of $\{X_j\}$ given by

$$V_{j,k} = E\{(X_j - m_j)(X_k - m_k)\}, \quad 1 \le j, k \le n.$$

Suppose that the rank of the matrix V is r. Then the distribution of $(X_1, X_2, \ldots, X_n)$ is concentrated on an r-dimensional hyperplane of $\mathbf{R}^n$

$$m + V\mathbf{R}^n, \quad m = (m_1, m_2, \ldots, m_n).$$

In case V is non-degenerate, that is, in the case $r = n$, the distribution is supported by the whole of $\mathbf{R}^n$ and has density function of the form

$$(2\pi)^{-n/2} |V|^{-1/2} \exp\left[-\frac{1}{2}(x-m)V^{-1}(x-m)'\right], \quad (1.51)$$

where $x = (x_1, x_2, \ldots, x_n) \in \mathbf{R}^n$, $|V|$ and V^{-1} are the determinant and inverse respectively, of V, and where $(x - m)'$ denotes the (column) vector transpose to the (row) vector $(x - m)$. The expression (1.51) is the general form of density of an n-dimensional Gaussian distribution. The characteristic function $\varphi(z)$ of this distribution is given by

$$\varphi(z) = \exp\left[i(m, z) - \frac{1}{2}(Vz, z)\right], \quad z = (z_1, z_2, \ldots, z_n) \in \mathbf{R}^n, \quad (1.52)$$

where $(\cdot, \cdot)$ is the inner product on $\mathbf{R}^n$.

We now return to a general Gaussian system $\mathbf{X} = \{X_\lambda : \lambda \in \Lambda\}$. For such a system we are given a

mean vector $m_\lambda = E(X_\lambda), \quad \lambda \in \Lambda$;

and

covariance matrix $V_{\lambda, \mu} = E\{(X_\lambda - m_\lambda)(X_\mu - m_\mu)\}, \quad \lambda, \mu \in \Lambda$.

The latter is positive definite, that is, for any $n \ge 1$, complex numbers $\alpha_1, \alpha_2, \ldots, \alpha_n \in \mathbf{C}$, and $\lambda_1, \lambda_2, \ldots, \lambda_n \in \Lambda$

$$\sum_{j,k=1}^{n} \alpha_j \bar{\alpha}_k V_{\lambda_j, \lambda_k} \ge 0. \quad (1.53)$$

Theorem 1.10. *Given $(m_\lambda : \lambda \in \Lambda)$, and a real positive definite $V = (V_{\lambda, \mu} : \lambda, \mu \in \Lambda)$, there exists a Gaussian system $\mathbf{X} = \{X_\lambda : \lambda \in \Lambda\}$, the mean vector and covar-*

1.6 Gaussian Systems

iance matrix of which coincide with (m_λ) and $V = (V_{\lambda,\mu})$ respectively. If there exists another system $\mathbf{X}'$ with the same property as $\mathbf{X}$, then $\mathbf{X}$ and $\mathbf{X}'$ have the same distribution.

OUTLINE OF PROOF. The proof begins with the construction of a probability space, by putting

$$\Omega = \mathbf{R}^\Lambda = \{\omega = (\omega_\lambda): \omega_\lambda \in \mathbf{R}, \lambda \in \Lambda\}.$$

For any finite number of $\lambda_j, j = 1, 2, \ldots, n$, we now take a Gaussian distribution with characteristic function

$$\exp\left[i \sum_{j=1}^n m_{\lambda_j} z_j - \frac{1}{2} \sum_{j,k=1}^n V_{\lambda_j, \lambda_k} z_j z_k\right], \quad (z_1, z_2, \ldots, z_n) \in \mathbf{R}^n, \quad (1.54)$$

and obtain a probability distribution $P_{\lambda_1, \lambda_2, \ldots, \lambda_n}$ on Ω in a manner similar to that found in §1.3 (ii). In other words, we introduce a probability measure in such a way that by setting $X_\lambda(\omega) = \omega_\lambda$, the distribution of $(X_{\lambda_1}, X_{\lambda_2}, \ldots, X_{\lambda_n})$ is Gaussian with characteristic function (1.54). The σ-field $\mathbf{B}$ is defined to be that generated by all cylinder sets of $\mathbf{R}^\Lambda$, and then the system $\{P_{\lambda_1, \lambda_2, \ldots, \lambda_n}\}$ of probability measures obtained as $(\lambda_1, \lambda_2, \ldots, \lambda_n)$ runs over all finite subsets of Λ defines a probability measure P on $(\Omega, \mathbf{B})$. This last step uses the Kolmogorov extension theorem (§1.3) and the Gaussian system $\mathbf{X} = \{X_\lambda(\omega): \lambda \in \Lambda\}$ so obtained on $(\Omega, \mathbf{B}, P)$ is precisely that which we wished to construct.

Suppose now that we are given another Gaussian system $\mathbf{X}' = \{X'_\lambda(\omega): \lambda \in \Lambda\}$ with

$$E(X'_\lambda) = m_\lambda, \qquad E\{(X'_\lambda - m_\lambda)(X'_\mu - m_\mu)\} = V_{\lambda, \mu}.$$

Since the distribution of $(X'_{\lambda_1}, X'_{\lambda_2}, \ldots, X'_{\lambda_n})$ is Gaussian, it is uniquely determined by $m_{\lambda_j}, j = 1, 2, \ldots, n$, and $V_{\lambda_j, \lambda_k}, 1 \le j, k \le n$, and hence must coincide with $P_{\lambda_1, \lambda_2, \ldots, \lambda_n}$. But this means that the distribution of $\mathbf{X}'$ is also the measure P just introduced on $(\Omega, \mathbf{B})$, and so our second assertion is proved. □

The distribution of $\mathbf{X}$ obtained above will be denoted by $N(m, V)$ where $m = (m_\lambda: \lambda \in \Lambda)$ and $V = (V_{\lambda, \mu}: \lambda, \mu \in \Lambda)$.

Let Λ be a finite set, say $\Lambda = \{1, 2, \ldots, n\}$. If the Gaussian random variables $X_j, 1 \le j \le n$, are independent, then the system $\mathbf{X} - \{X_j: 1 \le j \le n\}$ is obviously Gaussian. In particular, if $m_j = 0, 1 \le j \le n$, and $V_{j,k} = \delta_{jk}, 1 \le j, k \le n$, then the distribution is $N(0, E)$ where E is the identity matrix, and is called the n-dimensional standard Gaussian distribution.

A Gaussian system $\mathbf{X} = \{X_\lambda: \lambda \in \Lambda\}$ has many notable properties, and we will state some of the more important properties needed in later chapters, although the proofs are only given in outline.

Proposition 1.10. *Let* $\mathbf{X} = \{X_\lambda : \lambda \in \Lambda\}$ *be a Gaussian system. Then*

a. *a necessary and sufficient condition for* X_λ, $\lambda \in \Lambda$, *to be independent is that for every* $\lambda \neq \mu$

$$V_{\lambda, \mu} = 0; \tag{1.55}$$

b. *a necessary and sufficient condition for a member* X_{λ_0} *of the system to be independent of* $\{X_\lambda : \lambda \in \Lambda, \lambda \neq \lambda_0\}$ *is that for every* $\lambda \neq \lambda_0$,

$$V_{\lambda, \lambda_0} = 0. \tag{1.56}$$

OUTLINE OF PROOF

(a). Take any finite number of the X_λ, say X_{λ_1}, X_{λ_2}, ..., X_{λ_n}. Then the joint distribution of this set is Gaussian and (1.55) states that each covariance vanishes. Upon substituting V_{λ_j, λ_k} given by (1.55) into (1.54), we see that X_{λ_j}, $1 \leq j \leq n$, are mutually independent. We have therefore proved that X_λ, $\lambda \in \Lambda$, is an independent system. The converse is obvious.

(b). can be proved in a manner similar to that just given for (a). □

Proposition 1.11. *For a Gaussian system* $\mathbf{X} = \{X_n : n \geq 1\}$ *the convergence in probability of the sequence* $\{X_n\}$ *is equivalent to convergence in mean square (abbrev. m.s. convergence). The limit* X_∞ *of the sequence* $\{X_n\}$ *in this case is also a Gaussian random variable.*

OUTLINE OF PROOF. In general mean square convergence implies convergence in probability and so we need only prove the converse. Put

$$E(X_j - X_k) = m_{j,k}, \qquad V(X_j - X_k) = \sigma^2_{j,k}.$$

A necessary and sufficient condition for $\{X_n\}$ to converge in mean square is that

$$E\{(X_j - X_k)^2\} = \sigma^2_{j,k} + m^2_{j,k} \to 0 \quad \text{as} \quad j, k \to \infty,$$

and so if $\{X_n\}$ is not convergent in mean square, we would have

$$\limsup_{j,k \to \infty} (\sigma^2_{j,k} + m^2_{j,k}) > 0. \tag{1.57}$$

Now

$$P(|X_j - X_k| > \varepsilon) = \int_{|x| > \varepsilon} (2\pi\sigma^2_{j,k})^{-1/2} \exp\left[-\frac{(x - m_{j,k})^2}{2\sigma^2_{j,k}}\right] dx$$

and by (1.57), $\sigma^2_{j,k}$ and $m^2_{j,k}$ do not approach 0 simultaneously, so that for sufficiently small $\varepsilon > 0$ we have

$$\limsup_{j,k \to \infty} P(|X_j - X_k| > \varepsilon) \geq \frac{1}{2}.$$

But this means that $\{X_n\}$ does not converge in probability, a contradiction.

Convergence of $\{X_n\}$ in mean square is nothing but strong convergence in

1.6 Gaussian Systems

the Hilbert space $L^2(\Omega, \mathbf{B}, P)$ and the limit X_∞ can thus be found in $L^2(\Omega, \mathbf{B}, P)$. The inner product $E(X_n) = m_n$ of X_n and 1, and the squared norm $V(X_n) = \sigma_n^2$ of $X_n - m_n$ both converge to limits, say m and σ^2 respectively. Thus the characteristic function of X_n converges to that of $N(m, \sigma^2)$ and this limit function has to be the characteristic function of X_∞. This proves the second half of the proposition. □

Since it is generally true that almost sure convergence implies convergence in probability, we obtain

Corollary 1. *For a Gaussian system* $\{X_n, n \geq 1\}$, *the almost sure convergence of the sequence implies its convergence in mean square.*

Given a Gaussian system $\mathbf{X}$ it follows from the definition that the union of $\mathbf{X}$ together with any linear combination of elements of $\mathbf{X}$ is again a Gaussian system. Moreover we have

Corollary 2. *Let* $\mathbf{X}$ *be a Gaussian system. Then the closed linear subspace* $\bar{\mathbf{X}}$ *of* $L^2(\Omega, \mathbf{B}, P)$ *spanned by* $\mathbf{X}$ *is also a Gaussian system.*

OUTLINE OF PROOF. We may suppose that the given Gaussian system $\mathbf{X}$ forms a vector space, for if not we may add to it all necessary finite linear combinations and still preserve its Gaussian nature. Now take $X_j \in \bar{\mathbf{X}}$, $1 \leq j \leq n$, expressible as mean square limits

$$\lim_{n \to \infty} X_j^{(n)} = X_j, \qquad X_j^{(n)} \in \mathbf{X}.$$

Clearly any (finite) linear combination $\sum_{j=1}^m \alpha_j X_j$ is expressible as a limit

$$\lim_{n \to \infty} \sum_{j=1}^m \alpha_j X_j^{(n)} = \sum_{j=1}^m \alpha_j X_j, \qquad \sum_{j=1}^m \alpha_j X_j^{(n)} \in \mathbf{X},$$

and the second half of Proposition 1.11 tells us that the above limit is again a Gaussian random variable. This shows $\bar{\mathbf{X}}$ is Gaussian. □

Now we take a subset $\Lambda' \subseteq \Lambda$ and consider the subsystem $\mathbf{X}' = \{X_{\lambda'} : \lambda' \in \Lambda'\}$ of the Gaussian system $\mathbf{X} = \{X_\lambda : \lambda \in \Lambda\}$. Of course $\mathbf{X}'$ is again a Gaussian system, and we let $\mathbf{B}'$ be the smallest σ-field with respect to which all members of $\mathbf{X}'$ are measurable. For simplicity we suppose that the closed linear subspace $\bar{\mathbf{X}}'$ of $L^2(\Omega, \mathbf{B}, P)$ is separable.

Proposition 1.12. *For any element* X_λ *of the Gaussian system* $\mathbf{X}$ *we have*

$$E(X_\lambda | \mathbf{B}') = \text{the orthogonal projection of } X_\lambda \text{ onto } \bar{\mathbf{X}}'. \quad (1.58)$$

Remark. The conditional expectation $E(X_\lambda | \mathbf{B}')$ was shown in Proposition 1.4 to be the orthogonal projection of X_λ onto $L^2(\Omega, \mathbf{B}', P)$, but (1.58) asserts that this projection is onto a much smaller subspace, namely $\bar{\mathbf{X}}'$.

OUTLINE OF PROOF. Since $\bar{\mathbf{X}}'$ is separable, we may choose a sequence $\{X_{\lambda_n}\} \subset \bar{\mathbf{X}}'$ such that if $\mathbf{B}_n$ is the σ-field generated by $X_{\lambda_1}, X_{\lambda_2}, \ldots, X_{\lambda_n}$, then

$$\mathbf{B}_n \to \mathbf{B}' \quad \text{as} \quad n \to \infty.$$

For any $X_\lambda \in \mathbf{X}$ there exists a Gaussian random variable Y_n such that Y_n is independent of $\mathbf{B}_n$ and X_λ is expressed in the form

$$X_\lambda = \sum_{j=1}^n a_j X_{\lambda_j} + Y_n,$$

this following from the fact that $(X_\lambda, X_{\lambda_1}, X_{\lambda_2}, \ldots, X_{\lambda_n})$ is Gaussian. Let us write the sum $\sum_1^n a_j X_{\lambda_j}$ above by $X_\lambda^{(n)}$, and observe that it is $\mathbf{B}_n$-measurable and satisfies

$$E(X_\lambda | \mathbf{B}_n) = X_\lambda^{(n)}.$$

However $X_\lambda^{(n)}$ may be viewed as the orthogonal projection of X_λ onto the subspace spanned by $\{X_{\lambda_j}, j = 1, 2, \ldots, n\}$ and we may now let $n \to \infty$, use the corollary to Proposition 1.4 and the facts above, to complete the proof of our assertion. □

Continuing the above notation, we can easily use Proposition 1.12 to prove the following.

Corollary. *The collection $\{E(X_\lambda | \mathbf{B}'): \lambda \in \Lambda\}$ is also a Gaussian system on $(\Omega, \mathbf{B}, P)$.*

If the stochastic process $\{X(t): t \in T\}$ is a Gaussian system, then it is also a Gaussian process in the sense of §1.3. Paraphrasing the second half of Theorem 1.10, we can say that the distribution of a Gaussian process is uniquely determined by its mean vector and covariance function. This leads us to

Proposition 1.13. *If a Gaussian process is weakly stationary, then it is strictly stationary.*

1.7 Characterisations of Gaussian Distributions

As we have seen in the last section Gaussian distributions and Gaussian random variables possess many interesting properties. This section presents a converse approach, offering several kinds of characterisations of Gaussian random variables and distributions. Such characterisations are of interest in themselves, and in addition have important applications. We begin with one-dimensional Gaussian distributions.

a. In taking the logarithm log $\varphi(z)$ of a characteristic function we choose

1.7 Characterisations of Gaussian Distributions

a branch such that $\log \varphi(z) = 0$ at $z = 0$. If such a logarithm is expressed by a Taylor series

$$\log \varphi(z) = \sum_{k=1}^{N} \frac{(iz)^k \gamma_k}{k!} + o(|z|^N)$$

for some N, then the coefficient γ_k is real and is called the *k-th order cumulant* (semi-invariant). In general if the k-th order moment m_k of a distribution exists, then so does γ_k, and it can be expressed in terms of moments of order up to and including k. For example

$$\gamma_1 = m_1 \ (= \text{the expectation}), \qquad \gamma_2 = m_2 - m_1^2 \ (= \text{the variance}),$$

$$\gamma_3 = m_3 - 3m_1 m_2 + 2m_1^3, \qquad \text{and so on}.$$

Proposition 1.14. *A necessary and sufficient condition for a distribution to be Gaussian is that cumulants γ_k of all orders exist, and satisfy*

$$\gamma_k = 0, \qquad k \geq 3. \tag{1.59}$$

Note. The Poisson distribution is also characterised by a simple condition on cumulants, namely that for all $k \geq 1$,

$$\gamma_k = \gamma \ (\text{constant}).$$

b. When a probability distribution is absolutely continuous with respect to Lebesgue measure we can introduce a type of information measure $H(p)$ defined in terms of its density $p(x)$ by

$$H(p) = -\int_{-\infty}^{\infty} p(x) \log p(x) \, dx, \tag{1.60}$$

where $0 \log 0$ is taken by convention to be 0. The function $H(p)$ is called the *differential entropy*, or just the *entropy* of the distribution with density $p(x)$.

Proposition 1.15. *If p is restricted to vary over the set of all densities of distributions on $\mathbf{R}$ having a fixed finite variance σ^2, then the maximum value of $H(p)$ is $\log \sigma + \frac{1}{2} \log 2\pi e$ and occurs for the Gaussian density with variance σ^2.*

PROOF. The following relation holds for any pair of densities $p(x)$ and $q(x)$, and is a consequence of Jensen's inequality

$$\int_{-\infty}^{\infty} p(x) \log p(x) \, dx \geq \int_{-\infty}^{\infty} p(x) \log q(x) \, dx. \tag{1.61}$$

Now let $p(x)$ be an arbitrary density with variance σ^2 and let $q(x)$ be the density of a Gaussian distribution with variance σ^2 and the same expectation m as $p(x)$, i.e.

$$q(x) = (2\pi\sigma^2)^{-1/2} \exp\left[-\frac{(x-m)^2}{2\sigma^2}\right].$$

By the inequality (1.61) we have

$$H(p) \leq -\int_{-\infty}^{\infty} p(x) \left\{ -\log(2\pi\sigma^2)^{1/2} - \frac{(x-m)^2}{2\sigma^2} \right\} dx$$

$$= \frac{1}{2} \log(2\pi\sigma^2) + \frac{1}{2},$$

whilst $H(q)$ actually coincides with the last expression on the right. □

Remark. It is not hard to see from the definition of $H(p)$ that it is invariant under translations of p and so, in particular, it does not depend on the expectation of the distribution.

c. Gaussian distributions have a self-reproducing property: if X_1 and X_2 are independent Gaussian random variables, then their sum $S = X_1 + X_2$ is also Gaussian (and the expectation and variance of S is the sum of the expectations and variances of X_1 and X_2 respectively). It is a surprising and striking fact that a converse to this property holds: if the sum S is Gaussian, then X_1 and X_2 are both Gaussian as well. The following theorem, which states this result more precisely, was originally conjectured by P. Lévy and was proved in 1936 by H. Cramér.

Theorem 1.11. *Let X_1 and X_2 be independent random variables. If their sum $S = X_1 + X_2$ is Gaussian, then both X_1 and X_2 are Gaussian.*

PROOF. Let F_1 and F_2 be the distribution functions of X_1 and X_2 respectively. Since S has a Gaussian distribution with expectation m and variance σ^2 say, by our assumptions the relation $S = X_1 + X_2$ implies that

$$\Phi\left(\frac{x-m}{\sigma}\right) = \int_{-\infty}^{\infty} F_1(x-y) \, dF_2(y),$$

where Φ is the standard Gaussian distribution function. Replacing x with $x + y$ and using the monotonicity of $F_1(x)$ in x we get

$$\Phi\left(\frac{x+y-m}{\sigma}\right) \geq \int_{-\infty}^{y} F_1(x+y-u) \, dF_2(u)$$

$$\geq F_1(x) \int_{-\infty}^{y} dF_2(u) = F_1(x) F_2(y).$$

Fix y such that $F_2(y) > 0$, and use the inequality

$$1 - \Phi(x) + \Phi(-x) < \left(\frac{2}{\pi}\right)^{1/2} x^{-1} \exp\left(-\frac{1}{2} x^2\right)$$

valid for sufficiently large x, and we see that there exists positive constants A and B such that for all x

$$1 - F_1(x) + F_1(-x) < A \exp\left(-\frac{x^2}{2\sigma^2} + B|x|\right).$$

Integrating this inequality by parts proves that

$$J = \int_{-\infty}^{\infty} \exp\left(\frac{x^2}{4\sigma^2}\right) dF_1(x) < \infty$$

and thus that the characteristic function $\varphi_1(z)$ of F_1 is an entire function. Further, the evaluation

$$|\varphi_1(z)| \le \int \exp\left[\sigma^2|z|^2 + \frac{x^2}{4\sigma^2}\right] dF_1(x)$$

$$= J \exp(\sigma^2|z|^2)$$

tells us that the order of φ_1 is 2. Exactly the same results for the characteristic function $\varphi_2(z)$ of $F_2(x)$ follow by symmetry. Now our assumption implies that

$$\varphi_1(z)\varphi_2(z) = \exp\left(imz - \frac{1}{2}\sigma^2 z^2\right)$$

so that neither φ_1 nor φ_2 have any zeros. We can thus appeal to the Hadamard theorem on entire functions and obtain the following expressions for φ_j, $j = 1, 2$:

$$\varphi_j(z) = \exp[\alpha_j z^2 + \beta_j z + \gamma_j], \quad j = 1, 2. \quad (1.62)$$

At this point we recall the general properties of characteristic functions,

$$\overline{\varphi_j(z)} = \varphi_j(-z), \quad |\varphi_j(z)| \le 1, \quad j = 1, 2, \quad z \in \mathbf{R},$$

and the resulting restrictions on α_j, β_j, and γ_j lead us to the expressions

$$\varphi_j(z) = \exp\left[im_j z - \frac{1}{2}\sigma_j^2 z^2\right], \quad j = 1, 2,$$

which were to be proved. It is clear that $m = m_1 + m_2$ and $\sigma^2 = \sigma_1^2 + \sigma_2^2$.

□

d. Let X_1 and X_2 be independent Gaussian random variables. For a suitable choice of constants a, b, c and d the two linear forms

$$Y_1 = aX_1 + bX_2$$
$$Y_2 = cX_1 + dX_2$$

become independent. Such a property is typical of a Gaussian system, and with this background we turn to the following proposition due to M. Kac (1940).

Proposition 1.16. *Assume that the random variables X_1 and X_2 are independent. If the random variables Y_1 and Y_2 are independent for all θ, where*

$$Y_1 = X_1 \cos \theta + X_2 \sin \theta$$
$$Y_2 = X_1 \sin \theta - X_2 \cos \theta, \quad (1.63)$$

then X_1 and X_2 have the same Gaussian distribution with expectation 0 (and so also do Y_1 and Y_2).

Remark 1. Even if Y_1 and Y_2 are independent only for some θ not a multiple of $\frac{1}{2}\pi$ we can still prove that X_1 and X_2 are Gaussian (see Theorem 1.12 below).

Remark 2. In his investigation of the velocity distribution of molecules moving in $\mathbf{R}^3$, Maxwell assumed that the three velocity components behave like Gaussian random variables with expectation 0 for all orthogonal coordinate systems. If we use the above proposition, suitably generalised to three variables, we see that the three components must follow the same Gaussian distribution. This is also consistent with the conclusion in Example 3 of §1.2.

Theorem 1.12. *Suppose that X_1 and X_2 are independent random variables. If the random variables Y_1 and Y_2 given by*

$$Y_1 = aX_1 + bX_2$$
$$Y_2 = cX_1 + dX_2, \quad ad - bc \neq 0, \quad (1.64)$$

are independent, then, apart from the trivial case $b = c = 0$ or $a = d = 0$, both X_1 and X_2 are Gaussian.

Before proving this theorem we need a lemma.

Lemma 1.6. *Let f_1 and f_2 be non-constant continuous real-valued functions, and let a, b, c and d be constants such that $ad - bc \neq 0$. If there exist functions g_1 and g_2 such that*

$$f_1(ax + by)f_2(cx + dy) = g_1(x)g_2(y) \quad x, y \in \mathbf{R} \quad (1.65)$$

then $\log |f_j|$, $\log |g_j|$, $j = 1, 2$, are all polynomials of degree at most two.

PROOF. The functions g_1 and g_2 are clearly continuous, and since $ad - bc \neq 0$ we can think of the linear forms $ax + by$ and $cx + dy$ as independent variables. We further see that none of the functions, f_1, f_2, g_1 and g_2 vanish, and hence may be supposed always positive. Let us put

$$\log f_j = v_j, \quad \log g_j = w_j, \quad j = 1, 2.$$

Then (1.65) becomes

$$v_1(ax + by) + v_2(cx + dy) = w_1(x) + w_2(y). \quad (1.66)$$

1.7 Characterisations of Gaussian Distributions

Since v_1, v_2 and w_2 are locally integrable they can be convolved, viewed as functions of y, with a function in the space $\mathscr{D}$, and so we see that w_1 is a $\mathscr{C}^\infty$-function. Similarly for w_2. Now the assumption $ad - bc \neq 0$ shows that $ax + by$ and $cx + dy$ may be taken as new independent variables, and so the relation (1.66) is in fact symmetrical in the v_j and w_j ($j = 1, 2$). Thus we have proved that v_1 and v_2 are also $\mathscr{C}^\infty$-functions.

Returning to (1.66) a final time, we now take partial derivatives with respect to x and y and obtain the equation:

$$abv_1''(ax + by) + cdv_2''(cx + dy) = 0.$$

Assuming that $ab \neq 0$ and $cd \neq 0$ we see that v_1 and v_2 are polynomials of degree at most 2 and the lemma is proved. □

PROOF OF THEOREM 1.12.

1. We first prove the theorem under the additional assumption that the distributions of X_1 and X_2 are both symmetric with respect to the origin. Since Y_1 and Y_2 are independent we have

$$E\{\exp[i(z_1 Y_1 + z_2 Y_2)]\} = E\{\exp[iz_1 Y_1]\} E\{\exp[iz_2 Y_2]\}.$$

If we now let the characteristic functions of X_1, X_2, Y_1, Y_2 be denoted by φ_1, φ_2, ψ_1, ψ_2 respectively, we can use (1.64) and see that the above equation becomes

$$\varphi_1(az_1 + cz_2)\varphi_2(bz_1 + dz_2) = \psi_1(z_1)\psi_2(z_2).$$

Under the assumption that the distributions are symmetric, these characteristic functions are real-valued and, of course, continuous. Under the further assumption $abcd \neq 0$ we can use Lemma 1.6 (with b and c interchanged) and obtain

$$\varphi_j(z) = \exp[\alpha_j z^2 + \beta_j z + \gamma_j], \quad j = 1, 2.$$

The general condition $\varphi_j(0) = 1$ implies that $\gamma_j = 0$ whilst $\beta_j = 0$ follows from the symmetry property. Furthermore $|\varphi_j(z)| \leq 1$ implies that we can write $\alpha_j = -\frac{1}{2}\sigma^2$, and thus it has been proved that X_1 and X_2 have Gaussian distributions with expectations 0. In the case $ab = 0$ (resp. $cd = 0$), X_2 (resp. X_1) is trivial, that is, identically zero.

2. The general case, in which the distributions of X_1 and X_2 are not necessarily symmetric, can be reduced to the symmetric case by a trick explained below.

Let X_1^- and X_2^- be independent random variables with distributions the same as $-X_1$ and $-X_2$ respectively, such that the pairs (X_1, X_2) and (X_1^-, X_2^-) are independent. Setting

$$X_j^0 = X_j + X_j^-, \quad j = 1, 2,$$

we see that the distribution of X_j^0 is symmetric ($j = 1, 2$). Since Y_1 and Y_2 are independent, so also are $aX_1^- + bX_2^-$ and $cX_1^- + dX_2^-$. Therefore the symmetrised random variables

$$Y_1^0 = aX_1^0 + bX_2^0 = (aX_1 + bX_2) + (aX_1^- + bX_2^-)$$
$$Y_2^0 = cX_1^0 + dX_2^0 = (cX_1 + dX_2) + (cX_1^- + dX_2^-)$$

are independent. We can now apply part (1) above to deduce that X_1^0 and X_2^0 are both Gaussian, and conclude the proof by applying Theorem 1.11 to the sums X_1^0 and X_2^0 of independent random variables, finally deducing that X_1 and X_2 are both Gaussian. □

There have been several generalisations of this theorem, two of which we present below.

Proposition 1.17 (see Yu. V. Linnik, 1964). *Let $X_1, X_2, \ldots, X_n$ be an independent system of random variables. If the random variables Y_1 and Y_2 given by*

$$Y_1 = \sum_{j=1}^{n} a_j X_j$$

$$Y_2 = \sum_{j=1}^{n} b_j X_j$$

are independent, then each of the X_j such that $a_j b_j \neq 0$ must be subject to a Gaussian distribution.

Proposition 1.18 (see E. Lukacs, 1970). *Let $\{X_n\}$ be an independent sequence of identically distributed random variables and let $\{a_n\}$ and $\{b_n\}$ be two sequences of real numbers such that $\sum_{n=1}^{\infty} |a_n b_n| \neq 0$ and that one of the sums $\sum_{n=1}^{\infty} a_n^2/b_n^2$ or $\sum_{n=1}^{\infty} b_n^2/a_n^2$ converges. Suppose further that the sums $Y_1 = \sum_{n=1}^{\infty} a_n X_n$ and $Y_2 = \sum_{n=1}^{\infty} b_n X_n$ both exist and the Y_1 and Y_2 are independent. Then the common distribution of the X_j is Gaussian.*

e. The next theorem was proved by P. Lévy in 1957.

Theorem 1.13. *Let X and Y be two random variables such that there exist random variables U independent of X and V independent of Y satisfying*

$$Y = aX + U \tag{1.67}$$
$$X = bY + V,$$

for constants a and b. Then we may conclude
  i. *(X, Y) is Gaussian.*
  ii. *X and Y are independent.*
  iii. *There exists an affine relation between X and Y.*

The proof of this theorem is like that in Theorem 1.12 in that it involves solving a functional equation similar to (1.66). To avoid repeating such arguments we omit the details.

1.7 Characterisations of Gaussian Distributions

Although this theorem is similar in form to Theorem 1.12, it does have a deeper significance for the linear theory of stochastic processes, for example in the theory of linear prediction for stationary processes. Indeed the theorem determines a class of stochastic processes to which the linear operations can effectively be applied.

f. Finally we note an easy assertion which, although started under rather restrictive assumptions, does suggest an approach to certain distributions in infinite-dimensional spaces (see §5.6).

Proposition 1.19. *If the distribution of an n-dimensional random variable $(X_1, X_2, \ldots, X_n)$ is invariant under all rotations $(u_{j,k}: 1 \leq j, k \leq n)$ about the origin, [i.e. $(u_{j,k}) \in SO(n)$], and if the n random variables*

$$Y_j = \sum_{k=1}^{n} u_{j,k} X_k, \qquad j = 1, 2, \ldots, n,$$

are mutually independent, then for some constant c the random vector $(cX_1, cX_2, \ldots, cX_n)$ has an n-dimensional standard Gaussian distribution (cf. Proposition 1.16).

2 Brownian Motion

We begin this chapter with the definition of Brownian motion and a proof that its distribution is supported by the space of continuous functions (§2.1), and then go on to deal with important aspects of Brownian motion such as its sample path (§2.2) and Markov properties (§2.4). It is through these discussions that we can appreciate the place of Brownian motion within the class of all stochastic processes and, in particular, Gaussian processes. Two methods of constructing Brownian motion will be presented (§2.3), each of which is significant in its own right, and which also exhibits the ideas underlying constructions relevant to later chapters. Markov properties will only be touched upon briefly (§§2.4–2.6), but, hopefully, enough to enable a close connection with analysis to be seen.

2.1 Brownian Motion. Wiener Measure

Let us begin with the definition of Brownian motion.

Definition 2.1. A stochastic process $\{B(t, \omega): t \geq 0\}$ defined on a probability space $(\Omega, \mathbf{B}, P)$ is called a *Brownian motion* or a *Wiener process* if it satisfies:

a. $B(0, \omega) = 0$ for almost all ω; and
b. the system $\{B(t, \omega): t \geq 0\}$ is Gaussian on $(\Omega, \mathbf{B}, P)$, and for any t and h with $t + h > 0$, $B(t + h, \omega) - B(t, \omega)$ has expectation 0 and variance $|h|$.

The parameter t denotes time and usually extends over the interval $[0, 1]$ or $[0, \infty)$. Postponing a discussion of the existence of such a process for the moment, we state some immediate consequences of the definition.

1. In order to determine the distribution of $\{B(t): t \geq 0\}$ it suffices to know only the covariance function. This is because the process is Gaussian and has identically zero mean. Taking expectations of both sides of the identity

$$B(t)B(s) = \frac{1}{2}\{B(t)^2 + B(s)^2 - [B(t) - B(s)]^2\}$$

and using (b) gives us

$$E\{B(t)B(s)\} = \min\{t, s\}(= t \wedge s). \tag{2.1}$$

On the other hand, (2.1) together with $E\{B(t)\} \equiv 0$ implies the second half of (b) in Definition 2.1.

2. For $a < b \leq c < d$ the random vector $(B(b) - B(a), B(d) - B(c))$ is governed by a bivariate Gaussian distribution, by (b), with mean $(0, 0)$; moreover

$$E\{[B(b) - B(a)][B(d) - B(c)]\} = b \wedge d - b \wedge c - a \wedge d + a \wedge c = 0,$$

and so $B(b) - B(a)$ and $B(d) - B(c)$ are independent. More generally for any finite number of time instants $t_1, t_2, \ldots, t_n$ all prior to a, i.e. $t_i \leq a$, $i = 1, 2, \ldots, n$, $B(b) - B(a)$ is independent of $\{B(t_i): 1 \leq i \leq n\}$. Intuitively speaking, the increment $B(b) - B(a)$ of $B(t)$ over $[a, b]$ is independent of its past prior to a, where $\{B(s): s \leq a\}$ is the set of past values of $B(t)$ relative to the present instant $t = a$. In this sense $\{B(t): t \geq 0\}$ is an *additive process*. This fact, together with (b) of Definition 2.1, allows us to determine the distribution of the random vector $(B(t_1), B(t_2), \ldots, B(t_n))$ for any $t_1, t_2, \ldots, t_n$ with $0 < t_1 < t_2 < \cdots < t_n$. [Alternatively the fact that this distribution is Gaussian means that it is determined by the covariance function (2.1) above]. In any event the distribution has density function

$$\prod_{k=1}^{n} \{2\pi(t_k - t_{k-1})\}^{-1/2} \exp\left[\frac{-(x_k - x_{k-1})^2}{2(t_k - t_{k-1})}\right] \tag{2.2}$$

where we set $t_0 = x_0 = 0$. We write this density function

$$g(t_1, t_2 - t_1, \ldots, t_n - t_{n-1}; x_1, x_2 - x_1, \ldots, x_n - x_{n-1})$$

and call the function g the *Gauss kernel*.

3. Using the above results and recalling Definition 2.1 we can easily prove the following proposition.

Proposition 2.1. *Let $\{B(t): t \geq 0\}$ be a Brownian motion. Then*

i. *for arbitrary but fixed real numbers $a > 0$ and $\lambda \neq 0$ the processes given by (A) and (B) below are both Brownian motions:*

A. $\{B(t + a) - B(a): t \geq 0\}$,
B. $\{\lambda^{-1}B(\lambda^2 t): t \geq 0\}$ (*the space-time transformation*), *in particular* $\{-B(t): t \geq 0\}$; *and*

ii. $\{B(t): t > 0\}$ *and* $\{tB(1/t): t > 0\}$ *have the same distribution.*

Remark. Although assertion (ii) can easily be proved by noting that the covariance functions of the two processes coincide, this fact is worth recording because it becomes generalised in the property called the "projective invariance of Brownian motion" to be demonstrated in §5.4.

The existence of a Brownian motion as defined in Definition 2.1 is guaranteed by the Kolmogorov extension theorem [see §1.3(ii)] using finite dimensional distributions given by (2.2). In addition we will prove that for almost all ω the sample function $B(t, \omega)$ is continuous in t, and thus all distribution of $\{B(t): t \geq 0\}$, μ^W say, can be introduced onto the space $\mathscr{C} = \mathscr{C}[0, \infty)$ of all continuous functions on $[0, \infty)$. In this way the Wiener measure μ^W appears, and the remainder of this section is devoted to aspects of this measure.

We begin by constructing a measurable space over $\mathscr{C}$. A cylinder subset A of $\mathscr{C}$ [cf. §1.3, (ii)] is any subset of the form

$$A = \{w: (w(t_1), \ldots, w(t_n)) \in B_n\} \tag{2.3}$$

where $\{t_j\}$ are n time points with $0 \leq t_1 < t_2 < \cdots < t_n$, B_n is an n-dimensional Borel set, and a typical member of $\mathscr{C}$, that is, a continuous function on $[0, \infty)$, is denoted by w. As we vary n, the choice of points $t_1, t_2, \ldots, t_n$ and the set B_n, we obtain a collection $\mathfrak{U}$ of subsets of $\mathscr{C}$ which is easily seen to be an algebra. If we let $\mathfrak{B}$ denote the smallest σ-algebra containing $\mathfrak{U}$, then we have the measurable space $(\mathfrak{C}, \mathfrak{B})$ on which the measure μ^W will be defined.

In order that μ^W be identified as the distribution of a Brownian motion $\{B(t): t \geq 0\}$ we must prove that (1) μ^W is countably additive, and (2) for any $0 \leq t_1 < \cdots < t_n$ and n-dimensional Borel set B_n, we have

$$P((B(t_1), \ldots, B(t_n)) \in B_n) = \mu^W(A), \tag{2.4}$$

where A is the cylinder set (2.3). The following theorem proves that the distribution of a Brownian motion is concentrated upon $(\mathscr{C}, \mathfrak{B})$ and does coincide with μ^W [Yu. V. Prokhorov (1956), K. Itô, Lectures on stochastic processes, 1961].

Theorem 2.1. *There exists a unique probability measure μ^W on $(\mathscr{C}, \mathfrak{B})$ satisfying (2.4).*

PROOF. The assertion is proved in two steps. First we introduce a finitely additive measure m on $(\mathscr{C}, \mathfrak{U})$, and then m is extended to a (countably additive) measure μ^W on $(\mathscr{C}, \mathfrak{B})$.

1. If a cylinder set A is of the form (2.3), then we define $m(A)$ by

$$m(A) = \int_{B_n} \cdots \int g(t_1, t_2 - t_1, \ldots, t_n - t_{n-1}; x_1, x_2 - x_1, \ldots, x_n - x_{n-1})$$
$$\times dx_1 \, dx_2 \cdots dx_n \tag{2.5}$$

2.1 Brownian Motion. Wiener Measure

where g is the Gauss kernel defined by (2.2). A set-function m on $\mathfrak{A}$ is well-defined by this relation, that is, $m(A)$ is independent of the particular representation of A in terms of $t_1, \ldots, t_n$ and B_n. For example, we note that if a new time point $t_{n+1} > t_n$ is added, and the Borel set B_n is replaced by $B_n \times R$, the value of the corresponding integral would agree with that on the right-hand side of (2.5).

We now see that m is a finitely additive measure on $\mathfrak{A}$. Let $A_1, A_2, \ldots, A_n$ be pairwise disjoint members of $\mathfrak{A}$ and arrange the time points in the definition of the A_i in increasing order as $t_1 < t_2 < \cdots < t_p$. We may now assume that each of the A_i is determined by these p time points and a certain p-dimensional Borel set, and that the Borel sets are pairwise disjoint. But then it follows immediately from (2.5) that

$$m\left(\sum_{i=1}^n A_i\right) = \sum_{i=1}^n m(A_i),$$

and so m is finitely additive. This fact, coupled with the obvious relations $0 \le m(A) \le 1$, $A \in \mathfrak{A}$, and $m(\mathscr{C}) = 1$, shows that m is a finitely additive measure on $(\mathscr{C}, \mathfrak{A})$.

2. In order to extend the measure m just introduced to a countably additive measure on $(\mathscr{C}, \mathfrak{B})$ we will use the following lemma.

Lemma 2.1. *Suppose that a finitely additive measure m on $(\mathscr{C}, \mathfrak{A})$ satisfies the property:*

(*) *For every decreasing sequence $A_1 \supset A_2 \supset \cdots \supset A_n \supset \cdots$ with $A_n \in \mathfrak{A}$ for all n, the assumption $m(A_n) > \varepsilon \ (>0)$ for every n implies that*

$$\bigcap_n A_n \ne \emptyset.$$

Then m is uniquely extendable to a countably additive measure on $(\mathscr{C}, \mathfrak{B})$, where $\mathfrak{B}$ is the σ-field of subsets of $\mathscr{C}$ generated by $\mathfrak{A}$.

Now let us check (*) above. Assume that the set A_n can be written

$$A_n = \{w : (w(t_1^{(n)}), \ldots, w(t_{r_n}^{(n)})) \in B_n\},$$

and, adding some new time points if necessary, we may assume that the $\{t_k^{(n)}\}$ have the following property: there exists an increasing sequence $\{q_n\}$ of positive integers such that for every n,

i. $t_i^{(n)} \le q_n$, $1 \le i \le r_n$;
ii. for all k such that $1 < k < q_n 2^{q_n+1}$, there exists one and only one $t_i^{(n)}$ in the interval $[(k-1)2^{-q_n}, k2^{-q_n}]$;
iii. the time points $k2^{-q_n}$, $0 \le k \le 2^{q_n}$ are all included in $\{t_1^{(n)}, \ldots, t_{r_n}^{(n)}\}$.

By using further devices such as letting the number of time points increase and repeating the same A_n as many times as is necessary, we may suppose that

$$q_n = n, \ t_{2k}^{(n)} = k2^{-n}, \ r_n = n2^{n+1}.$$

Introducing the notation $\varphi(n) = n2^{n+1}$ we can recapitulate as follows:

$$A_n = \{w \in \mathscr{C} : (w(t_1^{(n)}), \ldots, w(t_{\varphi(n)}^{(n)})) \in B_n\},$$

B_n is a Borel subset of $\mathbf{R}^{\varphi(n)}$, \hfill (2.6)

$$t_{2k}^{(n)} = k2^{-n}, \{t_1^{(n)}, \ldots, t_{\varphi(n)}^{(n)}\} \subseteq \{t_1^{(n+1)}, \ldots, t_{\varphi(n+1)}^{(n+1)}\},$$

$$m(A_n) > \varepsilon \; (> 0).$$

A standard technique in measure theory allows us to take the B_n defining A_n to be a compact set, possibly requiring us to choose an $\varepsilon > 0$ smaller than the given one.

Now let us return to m. Fix $t_1, t_2, \ldots, t_n$ and form the σ-field $\mathfrak{A}(t_1, t_2, \ldots, t_n)$ consisting of all cylinder sets of the form (2.3) obtained as B_n ranges over the Borel subsets of $\mathbf{R}^n$. Then we have a measure space $(\mathscr{C}, \mathfrak{A}(t_1, t_2, \ldots, t_n), m)$, and $w \in \mathscr{C}$ may be regarded as a random parameter with $(w(t_1), w(t_2), \ldots, w(t_n))$ being a random vector with distribution given by (2.5). Consider the special case in which $n = 2$ and $t_1 = s, t_2 = t$. Then we have a Gaussian random variable $w(t) - w(s)$ with mean 0 and variance $|t - s|$, so that

$$\int_{\mathscr{C}} |w(t) - w(s)|^4 \, dm(w) = 3|t - s|^2. \tag{2.7}$$

Using this in Tchebychev's inequality we obtain

$$m(w: |w(t_i^{(n)}) - w(t_{i-1}^{(n)})| \geq |t_i^{(n)} - t_{i-1}^{(n)}|^{1/5}) \leq 3 \cdot 2^{-6n/5}.$$

We therefore have the result

$$m\left(\bigcup_{i=1}^{\varphi(n)} (w: |w(t_i^{(n)}) - w(t_{i-1}^{(n)})| \geq |t_i^{(n)} - t_{i-1}^{(n)}|^{1/5})\right) \leq 6n2^{-n/5}.$$

Now the series $\sum_n n2^{-n/5}$ is convergent, and so there exists an integer m_0 such that $6 \sum_{n \geq m_0} n2^{-n/5} < \frac{1}{2}\varepsilon$. With this choice of m_0 we have for any $l > m_0$:

$$m\left(\bigcup_{n=m_0}^{l} \bigcup_{i=1}^{\varphi(n)} (w: |w(t_i^{(n)}) - w(t_{i-1}^{(n)})| \geq |t_i^{(n)} - t_{i-1}^{(n)}|)\right) < \tfrac{1}{2}\varepsilon.$$

Denoting by C_l the intersection of A_l with the complement of the w-set in the expression just above, the assumption $m(A_l) > \varepsilon$ implies that

$$m(C_l) > \frac{1}{2}\varepsilon.$$

The sequence of sets C_l obviously satisfies

$$C_l \neq \varnothing, \; C_l \supseteq C_{l+1}, \; A_l \supseteq C_l,$$

and so to complete the proof it suffices to show that

$$\bigcap_l C_l \neq \varnothing \quad \left(\text{and hence } \bigcap_l A_l \neq \varnothing\right).$$

2.1 Brownian Motion. Wiener Measure

We now proceed to prove this fact by constructing a continuous function belonging to every C_l.

Let w_l be a member of C_l that varies linearly over each interval $[t_{i-1}^{(l)}, t_i^{(l)}]$ and satisfies $w(t_1^{(l)}) = 0$; such a function clearly exists. Since it satisfies

$$|w_l(t_i^{(n)}) - w_l(t_{i-1}^{(n)})| < |t_i^{(n)} - t_{i-1}^{(n)}|^{1/5} \leq 2^{-n/5}$$

for $m_0 \leq n \leq l$, $1 \leq i \leq n2^{n+1}$, we have

$$|w_l(k2^{-n}) - w_l((k-1)2^{-n})| < 2 \cdot 2^{-n/5}, \qquad m_0 \leq n \leq l, 1 \leq k \leq n2^{n+1}.$$

For any pair k, k' of integers such that

$$k' < k, \quad k2^{-l} - k'2^{-l} < 2^{-m_0},$$

there exists an integer q such that

$$q < l, \quad 2^{-q} \leq k2^{-l} - k'2^{-l} < 2^{-q+1}.$$

We can therefore find an integer j such that $k'2^{-l} \leq j2^{-q} < (j+1)2^{-q} \leq k2^{-l}$ and with this j

$$|w_l(j2^{-q}) - w_l((j+1)2^{-q})| < 2 \cdot 2^{-q/5},$$

since $w_l \in C_q$. By repeating similar arguments we can find a positive number μ for which the following holds:

$$|w_l(k2^{-l}) - w_l(k'2^{-l})| < \mu |k2^{-l} - k'2^{-l}|^{1/5}.$$

Hence there exists another positive number μ' such that if $|t_i^{(l)} - t_j^{(l)}| \leq 2^{-m_0}$ we have

$$|w_l(t_i^{(l)}) - w_l(t_j^{(l)})| < \mu' |t_i^{(l)} - t_j^{(l)}|^{1/5}.$$

Recalling that w_l is piecewise linear we see that for any $s, t \in [t_i^{(l)}, t_j^{(l)}]$

$$|w_l(t) - w_l(s)| \leq 4\mu' |t_i^{(l)} - t_j^{(l)}|^{1/5}.$$

Now we know that $w_{l+p} \in C_l$ for any $p \geq 0$, i.e.

$$(w_{l+p}(t_1^{(l)}), \ldots, w_{l+p}(t_{l2^l+1}^{(l)})) \in B_l,$$

and since B_l is assumed to be compact, the sequence of points in the preceding expression has a limit within B_l as $p \to \infty$. By using the diagonal method we may pass to a subsequence, $\{w_n\}$ say, for which $\{w_n(t_i^{(l)})\}$ converges as $n \to \infty$ for every i and l.

For any given t_0 and $\eta > 0$ there exists a sufficiently large n_0 for which we have

$$t_i^{(n_0)} \leq t_0 < t_{i+1}^{(n_0)}, \quad t_{i+1}^{(n_0)} - t_i^{(n_0)} < 2^{-n_0} < \eta^5,$$

$$|w_l(t_i^{(n_0)}) - w_m(t_i^{(n_0)})| < \eta \text{ for large } l, m \geq n_0.$$

Assuming $t_j^{(l)}, t_k^{(m)} \in [t_i^{(n_0)}, t_0]$, we now have,

$$\begin{aligned}
|w_l(t_0) - w_m(t_0)| &\leq |w_l(t_0) - w_l(t_j^{(l)})| + |w_l(t_j^{(l)}) - w_l(t_i^{(n_0)})| \\
&\quad + |w_l(t_i^{(n_0)}) - w_m(t_i^{(n_0)})| + |w_m(t_i^{(n_0)}) - w_m(t_k^{(m)})| \\
&\quad + |w_m(t_k^{(m)}) - w_m(t_0)| \\
&\leq |t_0 - t_j^{(l)}|^{1/5} + \mu' |t_j^{(l)} - t_i^{(n_0)}|^{1/5} + \eta \\
&\quad + \mu' |t_i^{(n_0)} - t_k^{(m)}|^{1/5} + |t_k^{(m)} - t_0|^{1/5} \\
&< a\eta, \quad a \text{ constant.}
\end{aligned}$$

This evaluation is valid for every $t \in [t_i^{(n_0)}, t_{i+1}^{(n_0)}]$. In addition the inequality

$$|w_l(t) - w_l(s)| < 4\mu' |t_i^{(l)} - t_j^{(l)}|^{1/5}$$

assures us that the limit function w^* of the sequence $\{w_n\}$ is also a continuous function. Since we can prove

$$(w^*(t_1^{(l)}), \ldots, w^*(t_{l2^l+1}^{(l)})) \in B_l$$

for every l, we have proved that $\bigcap_{l \geq n_0} C_l \neq \varnothing$, as desired. $\square$

Definition 2.2. The measure μ^W on $(\mathscr{C}, \mathfrak{B})$ whose existence and uniqueness has been guaranteed by Theorem 2.1 is called the *Wiener measure*.

Remark 1. Although the general theory (Kolmogorov's extension theorem) introduces the distribution of Brownian motion onto the space $\mathbf{R}^{[0, \infty)}$, Theorem 2.1 implies that, taking a suitable version of $\{B(t)\}$, it is supported by $\mathscr{C}$ and so the measure μ^W on $(\mathscr{C}, \mathfrak{B})$ can be viewed as the distribution of a Brownian motion. In other words, for almost all ω the sample path $B(\cdot, \omega)$ is a continuous function. We will discuss more precise sample path properties of Brownian motion in the next section.

Remark 2. As far as the author is aware, N. Wiener was the first person to succeed in introducing a Gaussian measure onto the space of continuous functions, and in view of this it seems fitting to call μ^W Wiener measure. Incidentally Wiener himself called the measure space $(\mathscr{C}, \mathfrak{B}, \mu^W)$ "differential space," the name coming from the additive property of $B(t)$. It is interesting to contrast this terminology with the fact that almost all sample paths are everywhere *non*-differentiable [see (i) of the next section].

If a general, not necessarily Gaussian, stochastic process $\{X(t): t \in T\}$ defined on an interval T satisfies the condition:

There exist constants $\alpha > 0$, $\beta > 1$ such that for some constant $C > 0$

$$E\{|X(t) - X(s)|^\alpha\} \leq C|t - s|^\beta, \quad t, s \in T, \tag{2.8}$$

then the distribution of the process, taking a suitable version, is known to be supported by $\mathscr{C}(T)$ [Kolmogorov-Prokhorov; see Yu. V. Prokhorov (1956) and (2.7)]. This result is of the type concerning the support of a stochastic process referred to in §1.3 (ii).

We shall now pause to give a brief discussion of multi-dimensional Brownian motion.

Definition 2.3. A system $\{B(t, \omega) = (B_1(t, \omega), B_2(t, \omega), \ldots, B_n(t, \omega)): t \geq 0\}$ of vector-valued random variables defined on a probability space $(\Omega, \mathbf{B}, P)$ is called an *n-dimensional Brownian motion* if

a. each component $\{B_i(t, \omega): t \geq 0\}$ is a (one-dimensional) Brownian motion; and
b. $\{B_i(t, \omega): t \geq 0;\ i = 1, 2, \ldots, n\}$ is an independent system of stochastic processes.

As one would anticipate from Theorem 2.1, the distribution of an n-dimensional Brownian motion is supported by the space $\mathscr{C}([0, \infty), \mathbf{R}^n)$, and is called the (*n*-dimensional) Wiener measure. Many of the sample path properties can be reduced to those in the one-dimensional case, although the number of dimensions will not be involved so much in the analysis of functionals of Brownian motion, the main topic of this text. However, when we view Brownian motion as a Markov process, we meet (§§2.4–6) phenomena which are more interesting or in which more fruitful results can be obtained in the higher dimensional cases. For example, when a Brownian motion is restricted to lie in a certain domain within the space, we will be able to speak of its behaviour on or near the boundary of this domain, and we can demonstrate intimate connections with areas in analysis such as potential theory or the theory of partial differential equations. These topics certainly become more interesting in higher dimensions, details being found in the famous books by W. Feller (1971) and by K. Itô and H. P. McKean (1965) which we recommend to readers.

Another direction in which Brownian motion can be generalised is by allowing the time parameter t to be a vector in $\mathbf{R}^n$ or a point in some Riemannian space, for example, a sphere. A short illustration of multiparameter Brownian motion can be found in Appendix §A.2.

2.2 Sample Path Properties

In the last section we observed the continuity, for almost all ω, of the sample function $t \to B(t, \omega)$, $t \geq 0$, of Brownian motion. Questions naturally arise concerning the more refined properties such as differentiability, asymptotic behaviour as $t \to \infty$ and so on of the Brownian sample path, and this section is devoted to answering some of these questions and presenting some typical results in the area.

(i) Non-differentiability

We first state the result.

Theorem 2.2. *Almost all sample paths of a Brownian motion $\{B(t, \omega): t \geq 0\}$ are nowhere-differentiable.*

Observation. Before giving the proof of the theorem we present the underlying idea. For small $h > 0$ the increment $B(t + h) - B(t)$ is a Gaussian random variable with mean 0 and variance h, whence $h^{-1/2}[B(t + h) - B(t)]$ is a standard (mean 0, variance 1) Gaussian random variable and so can be thought of as being of ordinary magnitude, however small the value of $h > 0$. If we then consider the ratio $h^{-1}[B(t + h) - B(t)]$ and let h tend to 0, we see that the variance of this ratio will become arbitrarily large, and so we would never expect the existence of a limit of the ratio for each ω, which would have to be the case to have a time derivative of $B(t, \omega)$.

PROOF OF THEOREM 2.2. By Proposition 2.1 (i) A it suffices to prove the theorem for t ranging over the interval $[0, 1]$, and we can then use the method of Dvoretzky, Erdös and Kakutani.

Suppose that $B(t, \omega)$ was differentiable at some point $s \in [0, 1)$. Then since $B(t, \omega)$ is differentiable from the right at s, there exists $\varepsilon > 0$ and an integer $l \geq 1$ such that for $0 < t - s < \varepsilon$,

$$|B(t, \omega) - B(s, \omega)| < l(t - s).$$

Now take a larger integer n, set $i = [ns] + 1$, and let j run over $i + 1, i + 2, i + 3$, successively (see Fig. 3). Then the above inequality gives us

$$\left| B\left(\frac{j}{n}, \omega\right) - B\left(\frac{j+1}{n}, \omega\right) \right| < \frac{7l}{n}, j = i + 1, i + 2, i + 3, \quad (2.9)$$

where it will become clear in the computations which follow why we take three successive j's.

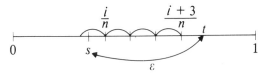

Figure 3

Let $A_{l,n}^{i,j}$ be the set of all ω satisfying (2.9), obviously a **B**-measurable set, and let us consider the **B**-measurable set

$$A = \bigcup_{l \geq 1} \bigcup_{m \geq 1} \bigcap_{n \geq m} \bigcup_{0 < i \leq n} \bigcap_{i < j \leq i+3} A_{l,n}^{i,j}. \quad (2.10)$$

This set is the event that there exists an integer l such that for all n sufficiently large the inequality (2.9) holds at some point i/n. Thus A includes

2.2 Sample Path Properties

every ω for which $B(\cdot, \omega)$ is differentiable at some point t, and so if we can prove that $P(A) = 0$, the proof will be complete. To this end we note that

$$P\left(\bigcap_{n \geq m} \bigcup_{0 < i \leq n} \bigcap_{i < j \leq i+3} A_{l,n}^{i,j}\right)$$

$$\leq \liminf_{n \to \infty} nP\left(\left|B\left(\frac{1}{n}\right)\right| < \frac{7l}{n}\right)^3$$

$$= \liminf_{n \to \infty} nP\left(|B(1)| < \frac{7l}{\sqrt{n}}\right)^3$$

$$\leq \lim_{n \to \infty} n\left\{(2\pi)^{-1/2} \frac{14l}{\sqrt{n}}\right\}^3 = 0.$$

The set A is thus the union of a countable number of sets of probability zero, as was to be proved. □

(ii) The Second Variation

For fixed ω we will define the (first) variation as well as the second variation of the continuous function $B(t, \omega)$, $0 \leq t \leq 1$, these serving to indicate how the Brownian sample path oscillates as the time t evolves.

To introduce the notation associated with the concept of variation, let $f(t)$ be an ordinary function and let $\pi = \{t_0, t_1, \ldots, t_n\}$, $0 = t_0 \leq t_1 \leq \cdots \leq t_n$, stand for a partition of the unit interval $[0, 1]$ into the sub-intervals $[t_j, t_{j+1}]$, $0 \leq j < n$. Define $\pi^k f$, $k > 0$, and $\delta \pi$ by the relations

$$\pi^k f = \sum_{j=0}^{n-1} |f(t_{j+1}) - f(t_j)|^k, \quad k > 0,$$

$$\delta \pi = \max_{0 \leq j < n} |t_{j+1} - t_j|.$$

We now take a sequence of finer and finer partitions π_n, $n \geq 0$;

$$\pi_0 = \{0, 1\} \subset \pi_1 \subset \pi_2 \subset \cdots. \tag{2.11}$$

If the limit

$$\lim_{n \to \infty} \pi_n^k f = V^k(f)$$

exists, then $V^k(f)$ is called the *variation* of f of *order* k relative to $\{\pi_n\}$.

Returning now to the Brownian sample path $B(t, \omega)$, $0 \leq t \leq 1$, we shall demonstrate a result on its variation originally pointed out by P. Lévy (1940).

Theorem 2.3. *Let $\{\pi_n : n \geq 0\}$ be a sequence of partitions of $[0, 1]$ satisfying (2.11). Then*

  i. *For almost all ω the limit*

$$\lim_{n \to \infty} \pi_n^2 B(\cdot, \omega) \equiv V^2(\omega) \tag{2.12}$$

exists and is finite.

ii. *If, in addition, $\{\pi_n\}$ satisfies the condition*

$$\lim_{n\to\infty} \delta\pi_n = 0,$$

then we also have

$$V^2(\omega) = 1, \quad \text{a.e.} \tag{2.13}$$

Before giving the proof of the theorem we introduce the notion of martingale, and state a limit theorem which provides a tool used in the proof.

Letting N denote the set of all positive integers N^+ or of all negative integers N^-, we suppose that $\{X_n: n \in N\}$ is a stochastic process on a probability space $(\Omega, \mathbf{B}, P)$. Suppose also that we have an increasing sequence $\mathbf{B}_n$, $n \in N$, of sub-σ-fields of $\mathbf{B}$. The system $\{X_n, \mathbf{B}_n: n \in N\}$ is called a *martingale* if it satisfies the following conditions:

a. X_n is $\mathbf{B}_n$-measurable, $n \in N$;
b. $E(|X_n|) < \infty, n \in N$;
c. $E(X_n|\mathbf{B}_m) = X_m$ a.e. for every $n \geq m$.

A martingale with continuous time parameter is defined similarly, and is written $\{X(t), \mathbf{B}(t): t \in T\}$ where T is taken to be $[0, \infty)$, $(-\infty, \infty)$ or $(-\infty, 0]$.

Remark. Recalling the sequence S_n introduced in Example 1, §1.4, we can show that $\{S_n, \mathbf{B}_n: n \in N^+\}$ is an example of a martingale.

The following lemma is due to J. L. Doob. For a proof of this and related results we refer to Appendix §A.1 and to Doob (1953) Chapter VII §4.

Lemma 2.2. i. *Let $\{X_n, \mathbf{B}_n: n \in N^+\}$ be a martingale. Then $E(|X_n|)$ is non-decreasing*:

$$E(|X_1|) \leq E(|X_2|) \leq \cdots \leq E(|X_n|) \leq \cdots.$$

If $\lim_{n\to\infty} E(|X_n|) = K < \infty$, then

$$\lim_{n\to\infty} X_n(\omega) \equiv X_\infty(\omega)$$

exists almost surely.

ii. *For a martingale $\{X_n, \mathbf{B}_n: n \in N^-\}$ the limit*

$$\lim_{n\to -\infty} X_n(\omega) \equiv X_{-\infty}(\omega)$$

exists almost surely, and if we write $\mathbf{B}_{-\infty} = \bigcap_{n \in N^-} \mathbf{B}_n$, then

$$\{X_n, \mathbf{B}_n: n \in N^- \cup \{-\infty\}\}$$

is also a martingale.

2.2 Sample Path Properties

PROOF OF THEOREM 2.3. i. Let $\{\pi_n: n \geq 0\}$ be a sequence of partitions of $[0, 1]$ satisfying (2.11) and for $n \geq 0$ set

$$V_n^2(\omega) = \pi_n^2 B(\cdot, \omega).$$

Our first task is to prove that for integers $k > 0$ and $K \geq 0$

$$E(\Delta_1^k B \, \Delta_2^k B \mid V_k^2, V_{k+1}^2, \ldots, V_{k+K}^2) = 0. \tag{2.14}$$

where Δ_i^k, $i = 1, 2$, are different sub-intervals of $[0, 1]$ from the partition π_k, and where $\Delta_i^k B$ is the increment of $B(t)$ over the interval Δ_i^k. We may suppose that the partition π_{k+1} of any sequence $\{\pi_n: n \geq 0\}$ satisfying (2.11) is obtained by dividing some interval of π_k into two, for if necessary new partitions can be inserted into the sequence $\{\pi_n\}$ without affecting its monotonic increasing nature. Suppose then that π_{k+1} is obtained from π_k by dividing an interval of π_k into two subintervals Δ_1^{k+1} and Δ_2^{k+1}. In this case

$$V_k^2 - V_{k+1}^2 = 2 \, \Delta_1^{k+1} B \, \Delta_2^{k+1} B. \tag{2.15}$$

Repeating this procedure K times, the left side of (2.14) becomes

$$E(\Delta_1^k B \, \Delta_2^k B \mid V_k^2, \Delta_1^{k+1} B \, \Delta_2^{k+1} B, \ldots, \Delta_1^{k+K} B \, \Delta_2^{k+K} B).$$

Further, V_k^2 can be written $V_k^2 = (\Delta_1^k B)^2 + (\Delta_2^k B)^2 + \tilde{V}_k^2$, the sum of three independent random variables, and so to prove (2.14) it suffices to show

$$E(\Delta_1^k B \, \Delta_2^k B \mid (\Delta_1^k B)^2, (\Delta_2^k B)^2, \tilde{V}_k^2, \Delta_1^{k+1} B \, \Delta_2^{k+1} B, \ldots, \Delta_1^{k+K} B \, \Delta_2^{k+K} B) = 0. \tag{2.16}$$

We note that for $0 < j \leq K$ the intervals Δ_1^{k+j} and Δ_2^{k+j} are included simultaneously in either Δ_1^k or Δ_2^k, or otherwise are both disjoint from $\Delta_1^k \cup \Delta_2^k$. In the second case $\Delta_1^{k+j} B \, \Delta_2^{k+j} B$ may be dropped from the conditioning above, since it is independent of $(\Delta_1^k B, \Delta_2^k B)$. Similarly $\tilde{V}_k^2$ can be dropped, and having done this we integrate the left-hand side of (2.16) over an ω-set determined by the conditioning random variables. Since increments over the intervals of π_{k+K} are mutually independent, the integral is expressed in terms of their distributions in the form

$$\int_A \left(\sum_{i=1}^m x_i\right)\left(\sum_{j=1}^n y_j\right)(2\pi)^{-(m+n)/2} \left(\prod_i c_i^{-1}\right)\left(\prod_j d_j^{-1}\right)$$
$$\times \exp\left[-\frac{1}{2}\sum_{i=1}^n \frac{x_i^2}{c_i^2} - \frac{1}{2}\sum_{i=1}^n \frac{y_j^2}{d_j^2}\right] dx_1 \cdots dx_m \, dy_1 \cdots dy_n.$$

In this expression the integration variables $\{x_i\}$ and $\{y_j\}$ correspond to the increments of $B(t)$ over the elements of π_{k+K} included in Δ_1^k and Δ_2^k, respectively, and A is the subset of $\mathbf{R}^{m+n}$ determined by conditions on the quadratic forms $(\sum_i x_i)^2$, $(\sum_j y_j)^2$, $x_i x_{i'}$, $y_j y_{j'}$, etc. As the set A is invariant under a simultaneous change of the signs of the $\{x_i\}$, whilst the integrand changes its sign, the value of the integral must be 0. This proves (2.16) and hence (2.14).

From (2.15) we have for any k
$$E(V_k^2 \mid V_{k+1}^2, V_{k+2}^2, \ldots, V_{k+K}^2) = V_{k+1}^2, \qquad k \geq 0$$
and so, letting $K \to \infty$, we obtain
$$E(V_k^2 \mid V_{k+1}^2, V_{k+2}^2, \ldots) = V_{k+1}^2.$$
If we now set $X_n = V_{-n}^2$, $n \in N^-$ and let $\mathbf{B}_n$ denote the σ-field generated by X_k, $k \leq n$, then the preceding equation states that $\{X_n, \mathbf{B}_n : n \in N^-\}$ is a martingale. From Lemma 2.2 (ii) we obtain the existence almost surely of
$$\lim_{n \to -\infty} X_n(\omega) = \lim_{k \to \infty} V_k^2(\omega) = V^2(\omega).$$
This proves assertion (i) of Theorem 2.3.

ii. Suppose that the partition π_n is determined by the points $\{t_j\}$. Then we have
$$E\{V_n^2\} = \sum_j E((\Delta_j B)^2) = \sum_j (t_{j+1} - t_j) = 1,$$
where $\Delta_j B = B(t_{j+1}) - B(t_j)$. Furthermore
$$E[(V_n^2 - 1)^2] = \sum_{j,k} E((\Delta_j B)^2 (\Delta_k B)^2) - 2\sum_j E((\Delta_j B)^2) + 1$$
$$= \sum_{j \neq k} (t_{j+1} - t_j)(t_{k+1} - t_k) + 3\sum_j (t_{j+1} - t_j)^2 - 2 + 1$$
$$= 1 - \sum_j (t_{j+1} - t_j)^2 + 3\sum_j (t_{j+1} - t_j)^2 - 1$$
$$\leq 2\delta\pi_n \sum_j (t_{j+1} - t_j)$$
$$= 2(\delta\pi_n) \to 0 \quad \text{as} \quad n \to \infty.$$

We have therefore proved that V_n^2 converges in mean square to 1 in the space $L^2(\Omega, \mathbf{B}, P)$. On the other hand, from (i) we know that $V_n^2(\omega)$ converges a.s. to $V(\omega)$, and so we finally obtain our result
$$V^2(\omega) = 1 \qquad \text{a.s.} \qquad \square$$

Let us replace the partition π_n of $[0, 1]$ in Theorem 2.3 by the partition $\pi_n(t)$ of $[0, t]$, $\pi_0(t)$ being tacitly defined as $\{0, t\}$, and set
$$\pi_n(t)^2 B(\cdot, \omega) = V_n^2(t, \omega).$$
Then we can prove

Corollary. i. *For almost all ω the limit*
$$\lim_{n \to \infty} V_n^2(t, \omega) = V^2(t, \omega) \tag{2.17}$$
exists, and $V^2(t, \omega)$ is increasing in t.

ii. If $\{\pi_n(t)\}$ satisfies the further condition

$$\lim_{n \to \infty} \delta\pi_n(t) = 0,$$

then

$$V^2(t, \omega) = t \quad \text{a.s.} \tag{2.18}$$

In order to state a result on the variation (of order 1) of a sample function $B(t, \omega)$, $0 \le t \le 1$, we need the following lemma. It is easily proved and so is presented without proof.

Lemma 2.3. *Let $\{\pi_n\}$ be a sequence of partitions of $[0, 1]$ such that $\delta\pi_n \to 0$. If a function f on $[0, 1]$ is of bounded variation, then $\pi_n^2 f \to 0$.*

With this lemma, Theorem 2.3 (ii) immediately implies the following:

Theorem 2.4. *For almost all ω the sample path $B(t, \omega)$ is of unbounded variation on any interval subset of $[0, 1]$.*

Remark. The non-differentiability of the sample paths can also be seen from this result.

(iii) The Law of the Iterated Logarithm and the Uniform Continuity of Sample Paths

From Proposition 2.1 (ii) we know that the Brownian motion process $\{B(t, \omega): t > 0\}$ and the process $\{tB(1/t, \omega): t > 0\}$ have the same distribution. Since $B(0+, \omega) = 0$ a.e., we have $\lim_{t \downarrow 0} tB(1/t, \omega) = 0$ a.e. and so, replacing t by $1/t$ we see that

$$\lim_{t \to \infty} t^{-1} B(t, \omega) = 0, \quad \text{a.e.} \tag{2.19}$$

A much finer result is due to A. Khintchine (1933).

Theorem 2.5. *For almost all ω we have*

$$\lim_{t \downarrow 0} \sup (2t \log \log t^{-1})^{-1/2} B(t, \omega) = 1, \tag{2.20}$$

and

$$\lim_{t \downarrow 0} \inf (2t \log \log t^{-1})^{-1/2} B(t, \omega) = -1. \tag{2.21}$$

Again martingales play an essential role in the proof, and we begin with a martingale inequality generalising Tchebychev's and Kolmogorov's inequality.

Lemma 2.4 (Doob's inequality). *Let $\{X_n, \mathbf{B}_n: n \in N^+\}$ be a martingale. Then for any $\lambda > 0$ we have*

$$P\left(\max_{k \leq n} X_k \geq \lambda\right) \leq \lambda^{-1} E(X_n^+), \tag{2.22}$$

where $X_n^+ = X_n \vee 0$.

The proof may be found in Appendix A§1; see also Doob (1953) Chapter VII, Theorem 3.2.

Remark. The system $\{X_n, \mathbf{B}_n: n \in N\}$ is said to be *submartingale* if it satisfies conditions (a) and (b) in the definition of a martingale, and instead of condition (c) there it satisfies

$$E(X_n | \mathbf{B}_m) \geq X_m \quad \text{a.e.,} \quad \text{for every } n \geq m. \tag{c'}$$

The inequality (2.22) is also valid for a submartingale. Similarly, for a continuous parameter martingale $\{X(t), \mathbf{B}_t: t \geq 0\}$ the inequality

$$P\left(\max_{s \leq t} X(s) \geq \lambda\right) \leq \lambda^{-1} E(X(t)^+) \tag{2.23}$$

holds whenever almost all sample functions of $X(t)$ are continuous.

PROOF OF THEOREM 2.5 [H. P. McKean Jr. (1969)]. Let $\mathbf{B}(t)$ be the σ-field generated by the random variables $B(s)$, $s \leq t$, and for a real constant a set

$$X(t, \omega) = \exp\left[aB(t, \omega) - \frac{1}{2}a^2 t\right], \quad t \geq 0. \tag{2.24}$$

Then each $X(t)$ is $\mathbf{B}_t$-measurable and integrable, for

$$E(X(t)) = \exp\left[-\frac{1}{2}a^2 t\right](2\pi t)^{-1/2} \int_{-\infty}^{\infty} \exp[ax] \cdot \exp\left[\frac{-x^2}{2t}\right] dx$$

$$= 1 \text{ for every real number } a.$$

For $t > s$ we may write

$$X(t) = X(s)\exp\left[a(B(t) - B(s)) - \frac{1}{2}a^2(t-s)\right],$$

and since $B(t) - B(s)$ is independent of any $\mathbf{B}_s$-measurable random variables, and has the same distribution as $B(t-s)$, we obtain

$$E(X(t)|\mathbf{B}_s) = X(s)E\left\{\exp\left[aB(t-s) - \frac{1}{2}a^2(t-s)\right]\right\}$$

$$= X(s).$$

Thus we have proved that $\{X(t), \mathbf{B}_t: t \geq 0\}$ is a martingale.

2.2 Sample Path Properties

Now let us take a positive constant b and use Lemma 2.4 to get

$$P\left(\max_{s \leq t}\left|B(s) - \frac{1}{2}as\right| \geq b\right) = P\left(\max_{s \leq t} X(s) \geq \exp(ab)\right)$$
$$\leq \exp[-ab]E(X(t))$$
$$= \exp[-ab].$$

Set $h(t) = (2t \log \log t^{-1})^{1/2}$ and choose constants $0 < \theta < 1$ and $\delta > 0$. Taking the constants a, b and the time t above to be $a_n = (1 + \delta)\theta^{-n}h(\theta^n)$, $b_n = \frac{1}{2}h(\theta^n)$ and $t_n = \theta^{n-1}$ respectively, we may use the relations

$$a_n b_n = \frac{1}{2}(1 + \delta)\theta^{-n}[h(\theta^n)]^2 = (1 + \delta)\log \log \theta^{-n}$$

and

$$\exp(-a_n b_n) = (\log \theta^{-1})^{-1-\delta}n^{-1-\delta}$$

to prove that

$$\sum_n P\left(\max_{s \leq t_n}\left|B(s) - \frac{1}{2}a_n s\right| \geq b_n\right) < \infty.$$

Thus the first part of the Borel-Cantelli Lemma implies that

$$P\left(\max_{s \leq t_n}\left|B(s) - \frac{1}{2}a_n s\right| < b_n \text{ for all but a finite number of } n\right) = 1.$$

Therefore for almost all ω there exists $n(\omega)$ such that for any $n \geq n(\omega)$ and t with $t_{n+1} < t \leq t_n$ we have

$$B(t, \omega) \leq \max_{s \leq t_n} B(s, \omega) < \frac{1}{2}a_n t_n + b_n = \left|\frac{1+\delta}{2\theta} + \frac{1}{2}\right|h(\theta^n)$$
$$< \left|\frac{1+\delta}{2\theta} + \frac{1}{2}\right|h(t_n).$$

Letting $\theta \to 1$ and $\delta \to 0$ we obtain the inequality

$$\limsup_{t \downarrow 0} \frac{B(t, \omega)}{h(t)} \leq 1. \tag{2.25}$$

We next work towards the converse inequality

$$\limsup_{t \downarrow 0} \frac{B(t, \omega)}{h(t)} \geq 1. \tag{2.26}$$

As before let $0 < \theta < 1$ and set

$$A_n = \{\omega : B(\theta^n, \omega) - B(\theta^{n+1}, \omega) \geq (1 - \theta^{1/2})h(\theta^n)\},$$

clearly an independent sequence of events. The well known inequalities for the Gaussian distribution function

$$(2\pi)^{-1/2}(a^{-1} - a^{-3})\exp\left(-\frac{1}{2}a^2\right)$$
$$< (2\pi)^{-1/2}\int_a^\infty \exp\left(-\frac{1}{2}x^2\right)dx < (2\pi)^{-1/2}a^{-1}\exp\left(-\frac{1}{2}a^2\right) \quad (2.27)$$

for $a > 0$ imply that

$$P(A_n) = (2\pi)^{-1/2}\int_{(1-\theta^{1/2})h(\theta n)/(\theta n - \theta n + 1)^{1/2}}^\infty \exp\left(-\frac{1}{2}x^2\right)dx$$
$$> c(\log n)^{-1/2}n^{-(1-2\theta^{1/2}+\theta)/(1-\theta)},$$

c a positive constant. The further inequality $(1 - 2\theta^{1/2} + \theta)/(1 - \theta) < 1$ proves that $\sum_n P(A_n) = \infty$ and so the second part of the Borel-Cantelli Lemma implies that

$$P\left(\limsup_{n\to\infty} A_n\right) = 1.$$

But this means that for infinitely many integers n we have

$$B(\theta^n, \omega) \geq (1 - \theta^{1/2})h(\theta^n) + B(\theta^{n+1}, \omega).$$

Now we proved above that for all integers n from some point onwards $B(\theta^{n+1}, \omega) < 2h(\theta^{n+1})$, and since the distribution of $B(t, \omega)$ is symmetric, we must also have $B(\theta^{n+1}, \omega) > -2h(\theta^{n+1})$ for all sufficiently large n. Therefore

$$B(\theta^n, \omega) > (1 - \theta^{1/2})h(\theta^n) - 2h(\theta^{n+1})$$
$$> (1 - \theta^{1/2} - 3\theta^{1/2})h(\theta^n)$$

for infinitely many n. Letting $\theta \to 0$ we obtain (2.26). □

The next result is a consequence of Proposition 2.1 (i) A and the fact that $\{B(a + h) - B(a): h \geq 0\}$ and $\{B(a - h) - B(a): h \geq 0\}$ have the same distribution around $h = 0$.

Corollary 1. *For almost all ω we have*

$$\limsup_{h\to 0} (2|h|\log\log|h|^{-1})^{-1/2}[B(a + h, \omega) - B(a, \omega)] = 1, \quad (2.28)$$

and

$$\liminf_{h\to 0} (2|h|\log\log|h|^{-1})^{-1/2}[B(a + h, \omega) - B(a, \omega)] = -1. \quad (2.29)$$

This result shows the so-called local continuity of the Brownian sample path. There is another approach which deals with the uniform continuity on

2.2 Sample Path Properties

a finite interval, and this will be discussed in Theorem 2.6 below. But before we come to this we state one more Corollary of Theorem 2.5, this one concerning the asymptotic behaviour of the Brownian sample path as $t \to \infty$.

Corollary 2. *For almost all ω we have*

$$\limsup_{t \to \infty} (2t \log \log t)^{-1/2} B(t, \omega) = 1, \quad (2.30)$$

$$\liminf_{t \to \infty} (2t \log \log t)^{-1/2} B(t, \omega) = -1. \quad (2.31)$$

This result is also proved from the fact that the processes $\{B(t, \omega): t > 0\}$ and $\{tB(1/t, \omega): t > 0\}$ have the same distribution.

The properties obtained so far, such as (2.20), (2.21), (2.28), (2.29), (2.30) and (2.31), are known generically as the *law of the iterated logarithm*.

Remark. The martingale $X(t)$ given by formula (2.24) in the proof of Theorem 2.4 was in fact defined by using the generating function of the Hermite polynomials [see Appendix A §5 (i)] where the variable x and parameter t are taken to be $B(t, \omega)/\sqrt{2t}$ and $a\sqrt{t/2}$, respectively. Functionals of Brownian motion of this type are extremely important and will be frequently used in the sequel.

We turn now to a discussion of the uniform continuity of the Brownian sample paths. The time variable t is here supposed to range over $[0, 1]$.

Theorem 2.6 [P. Lévy (1937)]. *Suppose that $c > 1$ is constant. Then for almost every ω there exists $\delta = \delta(\omega) > 0$ such that whenever $|t - t'| < \delta$*

$$|B(t, \omega) - B(t', \omega)| \le c[2|t - t'| \log |t - t'|^{-1}]^{1/2}. \quad (2.32)$$

PROOF. Let $h = 1/n$. Since $B(t + h) - B(t)$ is a zero mean Gaussian random variable with variance h, the second inequality of (2.27) gives the result that for large n

$$\alpha_n \equiv P(|B(t + h) - B(t)| > c(2h \log h^{-1})^{1/2})$$

$$= \left(\frac{2}{\pi}\right)^{1/2} \int_{c(2 \log n)^{1/2}}^{\infty} \exp\left(-\frac{1}{2} x^2\right) dx$$

$$< c^{-1}(\pi \log n)^{-1/2} n^{-c^2}.$$

The probability that at least one of the n increments $B((k + 1)h) - B(kh)$, $k = 0, 1, \ldots, n - 1$, exceeds $c(2h \log h^{-1})^{1/2}$ in absolute value is at most $n\alpha_n$ and, considering the particular case $n = 2^p$, we see that

$$\sum_p 2^p \alpha_{2^p} < \sum_p 2^p (c^2 \pi \log 2)^{-1/2} p^{-1/2} 2^{-c^2 p}$$

$$= (c^2 \pi \log 2)^{-1/2} \sum_p p^{-1/2} 2^{(1 - c^2)p} < \infty.$$

By the first part of the Borel-Cantelli lemma we see that for almost all ω there exists an integer $p_0 = p_0(\omega)$ such that whenever $p > p_0$

$$|B((k+1)h, \omega) - B(kh, \omega)| \leq c(2h \log h^{-1})^{1/2}, \qquad h = 2^{-p}. \quad (2.33)$$

This proves the inequality (2.32) for the values of t and t' of the form $(k+1)2^{-p}$ and $k2^{-p}$, $k = 0, 1, \ldots, 2^p - 1$, respectively.

We now turn to the case where $t = q2^{-p}$ and $t < t' < t + 2^{-p}$ for some $p > p_0$. Using the binary expansion of $t' - t$

$$t' - t = \sum_{v=1}^{\infty} \varepsilon_v 2^{-p-v}, \qquad \varepsilon_v = 0 \text{ or } 1,$$

inequality (2.33) implies that

$$|B(t', \omega) - B(t, \omega)| \leq c \sum_{v=1}^{\infty} \varepsilon_v \{2(p+v)\log 2\}^{1/2} 2^{-(p+v)/2}.$$

Letting v_0 denote the smallest integer v for which $\varepsilon_v = 1$ we set $v = v_0 + v' - 1$. Then from $p + v \leq (p + v_0)v'$ we have

$$|B(t', \omega) - B(t, \omega)| \leq c\{2(p+v_0)\log 2\}^{1/2} 2^{-(p+v_0)/2} \sum_{v'=1}^{\infty} (2v')^{1/2} 2^{-v'/2}$$

$$= cA\{2(p+v_0)\log 2\}^{1/2} 2^{-(p+v_0)/2}, \qquad A > 1.$$

The function $h \log h^{-1}$ is increasing on the interval $(0, e^{-1})$, a fortiori between 2^{-p-v_0} and $t' - t$, so that

$$|B(t', \omega) - B(t, \omega)| < c'\{2|t' - t|\log|t' - t|^{-1}\}^{1/2} \quad (2.34)$$

holds with some constant $c' > 1$. In a similar manner we can prove (2.34) in the case where $t = q2^{-p}$ and $t > t' > t - 2^{-p}$.

Further, we can prove (2.33) in the same way when t and t' are of the form $q2^{-p}$ and $(q+v)2^{-p}$, $1 \leq v \leq N$, N fixed. That is, for almost all ω there exists an integer $p = p_1(\omega)$ such that for every $p > p_1$

$$|B((q+v)2^{-p}, \omega) - B(q2^{-p}, \omega)| < c''(2h_v \log h_v^{-1})^{1/2}, \quad h_v = v2^{-p} \quad (2.33')$$

where $c'' > 1$.

Finally, we consider the case where t and t' are arbitrary but close together. We may suppose that $t' > t$ and that for some integer N

$$N2^{-p-1} < t' - t \leq N2^{-p}, \qquad p > \max\{p_0, p_1\}.$$

Then there exist q, q' such that

$$q2^{-p} < t \leq t_1 = (q+1)2^{-p} < t'_1 = q'2^{-p} \leq t' < (q'+1)2^{-p},$$

where $\frac{1}{2}N - 1 < q' - q < N + 1$. Take the inequality

$$|B(t', \omega) - B(t, \omega)| \leq |B(t', \omega) - B(t'_1, \omega)|$$
$$+ |B(t'_1, \omega) - B(t_1, \omega)| + |B(t_1, \omega) - B(t, \omega)|$$

and choose positive constants ε, c', c'' and N so that $c = 1 + 2\varepsilon$, $c'' = 1 + \varepsilon$, and $N \geq 16c'^2\varepsilon^{-2}$. As before the numbers $p_0(\omega)$ and $p_1(\omega)$ are suitably large. Since $h \log h^{-1}$ is monotonic increasing for small h, inequality (2.33′) implies

$$|B(t'_1, \omega) - B(t_1, \omega)| < (1 + \varepsilon)\{2|t' - t|\log|t' - t|^{-1}\}^{1/2}.$$

Also (2.34) implies

$$|B(t', \omega) - B(t'_1, \omega)| + |B(t_1, \omega) - B(t, \omega)|$$
$$< 2c'(2 \cdot 2^{-p} \log 2^p)^{1/2}$$
$$< 2c'\{4|t' - t|N^{-1} \log(N|t' - t|^{-1})\}^{1/2}.$$

Since we may assume that $|t' - t| < N^{-1}$, this last expression is less than

$$4c'\{2|t' - t|N^{-1} \log|t' - t|^{-1}\}^{1/2} < \varepsilon\{2|t' - t|\log|t' - t|^{-1}\}^{1/2}.$$

Collecting up these inequalities we complete the proof of the theorem. □

Many authors have improved Lévy's original theorem since 1937, and we mention in particular Chung, Erdös and Sirao who in 1959 obtained the definitive results concerning the uniform continuity of the Brownian sample path. Similar beautiful and definitive results have been obtained for local continuity as well, but these all require too much preparation for even the statements to be given here.

Before closing this section it is worth pointing out that properties involving the continuity of sample paths of Gaussian processes are being discussed even now, and the area continues to provide interesting problems for investigation.

2.3 Constructions of Brownian Motion

In this section we present two different methods of constructing a Brownian motion on a probability space $(\Omega, \mathbf{B}, P)$. One, due to P. Lévy, uses an approximating sequence of processes, and the other, introduced by R. E. A. C. Paley and N. Wiener, appeals to the Fourier series expansion. Each can be thought of as a descriptive method exhibiting certain important properties of Brownian motion.

(i) P. Lévy's Method (1948, §1)

Before we give the details of the construction, we point out the significant property which underlies the idea of the construction. Let $\{B(t, \omega): t \geq 0\}$ be a Brownian motion on $(\Omega, \mathbf{B}, P)$.

Proposition 2.2. *For any point t in the finite interval (a, b), $B(t)$ can be expressed in the form*

$$B(t) = \mu(t) + \sigma(t)X(t) \tag{2.35}$$

where

$$\mu(t) = (b - a)^{-1}\{(b - t)B(a) + (t - a)B(b)\}, \tag{2.36}$$

$X(t)$ *is a standard Gaussian random variable independent of* $\{B(s): s \in [0, a] \cup [b, \infty)\}$, *and $\sigma(t)$ is given by*

$$\sigma^2(t) = \frac{(t - a)(b - t)}{b - a} \tag{2.37}$$

PROOF. Since $\{B(t): t \geq 0\}$ is a Gaussian system, we can make use of the results of §1.6.

We begin by showing that $\mu(t)$ in (2.36) coincides with the conditional expectation $E(B(t) | B(s), s \in [0, a] \cup [b, \infty))$. Clearly $\mu(t)$ is a function of the $B(s)$, $s \in [0, a] \cup [b, \infty]$, and we know that $\{B(t): t \geq 0\}$ is a Gaussian system with expectation 0. From Proposition 1.10 (b) it suffices to prove that

$$E[(B(t) - \mu(t))B(s)] = 0$$

for any $s \in [0, a] \cup [b, \infty)$. Direct computations show that the left side equals $s - (b - t)s/(b - a) - (t - a)s/(b - a)$ or $t - (b - t)a/(b - a) - (t - a)b/(b - a)$ according as $s < a$ or $s > b$, which is 0 in either case. Thus $\mu(t)$ is the required conditional expectation and hence $B(t) - \mu(t)$ is independent of $\{B(s): s \in [0, a] \cup [b, \infty)\}$. By computing the variance of $B(t) - \mu(t)$ we obtain (2.37). □

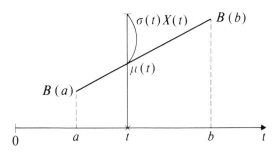

Figure 4

The content of this proposition may be paraphrased as follows: given the values of $B(t)$ outside a time interval (a, b), its value at a point t inside (a, b) is obtained by interpolating linearly between $B(a)$ and $B(b)$, and adding to this an independent amount $\sigma(t)X(t)$, where $\sigma(t)$ is given by (2.37) and $X(t)$ is a standard Gaussian random variable; see Fig. 4.

We note in passing a consequence of the proof of Proposition 2.2.

2.3 Constructions of Brownian Motion

Corollary. *The expression $\mu(t)$ in the decomposition (2.35) of $B(t)$ with $a < t < b$ can be expressed as a conditional expectation*

$$\mu(t) = E(B(t) | B(a), B(b)). \tag{2.38}$$

With Proposition 2.2 as our inspiration, we come now to Lévy's construction of Brownian motion, beginning with the case in which the time interval is $[0, 1]$.

We start with an independent sequence $\{Y_n(\omega): n \geq 1\}$ of standard Gaussian random variables defined on a probability space $(\Omega, \mathbf{B}, P)$. Let T_n denote the set of all binary numbers $k/2^{n-1}$, $k = 0, 1, \ldots, 2^{n-1}$, $n \geq 1$, and put $T_0 = \bigcup_{n \geq 1} T_n$.

A Gaussian process $\{X_1(t, \omega): t \in [0, 1]\}$ is defined by

$$X_1(0, \omega) = 0$$
$$X_1(1, \omega) = Y_1(\omega)$$
$$X_1(t, \omega) = tY_1(\omega), \qquad t \notin T_1,$$

and a sequence $\{X_n(t): t \in [0, 1]\}$, $n \geq 1$, can now be defined inductively. Suppose that $\{X_n(t): t \in [0, 1]\}$ is given; then $\{X_{n+1}(t): t \in [0, 1]\}$ is defined by

$$X_{n+1}(t, \omega) = \begin{cases} X_n(t, \omega), & t \in T_n; \\ \dfrac{1}{2}[X_n(t + 2^{-n}, \omega) + X(t - 2^{-n}, \omega)] + 2^{-(n+1)/2} Y_k(\omega), \\ \qquad\qquad\qquad\qquad\qquad\qquad\qquad\qquad\qquad t \in T_{n+1} \setminus T_n; \\ 2^n((k+1)2^{-n} - t)X_{n+1}(k2^{-n}, \omega) \\ \quad + 2^n(t - k2^{-n})X_{n+1}((k+1)2^{-n}, \omega), \\ \qquad\qquad\qquad\qquad\qquad\qquad t \in [k2^{-n}, (k+1)2^{-n}]; \end{cases}$$

where the integer k in the middle expression is determined by the relation $k = k(t) = 2^{n-1} + \frac{1}{2}(2^n t - 1)$, $t \in T_{n+1} \setminus T_n$.

If t is restricted to the set T_n we see that $\{B(t): t \in T_n\}$ and $\{X_n(t): t \in T_n\}$ have the same distribution; indeed the covariance functions of the two processes coincide. Thus we would expect that the sequence of Gaussian processes $\{X_n(t): t \in [0, 1]\}$, $n \geq 1$, converges in some sense to Brownian motion.

Take a time point $t \in T_0$. By the definition of the $X_n(t)$, there exists a number $N = N(t, \omega)$ such that for every $n > N$, $X_n(t, \omega) = X_N(t, \omega)$. Consequently, if we regard $X_n(t)$ as an element of $L^2(\Omega, \mathbf{B}, P)$, we certainly have a m.s. limit, $\lim_{n \to \infty} X_n(t)$. Now the collection $\{X(t): t \in T_0\}$ can be viewed as a uniformly continuous function on T_0 taking values in $L^2(\Omega, \mathbf{B}, P)$ since $E[|X(t) - X(t')|^2] \leq |t - t'|$, $t, t' \in T_0$. Therefore we can extend it to a continuous function $\{X(t): 0 \leq t \leq 1\}$ which, by construction, is necessarily Gaussian. Since $X_n(t)$ is obtained by successively interpolating independent Gaussian random variables we must, for any $t \in [0, 1]$, have

$$\lim_{n \to \infty} X_n(t) = X(t) \quad \text{in} \quad L^2(\Omega, \mathbf{B}, P).$$

Further, we have

$$E(X(t)) = 0, \qquad E(X(t)X(s)) = t \wedge s,$$

and so we have proved that $\{X(t, \omega): t \in [0, 1]\}$ is a Brownian motion.

The convergence of $\{X_n(t)\}$ to Brownian motion just demonstrated does not guarantee the convergence of the sequence $X_n(\cdot, \omega)$ of sample functions, but the next proposition resolves this question.

Proposition 2.3. *For almost all ω the limit*

$$\lim_{n \to \infty} X_n(t, \omega) = \bar{X}(t, \omega) \tag{2.39}$$

exists, and $\bar{X}(t, \omega)$, $0 \leq t \leq 1$, is a continuous function.

PROOF. We begin by setting

$$Z_n(t, \omega) = X_{n+1}(t, \omega) - X_n(t, \omega).$$

Clearly $Z_n(t, \omega) = 0$ for $t \in T_n$, and also

$$\max_{0 \leq t \leq 1} |Z_n(t, \omega)| = \max_{2^{n-1} \leq k < 2^n} 2^{-(n+1)/2} |Y_k(\omega)|.$$

Now we evaluate the probability

$$P_n = P\left(\max_{0 \leq t \leq 1} |Z_n(t, \omega)| > \lambda_n\right)$$

$$\leq 2^{n-1} P(|Y_k(\omega)| \geq 2^{(n+1)/2} \lambda_n)$$

$$\leq 2^{n-1} (2^n \pi)^{-1/2} \lambda_n^{-1} \exp\left[-\frac{1}{2}(2^{(n+1)/2} \lambda_n)^2\right]$$

by the second inequality of (2.27).

Setting $\lambda_n = (2 \, cn \log 2)^{1/2} 2^{-(n+1)/2}$ with $c > 1$, we can see that

$$\sum_{n \geq 1} P_n < \infty.$$

Applying the first part of the Borel–Cantelli Lemma, we get

$$\max_t |Z_n(t, \omega)| \leq (2 \, cn \log 2)^{1/2} 2^{-(n+1)/2} \qquad \text{a.e.}$$

except for finitely many n (depending on ω). This implies that, for almost all ω, the series $\sum_k Z_n(t, \omega)$ is uniformly absolutely convergent, a fact which implies the existence of limit (2.39). Furthermore, since $Z_n(t, \omega)$ is continuous in t, the continuity of $\bar{X}(t, \omega)$ is also proved. □

2.3 Constructions of Brownian Motion

The foregoing discussion also implies that for each fixed t we have the relation

$$\bar{X}(t, \omega) = X(t, \omega) \quad \text{a.e.}$$

Equivalently, $\{\bar{X}(t, \omega): t \in [0, 1]\}$ is a version of Brownian motion for which almost all sample paths are continuous. The existence of such a process is of course consistent with the fact that the Wiener measure on $\mathscr{C}$ discussed in §2.1 is the distribution of Brownian motion.

When the time index t runs over the half-line $[0, \infty)$ we form a Brownian motion in the following stages. Firstly take two independent Brownian motions $\{X^{(i)}(t): t \in [0, 1]\}$, $i = 1, 2$, on the same probability space $(\Omega, \mathbf{B}, P)$. This could be done by starting with an independent system of standard Gaussian random variables $\{Y_n^{(i)}(\omega): i = 1, 2, n \geq 1\}$ and forming $X^{(i)}(t)$ using $\{Y_n^{(i)}: n \geq 1\}$, $i = 1, 2$, as we did above. Then we connect the two Brownian motions in the following way:

$$B(t, \omega) = \begin{cases} X^{(1)}(t, \omega) & t \in [0, 1] \\ X^{(1)}(1, \omega) + tX^{(2)}(t^{-1}, \omega) - X^{(2)}(1, \omega) & t > 1 \end{cases}$$

and see that by Proposition 2.1 (i), we have obtained a Brownian motion with parameter space $[0, \infty)$.

(ii) The Paley-Wiener Method (1934, Chapter IX)

Since we will be using the technique of Fourier expansions, it is convenient to work with complex-valued functions and complex random variables. In any event we proceed to form a complex Brownian motion.

Let $X_k(\omega)$, $Y_k(\omega)$, $k = 0, \pm 1, \pm 2, \ldots$ be an independent sequence of standard Gaussian random variables, and define

$$Z_k(\omega) = 2^{-1/2}(X_k(\omega) + iY_k(\omega)), \quad i = \sqrt{-1}.$$

Each $Z_k(\omega)$ is a complex Gaussian random variable (see §6.1), and $2^{1/2}Z_k(\omega)$ has a standard two-dimensional Gaussian distribution, the complex plane being regarded as two-dimensional Euclidean space. From their definition the $Z_k(\omega)$, $k = 0, \pm 1, \pm 2, \ldots$ are seen to form an independent system of random variables with $E(Z_k) = 0$ and

$$E(Z_k Z_l) = 0, \; E(Z_k \bar{Z}_l) = \delta_{k,l}. \tag{2.40}$$

We now consider and *ideal* random function of t (see §6.2) in the sense that its spectrum $[= 2\pi \times \text{(frequencies)}]$ is distributed over the entire set of integers, and the associated amplitudes are the independent identically distributed random variables, $Z_k(\omega)$, $k \geq 1$. That is, we form

$$\sum_{k=-\infty}^{\infty} Z_k(\omega)e^{ikt}. \tag{2.41}$$

The series (2.41) unfortunately does not converge for any fixed ω, but nevertheless it is still possible to modify it by term-by-term integration in t and obtain a well-defined random function with convergent series in the form

$$Z_1(t, \omega) = tZ_0(\omega) + \sum_{n=1}^{\infty} \frac{Z_n(\omega)(e^{int} - 1)}{in}$$

$$+ \sum_{n=1}^{\infty} \frac{Z_{-n}(\omega)(e^{-int} - 1)}{-in} \qquad (2.42)$$

Since by (2.40) $\{Z_k\}$ forms an orthonormal sequence in the complex Hilbert space $L_c^2(\Omega, \mathbf{B}, P)$, we see that the two series on the right of (2.42) both converge strongly; indeed both are less than $4 \sum_n n^{-2}$ in norm.

We now show that $\{(2\pi)^{-1/2} Z_1(t): t \in [0, 2\pi]\}$ is a complex form of Brownian motion with parameter set $[0, 2\pi]$.

Lemma 2.5. *For every t, s in $[0, 2\pi]$ we have the equality*

$$\sum_n' \frac{(e^{int} - 1)(e^{-ins} - 1)}{n^2} = \begin{cases} s(2\pi - t), & s \leq t; \\ t(2\pi - s), & t \leq s; \end{cases} \qquad (2.43)$$

where $\sum_n'$ denotes the sum $\sum_{n=-\infty}^{-1} + \sum_{n=1}^{\infty}$.

PROOF. Denote the right side of (2.43) by $G(t, s)$, where we have $G(t, s) = (s \wedge t)[2\pi - (s \vee t)]$. We will take it to be the kernel of an integral operator acting on a space of functions defined on the unit circle. The eigenfunctions and eigenvalues are obtained by the formula

$$\lambda \int_0^{2\pi} G(t, s) \varphi(s) \, ds = \varphi(t),$$

or, equivalently, by the differential equation

$$\varphi''(t) = -2\pi \lambda \varphi(t), \qquad \varphi(0) = \varphi(2\pi) = 0, \qquad (2.44)$$

the well-known equation describing a simple harmonic oscillator.

Also we can view the left-hand side of (2.43) as an integral kernel and obtain an integral equation of the form

$$\sum_n' \frac{(e^{int} - 1)}{n^2} \int_0^{2\pi} (e^{-ins} - 1) \varphi(s) \, ds = \lambda^{-1} \varphi(t). \qquad (2.45)$$

Set $\int_0^s \varphi(u) \, du = \Phi(s)$. Then integration by parts gives

$$\int_0^{2\pi} (e^{-ins} - 1) \varphi(s) \, ds = in \int_0^{2\pi} e^{-ins} \Phi(s) \, ds.$$

With this formula we can differentiate (2.45) with respect to t and obtain

$$\sum_n' \frac{ie^{int}}{n} \cdot in \int_0^{2\pi} e^{-ins} \Phi(s) \, ds = \lambda^{-1} \varphi'(t).$$

2.3 Constructions of Brownian Motion

Replacing e^{-ins} there with $(2\pi)^{-1/2}e^{-ins}$, the left side of the above expression coincides, up to a constant multiple ($= -2\pi$), with the Fourier expansion of Φ without the constant term. Differentiating once more in t we obtain

$$-2\pi\Phi'(t) = \lambda^{-1}\varphi''(t),$$

in agreement with (2.44).

So far we have just proved (2.44) by regarding both sides as $L^2([0, 2\pi] \times [0, 2\pi])$ functions, but the proof is complete once we observe that both sides are in fact continuous functions. □

Proposition 2.4. *Set* $Z(t, \omega) = (2\pi)^{-1/2}Z_1(t, \omega)$. *Then we have*

$$E(Z(t)) = 0$$

and (2.46)

$$E(Z(t)\overline{Z(s)}) = t \wedge s.$$

PROOF. Since each Z_k has zero expectation, the assertion that $E(Z(t)) = 0$ is obvious. Turning to the covariance function we compute as follows:

$$E(Z_1(t)\overline{Z_1(s)}) = ts + \sum_{n=1}^{\infty} \frac{(e^{int} - 1)(e^{-ins} - 1)}{n^2}$$

$$+ \sum_{n=1}^{\infty} \frac{(e^{-int} - 1)(e^{ins} - 1)}{n^2}$$

$$= ts + \sum_{n}' \frac{(e^{int} - 1)(e^{-ins} - 1)}{n^2}$$

$$= ts + G(t, s)$$

$$= \begin{cases} 2\pi s & s \leq t, \\ 2\pi t & t \leq s, \end{cases} \quad \text{by Lemma 2.5}$$

$$= 2\pi(s \wedge t). \qquad \square$$

As in the case (i) we are also interested in the formation of each sample path and so we now ask whether the series (2.42) converges for a given ω. To this end we form the partial sums

$$Z_{m,n}(t, \omega) = \sum_{k=m+1}^{n} \frac{Z_k(\omega)e^{ikt}}{ik}, \quad m < n,$$

and prove our first result in this direction.

Proposition 2.5. *For almost all ω the series*

$$\sum_{n=1}^{\infty} |Z_{2^n, 2^{n+1}}(t, \omega)| \qquad (2.47)$$

converges uniformly in t.

PROOF. We evaluate $Z_{m,n}(t, \omega)$ using the notation Re(·) for real part, obtaining

$$|Z_{m,n}(t, \omega)|^2 = \sum_{k=m+1}^{n} \frac{|Z_k(\omega)|^2}{k^2} + 2\operatorname{Re}\left\{\sum_{j=1}^{n-m-1} e^{ijt} \sum_{k=m+1+j}^{n} \frac{Z_k(\omega)\overline{Z_{k-j}(\omega)}}{k(k-j)}\right\}$$

$$\leq \sum_{k=m+1}^{n} \frac{|Z_k(\omega)|^2}{k^2} + 2\sum_{j=1}^{n-m-1} \left|\sum_{k=m+1+j}^{n} \frac{Z_k(\omega)\overline{Z_{k-j}(\omega)}}{k(k-j)}\right|.$$

Set $T_{m,n}(\omega) = \sup_t |Z_{m,n}(t, \omega)|$. Then we have

$$E(T_{m,n}^2) = E\left\{\sup_t |Z_{m,n}(t, \omega)|^2\right\}$$

$$\leq \sum_{k=m+1}^{n} k^{-2} + 2\sum_{j=1}^{n-m-1} E\left(\left|\sum_{k=m+1+j}^{n} \frac{Z_k \overline{Z}_{k-j}}{k(k-j)}\right|\right) \quad (2.48)$$

and also

$$E\left\{\left|\sum_{k=m+1+j}^{n} \frac{Z_k \overline{Z}_{k-j}}{k(k-j)}\right|^2\right\}$$

$$= \sum_{k=m+1+j}^{n} E\left(\frac{|Z_k|^2 |Z_{k-j}|^2}{k^2(k-j)^2}\right)$$

$$+ 2\operatorname{Re} \sum_{m+1+j \leq l < k \leq n} E\left(\frac{Z_k \overline{Z}_{k-j} \overline{Z}_l Z_{l-j}}{k(k-j)l(l-j)}\right) \quad (2.49)$$

$$= \sum_{k=m+1+j}^{n} k^{-2}(k-j)^{-2},$$

the second sum vanishing because the $\{Z_k\}$ form an independent system. One should be careful in the case $k - j = l$, but the expectation still vanishes since Z_k is independent of $\overline{Z}_{k-j}\overline{Z}_l Z_{l-j}$, i.e.

$$E(Z_k \overline{Z}_{k-j} \overline{Z}_l Z_{l-j}) = E(Z_k)E(\overline{Z}_{k-j}\overline{Z}_l Z_{l-j}) = 0.$$

Upon substituting the outcome of (2.49) into (2.48) and using the Schwarz inequality in the form $(E(|X|))^2 \leq E(|X|^2)$, we obtain

$$E(T_{m,n}^2) \leq \sum_{k=m+1}^{n} k^{-2} + 2\sum_{j=1}^{n-m-1} \left\{\sum_{k=m+1+j}^{n} k^{-2}(k-j)^{-2}\right\}^{1/2}$$

$$\leq \frac{n-m}{m^2} + 2(n-m)\left|\frac{n-m}{m^4}\right|^{1/2}.$$

Setting $n = 2m$ and again using the Schwarz inequality we find that

$$E\left(\sup_t |Z_{m,2m}(t, \omega)|\right) \leq E(T_{m,2m}^2)^{1/2}$$

$$\leq (m^{-1} + 2m^{-1/2})^{1/2} \leq 2m^{-1/4}.$$

2.3 Constructions of Brownian Motion

Consequently

$$\sum_{n=1}^{\infty} E\left(\sup_t |Z_{2^n, 2^{n+1}}(t, \omega)|\right) \le 2 \sum_{n=1}^{\infty} 2^{-n/4} < \infty,$$

whence we have proved that

$$E\left(\sum_{n=1}^{\infty} \sup_t |Z_{2^n, 2^{n+1}}(t, \omega)|\right) < \infty$$

thus showing that, almost surely, the series (2.47) converges uniformly in t. □

The third term in the series (2.42) can be dealt with in a similar manner, and we can then go on to prove the almost sure convergence of the series (2.42) by turning it into a series of partial sums [and scaling by $(2\pi)^{-1/2}$]

$$Z(t, \omega) = (2\pi)^{-1/2} t Z_0(\omega)$$
$$+ \sum_{n=0}^{\infty} \left[\sum_{k=2^n+1}^{2^{n+1}} \left|\frac{Z_k(\omega)(e^{ikt} - 1)}{ik(2\pi)^{1/2}} + \frac{Z_{-k}(\omega)(e^{-ikt} - 1)}{-ik(2\pi)^{1/2}}\right|\right]. \quad (2.50)$$

Indeed if the almost sure convergence of the terms $\sum_{k=2^n+1}^{2^{n+1}} Z_k(\omega)/k$ is proved, then our assertion immediately follows from Proposition 2.5.

The convergence (in fact uniform in t) in question is a consequence of the convergence in $L_c^2(\Omega, \mathbf{B}, P)$ of the series

$$\sum_k \frac{Z_k(\omega)}{k}$$

and hence, since the terms are Gaussian, of its almost sure convergence. In Theorem 1.9 we discussed the real case of this implication, and this result applies now to the real and imaginary parts separately.

As a corollary to Proposition 2.5 we thus have

Corollary. *For almost all ω the function $Z(t, \omega)$ given by the series (2.50), $t \in [0, 2\pi]$, is a continuous function of t.*

We now have constructed a complex Brownian motion $\{Z(t, \omega): t \in [0, 2\pi]\}$ whose sample paths have been proved to be continuous functions expressible as sums of Fourier series. Returning again to (2.42) we write it as the sum of its real and imaginary parts in the form

$$(2\pi)^{-1/2} Z_1(t, \omega) = Z(t, \omega) = 2^{-1/2}[X(t, \omega) + iY(t, \omega)] \quad (2.51)$$

$X(t, \omega)$, $Y(t, \omega)$ real-valued, and recall that $Z_k(\omega) = 2^{-1/2}[X_k(\omega) + iY_k(\omega)]$. Then we can read off the expression

$$X(t, \omega) = \frac{tX_0(\omega)}{\sqrt{2\pi}} + \sum_n{}' \frac{X_n(\omega)\sin nt + Y_n(\omega)(\cos nt - 1)}{\sqrt{2\pi} n}, \quad (2.52)$$

$$Y(t, \omega) = \frac{tY_0(\omega)}{\sqrt{2\pi}} + \sum_n{}' \frac{-X_n(\omega)(\cos nt - 1) + Y_n(\omega)\sin nt}{\sqrt{2\pi} n}. \quad (2.53)$$

Proposition 2.6. *The two stochastic processes $\{X(t, \omega): t \in [0, 2\pi]\}$ and $\{Y(t, \omega): t \in [0, 2\pi]\}$ given by (2.52) and (2.53) respectively are independent Brownian motions.*

PROOF. Recall that $\{X_n, Y_n: n = 0, \pm 1, \pm 2, \ldots\}$ is a system of independent random variables and that the X_n and Y_n all have standard Gaussian distributions. The system if of course a Gaussian system so that the system given by

$$\{X_0, Y_0, X_n \sin nt + Y_n(\cos nt - 1),$$
$$- X_n(\cos nt - 1) + Y_n \sin nt: n = \pm 1, \pm 2 \cdots\} \quad (2.54)$$

is again a Gaussian system. In addition, the random variables in the system are all mutually independent. To see this it is enough to prove that the covariance between each pair of random variables in (2.54) vanishes, a fact which is obvious when the subscripts differ, and when they coincide we have

$$E\{[X_n \sin nt + Y_n(\cos nt - 1)][-X_n(\cos nt - 1) + Y_n \sin nt]\}$$
$$= -\sin nt(\cos nt - 1)E(X_n^2) + \sin nt(\cos nt - 1)E(Y_n^2)$$
$$= 0.$$

It follows from this fact and the formulae (2.52) and (2.53) that the two processes $\{X(t, \omega): t \in [0, 2\pi]\}$ and $\{Y(t, \omega): t \in [0, 2\pi]\}$ are independent, and that they are Gaussian processes with mean 0. We now compute their covariance functions, firstly obtaining the variance of the random variables in the system.

$$E(X_0^2) = E(Y_0^2) = 1,$$
$$E\{[X_n \sin nt + Y_n(\cos nt - 1)]^2\}$$
$$= E\{[-X_n(\cos nt - 1) + Y_n \sin nt]^2\}$$
$$= \sin^2 nt + (\cos nt - 1)^2$$
$$= 4 \sin^2 \tfrac{1}{2} nt, \quad n \neq 0.$$

With these formulae it can be proved that $\{X(t): t \in [0, 2\pi]\}$ and $\{Y(t): t \in [0, 2\pi]\}$ have the same (Gaussian) distribution, and in particular, that

$$E(X(t)X(s)) = E(Y(t)Y(s)).$$

On the other hand we can use the independence of $\{X(t)\}$ and $\{Y(t)\}$, and the fact that these processes have zero expectations, to rewrite formula (2.46)

$$E(Z(t)\overline{Z(s)}) = t \wedge s$$

in the form

$$\tfrac{1}{2}E\{[X(t) + iY(t)][X(s) - iY(s)]\} = \tfrac{1}{2}\{E(X(t)X(s)) + E(Y(t)Y(s))\} = t \wedge s.$$

Thus we have proved that both $\{X(t): t \in [0, 2\pi]\}$ and $\{Y(t): t \in [0, 2\pi]\}$ are Brownian motion processes. □

Remark 1. In the light of Proposition 2.6 we might start with independent normalised (real) Gaussian random variables

$$X'_n = \frac{X_n \sin nt + Y_n(\cos nt - 1)}{2 \sin \tfrac{1}{2} nt} \qquad n \neq 0$$

and obtain the expression (2.52) directly in the form

$$X(t, \omega) = \frac{t X_0(\omega)}{\sqrt{2\pi}} + \sum_n{}' \frac{\sqrt{2} X'_n(\omega) \sin(\tfrac{1}{2} nt)}{\sqrt{\pi} n}, \qquad t \in [0, 2\pi]. \quad (2.55)$$

Thus we have a method of constructing (real) Brownian motion starting from an independent system $\{X_n : n = 0, \pm 1, \pm 2, \ldots\}$ of standard Gaussian random variables [cf. K. Itô (1953) §39].

Remark 2. In order to prove that the series (2.52), (2.53) and (2.55) converge uniformly in t almost surely, it is necessary to pass to the series of partial sums as we did for the series (2.50). We have omitted the details of this to avoid further complex calculations.

Definition 2.4. A complex-valued stochastic process $\{Z(t, \omega): t \in T\}$ defined on a finite or infinite interval subset of $[0, \infty)$ is called a *complex Brownian motion* if there exist independent real-valued Brownian motions $\{X(t, \omega): t \in T\}$ and $\{Y(t, \omega): t \in T\}$ such that

$$Z(t, \omega) = 2^{-1/2}(X(t, \omega) + iY(t, \omega)). \quad (2.56)$$

Our discussion so far may be summarised by saying that we have constructed a complex Brownian motion, and have obtained two independent (real) Brownian motions as its real and imaginary parts. The reader may consult N. Wiener and E. J. Akutowicz (1957) for interesting results concerning complex Brownian motion.

Many of the properties of complex Brownian motion, such as its additivity properties (cf. §2.1), sample path behaviour, and so on, may be reduced to the corresponding ones for real Brownian motion through the decomposition (2.56). We will not state such obvious generalisations, but we are interested in nonlinear functionals of a complex Brownian motion, these being discussed in detail in Chapter 6.

Before we close this section we will explain why we started with the formal series (2.41) or with the series (2.42) to construct our complex Brownian motion. Of course this choice was not accidental, but was based upon a consideration of the nature of the Fourier expansion of a typical Brownian sample path. We make this plausible in what follows.

Let us suppose that a complex Brownian motion $\{Z(t, \omega): t \in [0, 2\pi]\}$ is given, and assume that all its sample paths are continuous. Modify $Z(t)$ to obtain a function on the unit circle, $Z_0(t)$ say, by identifying its values at $t = 0$ and $t = 2\pi$,

$$Z_0(t, \omega) = Z(t, \omega) - (2\pi)^{-1} t Z(2\pi, \omega). \tag{2.57}$$

Clearly $E(Z_0(t)) \equiv 0$ and

$$E(Z_0(t)\overline{Z_0(s)}) = t \wedge s - (2\pi)^{-1} st, \tag{2.58}$$

which equals $(2\pi)^{-1} G(s, t)$, G being the function defined by (2.43). Corresponding to the expansion (2.43) of G we would expect an expansion of $Z_0(t)$ itself.

For any fixed ω we regard $Z(t, \omega)$ as an $L^2([0, 2\pi])$-function of t and expand it in a Fourier series using $\{(2\pi)^{-1/2} \exp(int): n = 0, \pm 1, \pm 2, \ldots\}$ as a complete orthonormal system. We obtain the Fourier coefficients

$$A_n(\omega) = (2\pi)^{-1/2} \int_0^{2\pi} e^{-int} Z_0(t, \omega) \, dt,$$

and the formal Fourier series expansion is given by

$$Z_0(t, \omega) \sim \frac{1}{\sqrt{2\pi}} \sum_{n=-\infty}^{\infty} A_n(\omega) e^{int}. \tag{2.59}$$

We now proceed to show that the preceding expression is in fact an equality. For $m \neq 0$, $n \neq 0$ we make use of the elementary result

$$\int_0^{2\pi} e^{ins} G(t, s) \, ds = \frac{2\pi(e^{int} - 1)}{n^2},$$

and obtain

$$E(A_m \bar{A}_n) = (2\pi)^{-1} \int_0^{2\pi} \int_0^{2\pi} \frac{e^{-imt + ins} G(t, s) \, dt \, ds}{2\pi}$$

$$= (2\pi n^2)^{-1} \int_0^{2\pi} [e^{i(n-m)t} - e^{-imt}] \, dt \tag{2.60}$$

$$= \delta_{m,n} n^{-2}.$$

Now the definition of A_n readily implies a decomposition of the form $A_n = X_n + iY_n$, where X_n and Y_n are real-valued independent random variables with the same Gaussian distribution. The result (2.60) shows that the collection $\{A_n: n = 0, \pm 1, \pm 2, \ldots\}$ is an independent system; a general theory of such complex-valued random variables will be developed in §6.1.

Concerning the exceptional coefficient $A_0(\omega)$ we will prove the relation

$$-A_0(\omega) = \sum_n{}' A_n(\omega). \tag{2.61}$$

To see this we first note that the right-hand side converges almost surely, because the A_n are mutually independent and from (2.60) we see that the sum of their variances is finite. We now begin to prove the equality (2.61). Since

$G(t, s)$ is integrable over the square $[0, 2\pi] \times [0, 2\pi]$, $Z_0(t, \omega)$ is square-integrable in (t, ω) i.e.

$$Z_0(t, \omega) \in L^2([0, 2\pi] \times \Omega, m \times P), \quad m = \text{Lebesgue measure}.$$

The formulae (2.43) in Lemma 2.5 and $\sum_{n>0} n^{-2} = \pi^2/6$ then imply that

$$E\left\{ \int_0^{2\pi} \left| Z_0(\omega, t) - (2\pi)^{-1/2} \sum_{1 \leq |n| \leq N} A_n(\omega)(e^{int} - 1) \right|^2 dt \right\} \to 0$$

as $N \to \infty$ and so we deduce from Fubini's theorem that

$$(2\pi)^{-1/2} \sum_n' A_n(\omega)(e^{int} - 1) = Z_0(t, \omega) \quad \text{a.s. } (P).$$

Let the sum $\sum_n'$ on the left-hand side be expressed as a series of partial sums as in (2.50), and obtain a continuous (sample) function in $t \in [0, 2\pi]$. Then term-by-term integration with respect to t must coincide with the integral of $Z_0(t, \omega)$, which easily proves (2.61).

If we now put $Z_m = inA_n$ to normalise A_n, then we see that the series (2.42) arises quite naturally when considering an expansion in a Fourier series.

2.4 Markov Properties of Brownian Motion

This section is devoted to a discussion of the more basic and important properties of Brownian motion that arise when its development in time is taken into account. They are the Markov and strong Markov properties, and related topics.

For a time-interval T let $\{X(t, \omega): t \in T\}$ be a stochastic process such that $X(t, \omega)$ is jointly measurable in t and ω, and set $\mathbf{B}_t(X) = \mathbf{B}(X(s): s \leq t)$.

Definition 2.5. A stochastic process $\{X(t)\}$ is called a *Markov process* if for any s and $t > s$, and any Borel subset A of $\mathbf{R}$, we have

$$P(X(t) \in A | \mathbf{B}_s(X)) = P(X(t) \in A | X(s)) \quad \text{a.e.} \quad (2.62)$$

Remark. For the definitions and notation associated with conditional probabilities we refer to §1.4.

Associated with any Markov process $\{X(t): t \in T\}$ is a system $P(s, x, t, A)$ of functions defined for $s, t > s$ in T, $x \in \mathbf{R}$ and $A \in \mathfrak{B}(\mathbf{R})$ and satisfying the following conditions:

i. $P(s, x, t, A)$ is a Borel-measurable function of x, for fixed s, t and A, and a probability measure in A, for fixed s, t and x.
ii. $P(s, x, s, A) = \chi_A(x)$.
iii. On the set $\{\omega: X(s, \omega) = x\}$ we have

$$P(X(t) \in A | X(s)) = P(s, x, t, A) \quad \text{a.e.}$$

iv. The following Chapman-Kolmogorov equation holds for almost all x, this being with respect to the distribution of $X(s)$:

$$P(s, x, u, A) = \int_{-\infty}^{\infty} P(s, x, t, dy) P(t, y, u, A) \qquad s < t < u. \quad (2.63)$$

Any such function $P(s, x, t, A)$ is called a *transition probability* of the Markov process $\{X(t)\}$. If the transition probability depends on t and s only through $t - s$, i.e. if $P(s, t, x, A) = P(t - s, x, A)$ for all fixed x and A, then the Markov process $\{X(t)\}$ is said to be *temporally homogeneous*. In this case we may set

$$P_x(X(t) \in A) = P(t, x, A), \qquad A \in \mathfrak{B}, \quad (2.64)$$

and obtain a system $\{P_x : x \in \mathbf{R}\}$ of probability measures on $(\Omega, \mathbf{B})$ associated with $\{X(t)\}$.

For definiteness in what follows we will take $T = [0, \infty)$. We claim that the distribution of a Markov process $\{X(t): t \in T\}$ is uniquely determined by the distribution Φ_0 of $X(0)$, and the system $\{P(s, x, t, A)\}$ of transition probabilities. In fact any finite-dimensional joint distribution, say that of $(X(t_1), X(t_2), \ldots, X(t_n))$ where $t_1 < t_2 < \cdots < t_n$, can be obtained by using the Markov property (2.62) repeatedly from

$$P((X(t_1), \ldots, X(t_n)) \in A_n)$$

$$= \int_{-\infty}^{\infty} \Phi_0(dx_0) \int \cdots \int_{A_n} P(0, x_0, t_1, dx_1) P(t_1, x_1, t_2, dx_2)$$

$$\cdots P(t_{n-1}, x_{n-1}, t_n, dx_n)$$

where A_n is an n-dimensional Borel set.

Now we change the point of view and start with a system $\{P(s, x, t, A)\}$ satisfying (i), (ii) and (2.63) above for all x, A, and construct a Markov process whose transition probability coincides with the given $P(s, x, t, A)$ as in (iii) above. The construction of $\{X(t)\}$ is the same as that illustrated in §1.3 (ii) with $\Omega = \mathbf{R}^T$, the Markov property deriving from the Chapman-Kolmogorov equation, which relates the finite-dimensional distributions to each other in the appropriate manner. In this case where $\Omega = \mathbf{R}^T$, the initial distribution Φ_0 and the system of transition probabilities uniquely determine the Markov process $\{X(t)\}$.

If the transition probability $P(s, x, t, A)$ is absolutely continuous with respect to Lebesgue measure when viewed as a measure in A, then we denote the density function by $p(s, x, t, y)$:

$$P(s, x, t, A) = \int_A p(s, x, t, y) \, dy$$

and $p(s, x, t, y)$ is called the *transition probability density*.

The general theory of Markov processes may be found in several standard texts [e.g. K. Itô and H. P. McKean Jr. (1965), W. Feller (1966), and

2.4 Markov Properties of Brownian Motion

E. B. Dynkin and A. A. Yushkevich (1969)]. Therefore we make no attempt to develop a general theory, but centre our attention on Brownian motion, for the Markov property is one of its most important properties. In addition to this we discuss some related topics.

Theorem 2.7. *Any Brownian motion $\{B(t, \omega): t \geq 0\}$ is a temporally homogeneous Markov process, and its transition probability has a density $p(s, x, t, y)$ which coincides with the Gauss kernel $g(t - s, x, y)$.*

PROOF. To prove the Markov property it is enough to prove for any $t > s$ and cylinder set $C \in \mathbf{B}_s = \mathbf{B}_s(B)$ that we have

$$\int_C P(B(t) \in A \mid \mathbf{B}_s) \, dP(\omega) = \int_C P(B(t) \in A \mid B(s)) \, dP(\omega). \tag{2.65}$$

Let us suppose that C is given by

$$C = \{B(s_n) \in A_n, \ldots, B(s_1) \in A_1, B(s) \in A_0\}$$

where $0 \leq s_n < s_{n-1} < \cdots < s_1 < s$, and $A_n, \ldots, A_1, A_0$ are all Borel sets in $\mathbf{R}$. (Since A_0 may always be taken to be $\mathbf{R}$, we may always include s as one of the s_j which help determine C.) With such a C the left-hand side of (2.65) is

$$\int_C \chi_{\{B(t) \in A\}}(\omega) \, dP(\omega) = P(C \cap \{B(t) \in A\})$$

$$= \int_{A_n} \cdots \int_{A_1} \int_{A_0} \int_A g(s_n, s_{n-1} - s_n, \ldots, s - s_1, t - s;$$

$$x_n, x_{n-1} - x_n, \ldots, x - x_1, y - x) \, dx_n \cdots dx_1 \, dx \, dy$$

$$= \int_{A_n} \cdots \int_{A_1} \int_{A_0} g(s_n, s_{n-1} - s_n, \ldots, s - s_1;$$

$$x_n, x_{n-1} - x_n, \ldots, x - x_1)$$

$$\times \left\{ \int_A g(t - s; y - x) \, dy \right\} dx_n \cdots dx_1 \, dx.$$

We now recall that $(B(s), B(t))$ has a two-dimensional Gaussian distribution with density $g(s, t - s; x, y - x)$ and so the expression inside the braces $\{\cdot\}$ of the last line above is the conditional probability $P(B(t) \in A \mid B(s))$ under the condition $B(s) = x$. Having made this observation we see that the last line above may be written as

$$\int_C P(B(t) \in A \mid B(s)) \, dP(\omega).$$

Furthermore we have seen that the transition probability $P(s, x, t, A)$ is expressible as $\int_A g(t - s; y - x) \, dy$ and, in particular, is a function depending on t and s only through $t - s$. Thus $\{B(t)\}$ a temporally homogeneous Markov process and its transition probability density is given by $g(t - s; y - x)$ where g is the Gauss kernel. □

Assuming that $B(s) = a$, $0 < t_1 < t_2 < \cdots < t_n$ and B_n is an n-dimensional Borel set, the Markov property of Brownian motion and a general relation between joint and conditional probabilities readily give us the following formula:

$$P((B(s + t_1), B(s + t_2), \ldots, B(s + t_n)) \in B_n | B(s))$$
$$= \int_{B_n} g(t_1; y_1 - a)g(t_2 - t_1; y_2 - y_1) \cdots g(t_n - t_{n-1}; y_n - y_{n-1}) \, dy_1 \cdots dy_n$$
$$= \int_{B_n} g(t_1, t_2 - t_1, \ldots, t_n - t_{n-1}; y_1 - a, y_2 - y_1, \ldots, y_n - y_{n-1}) \, dy_1 \cdots dy_n.$$
(2.66)

Setting $B_a(t, \omega) = B(t, \omega) + a$, $t \geq 0$ we see that the probability (2.66) is the distribution of $(B_a(t_1), \ldots, B_a(t_n))$. By following the same method as was used in §2.1 in introducing the Wiener measure μ^W, we can obtain a probability measure μ_a^W on the space $\mathscr{C}([0, \infty))$ of continuous functions, and this measure will be the distribution of $\{B_a(t, \omega): t \geq 0\}$. This measure may also be thought of as the distribution on $\mathscr{C}([0, \infty))$ of the process $\{B(t, \omega): t \geq 0\}$, defined on the space $(\Omega, \mathbf{B})$ equipped with the probability measure P_a introduced in (2.64). Because of this equivalence μ_a^W, $a \in \mathbf{R}$, is called *conditional Wiener measure*. Also this observation leads us to view the Markov property is implying that once the process arrives at the point a at some instant, it behaves thereafter as if it had started movement afresh from the point a.

Let $\{X(t, \omega): t \geq 0\}$ be a general Markov process being jointly measurable in (t, ω). A measurable function $\sigma(\omega)$ on $(\Omega, \mathbf{B})$ with $0 \leq \sigma(\omega) \leq \infty$ which satisfies the condition: for all t

$$\{\omega: \sigma(\omega) < t\} \in \mathbf{B}_t(X), \tag{2.67}$$

is called a *Markov time* [relative to $\{X(t)\}$].

Remark. Occasionally a more general definition of Markov time is desirable. Instead of the increasing system $\{\mathbf{B}_t(X): t \geq 0\}$ of sub-σ-fields, it is often useful to consider a system $\{\mathbf{A}_t: t \geq 0\}$ of sub-σ-fields such that (i) $\mathbf{A}_t \supseteq \mathbf{B}_t(X)$; (ii) $\mathbf{A}_t \supseteq \mathbf{A}_s$ when $t > s$; and (iii) $\mathbf{A}_s$ is independent of $\{X(t + s) - X(s): s \geq 0\}$. In this case σ would be called a Markov time relative to $\{\mathbf{A}_t\}$ if for all t

$$\{\omega: \sigma(\omega) < t\} \in \mathbf{A}_t.$$

We do not need this wider definition in what follows, and so content ourselves with the narrow definition using $\mathbf{B}_t(X)$ as in (2.67).

2.4 Markov Properties of Brownian Motion

EXAMPLE 1. Setting $G = [-1, 1]^c$, we define σ by

$$\sigma(\omega) = \sigma_G(\omega) = \inf\{t: B(t, \omega) \in G\} \tag{2.68}$$

where $\{B(t)\}$ is a Brownian motion. Since we have the relations

$$\{\omega: \sigma_G(\omega) < t\} = \{\omega: B(r, \omega) \in G \text{ for some rational } r < t\}$$
$$= \bigcup_{r<t} \{\omega: B(r, \omega) \in G\} \in \mathbf{B}_t,$$

we have proved that σ is a Markov time.

Using the notion of Markov time we are now able to introduce the strengthening of the usual Markov property known as the strong Markov property. Since we have been assuming all along that $X(t, \omega)$ is jointly measurable in (t, ω), the function $X(t \wedge \sigma(\omega), \omega)$ is **B**-measurable for every $t > 0$, and thus we have a new stochastic process $\{X(t \wedge \sigma(\omega), \omega): t \geq 0\}$. Let $\mathbf{B}_\sigma$ be the smallest sub-σ-field with respect to which all the random variables $X(t \wedge \sigma)$, $t \geq 0$, are measurable, and set $\mathbf{B}_{\sigma+} = \bigcap_{\varepsilon > 0} \mathbf{B}_{\sigma+\varepsilon}$.

Definition 2.6. A temporally homogeneous Markov process $\{X(t)\}$ is called a *strong Markov process* if for any Markov time σ, $A \in \mathbf{B}_{\sigma+}$, times $t_1, \ldots, t_n \geq 0$, and Borel subsets $B_1, \ldots, B_n$ of **R**, we have:

$$P_a\{A \cap (X(\sigma + t_j) \in B_j: 1 \leq j \leq n) \cap (\sigma < \infty)\}$$
$$= E_a\{\chi_A \cdot P_{X(\sigma)}(X(t_j) \in B_j: 1 \leq j \leq n); \sigma > \infty\}. \tag{2.69}$$

In this expression E_a denotes the expectation with respect to the measure P_a and $E(\cdot; \sigma < \infty)$ means the integral is taken over the set $\{\omega: \sigma(\omega) < \infty\}$.

Brownian motion is a strong Markov process in the sense of this definition and we now proceed to prove this in several steps. Firstly we recall that by removing P-null sets from Ω if necessary, we may suppose that the sample functions $B(t, \omega)$ are continuous in t. Following this **B** and $\mathbf{B}_t$ should be somewhat modified, but we continue to use the same symbols. Let σ be a Markov time and set

$$B_\sigma(t, \omega) = B(t \wedge \sigma(\omega), \omega), \qquad t \geq 0. \tag{2.70}$$

The sample functions of the stochastic process $\{B_\sigma(t): t \geq 0\}$ thus obtained are all continuous. For example let us consider the Markov time σ_G introduced in Example 1 above. We see that $B_\sigma(t, \omega)$ behaves in exactly the same manner as the Brownian motion $B(t, \omega)$ whilst $B(t, \omega)$ remains within the interval $[-1, 1]$, i.e. up until the time $\sigma(\omega)$, but as soon as $B(t, \omega)$ arrives at either 1 or -1, the arrival time $\sigma(\omega)$ depending on ω, $B_\sigma(t, \omega)$ stops moving and stays where it is forever.

Proposition 2.7. *A Markov time $\sigma(\omega)$ is $\mathbf{B}_\sigma$-measurable.*

PROOF. We first note that it is a consequence of the continuity of the sample paths and the additive property of Brownian motion (which implies the 0–1 law, see K. Itô and H. P. McKean Jr. (1965), Chap. 1) that we have

$$\mathbf{B}_t = \mathbf{B}_{t+} \left(= \bigcap_{\varepsilon > 0} \mathbf{B}_{t+\varepsilon} \right).$$

To prove the proposition it suffices to show that $\sigma(\omega)$ is a measurable function of $B(t \wedge \sigma(\omega), \omega)$, $t \geq 0$. When $\sigma(\omega) = \infty$ the assertion is obvious. Assuming now that $\sigma(\omega) < \infty$, then

$$(\sigma \leq t) = \bigcap_n (\sigma < t + n^{-1}) \in \mathbf{B}_{t+} = \mathbf{B}_t$$

follows. Thus the assertion $\sigma(\omega) \leq t$ is equivalent to one restricting the continuous function $B(\cdot, \omega)$ on $[0, t]$ to belong to some Borel subset C_t of the space of continuous functions $\mathscr{C}([0, t])$. Since $B(s, \omega) \equiv B_\sigma(s, \omega)$ on $[0, t]$ for $\omega \in (\sigma > t)$ our proposition has been proved. □

Remark. For a general Markov process $\{X(t)\}$ it is not always the case that σ is $\mathbf{B}_\sigma(X)$-measurable.

Corollary. *For any Markov time $\sigma(\omega)$, $B(\sigma(\omega), \omega)$ is $\mathbf{B}_\sigma$-measurable.*

Theorem 2.8. *Let $\sigma(\omega)$ be a Markov time such that $P(\sigma < \infty) = 1$. Then $\{B(\sigma + t) - B(\sigma) : t \geq 0\}$ is a Brownian motion which is independent of $\mathbf{B}_\sigma$.*

PROOF. Let A be in $\mathbf{B}_\sigma$ and take non-negative time points $t_1, t_2, \ldots, t_n$. Denoting a bounded continuous function $f(x_1, x_2, \ldots, x_n)$ on $\mathbf{R}^n$ by $f(x_j : 1 \leq j \leq n)$ we can prove that for any Markov time σ

$$\int_A f(B(\sigma + t_j) - B(\sigma) : 1 \leq j \leq n) \, dP \quad (2.71)$$

$$= P(A) \int_\Omega f(B(t_j) : 1 \leq j \leq n) \, dP.$$

In the proof we approximate σ by the sequence $\{\sigma_m\}$ where $\sigma_m = \min_j \{j 2^{-m} : j 2^{-m} > \sigma\}$. Setting $A_k = A \cap (\sigma_m = k 2^{-m})$ we have $A = \bigcup_1^n A_k$ (disjoint union) and we obtain

$$\int_A f(B(\sigma_m + t_j) - B(\sigma_m) : 1 \leq j \leq n) \, dP$$

$$= \sum_k \int_{A_k} f(B(k 2^{-m} + t_j) - B(k 2^{-m}) : 1 \leq j \leq n) \, dP.$$

2.4 Markov Properties of Brownian Motion

After noting that $A_k \in \mathbf{B}_{k2^{-m}}$, we see that each term in the last expression becomes

$$\int_{A_k} E\{f(B(k2^{-m} + t_j) - B(k2^{-m})): 1 \le j \le n) | \mathbf{B}_{k2^{-m}}\} \, dP$$

$$= \int_{A_k} E\{f(B(k2^{-m} + t_j) - B(k2^{-m})): 1 \le j \le n)\} \, dP$$

(by the additive property of Brownian motion)

$$= P(A_k) \cdot E\{f(B(t_j)): 1 \le j \le n)\}.$$

By letting $m \to \infty$ we get $\sigma_m \to \sigma$ and the equality (2.71) is proved.

For a finite number of finite intervals I_j, $1 \le j \le n$, a simple function of the form $\prod_1^n \chi_{I_j}(x_j)$ can be approximated by a sequence of bounded continuous functions, and thus (2.71) still holds for such a simple function. This means that $(B(\sigma + t_1) - B(\sigma), \ldots, B(\sigma + t_n) - B(\sigma))$ is independent of $\mathbf{B}_\sigma$, and thus $\{B(\sigma + t) - B(\sigma): t \ge 0\}$ is also independent of $\mathbf{B}_\sigma$.

Finally we note that the above computations have already shown $\{B(\sigma + t) - B(\sigma)\}$ to be a Brownian motion for a fixed Markov time σ. □

The next assertion is an immediate consequence of the theorem just proved since the relation (2.69) follows as a corollary. We emphasize that the result itself is a key one.

Theorem 2.9. *Brownian motion is a strong Markov process.*

The topic we now turn to was first discussed by P. Lévy (1948, Chapter VI). By using the strong Markov property of Brownian motion we can gain a deep insight into the complex behaviour of the Brownian sample path, and find intuitively satisfying proofs of important results [see also K. Itô and H. P. McKean Jr. (1965), D. Freedman (1971) etc.] We first explain the reflection principle for a Brownian motion. Given a Brownian motion $\{B(t)\}$ and a Markov time σ, we can form a stochastic process $\{X(t): t \ge 0\}$ as follows: for $\omega \, \varepsilon \, \Omega$

if $\sigma(\omega) = \infty$, then $X(t, \omega) = B(t, \omega)$ \qquad $t \ge 0$;

if $\sigma(\omega) < \infty$, then $X(t, \omega) = \begin{cases} B(t, \omega) & 0 \le t \le \sigma(\omega); \\ 2B(\sigma(\omega), \omega) - B(t, \omega) & t \ge \sigma(\omega). \end{cases}$

(2.72)

Obviously $\{X(t): t \ge 0\}$ is a stochastic process whose sample paths are all continuous. The idea behind the definition of $X(t)$ is as follows: prior to the instant σ, $X(t)$ is exactly the same as Brownian motion $B(t)$. As soon as t reaches the value σ, $X(t)$ becomes the value $B(t)$ reflected about the line $y = B(\sigma)$, see Fig. 5.

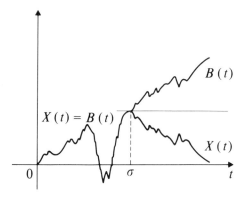

Figure 5

Theorem 2.10. *The process* $\{X(t, \omega): t \geq 0\}$ *is a Brownian motion.*

PROOF. On the set $(\sigma = \infty)$ we have $X(t, \omega) = B(t, \omega)$ for all $t \geq 0$ so that nothing needs to be proved. We may therefore suppose that $P(\sigma < \infty) = 1$. Let $B_\sigma(t)$ be given by (2.70) and consider the following two triples:

$$(\sigma(\omega), \{B_\sigma(t, \omega): t \geq 0\}, \{B(\sigma(\omega) + t, \omega) - B(\sigma(\omega), \omega): t \geq 0\}) \quad (2.73)$$

$$(\sigma(\omega), \{B_\sigma(t, \omega): t \geq 0\}, \{-[B(\sigma(\omega) + t, \omega) - B(\sigma(\omega), \omega)]: t \geq 0\}). \quad (2.74)$$

From the definition and Proposition 2.7 we prove that $\{B_\sigma(t)\}$ and σ are both $\mathbf{B}_\sigma$-measurable. Thus, by Theorem 2.8, both $\{B_\sigma(t)\}$ and σ are independent of the Brownian motion $\{B(\sigma + t) - B(\sigma): t \geq 0\}$, and hence also independent of $\{-[B(\sigma + t) - B(\sigma)]: t \geq 0\}$. Furthermore we can easily show that the two triples above have the same distribution. We now note that the Brownian motion $\{B(t): t \geq 0\}$ is obtained by connecting the two processes in (2.73), i.e. there is a measurable function Φ such that

$$\{B(t)\} = \Phi(\sigma, \{B_\sigma(t)\}, \{B(\sigma + t) - B(\sigma)\}).$$

By using this same Φ on (2.74) we obtain the process $\{X(t)\}$, i.e. from (2.72) we have

$$\{X(t)\} = \Phi(\sigma, \{B_\sigma(t)\}, \{-[B(\sigma + t) - B(\sigma)]\}).$$

Thus $\{B(t)\}$ and $\{X(t)\}$ have the same distribution, which means that $\{X(t)\}$ is also a Brownian motion. □

This result is nothing but a symmetry principle for Brownian motion, and is usually called the *reflection principle*.

Again following P. Lévy we introduce a new stochastic process $\{M(t, \omega): t \geq 0\}$ defined in terms of Brownian motion by

$$M(t, \omega) = \max_{0 \leq s \leq t} B(s, \omega), \quad t \geq 0. \quad (2.75)$$

2.4 Markov Properties of Brownian Motion

Since $M(t, \omega) = \sup\{B(r, \omega) : r \text{ rational}, 0 \leq r \leq t\}$, we see that $M(t)$ is $\mathbf{B}_t$-measurable.

Proposition 2.8. *For $y \geq 0$ and $z \geq 0$ we have the following equality:*

$$P(B(t) < z - y, M(t) \geq z) = P(B(t) > y + z). \tag{2.76}$$

PROOF. Set $\sigma_z = \min\{s : B(s) = z\}$. Then σ_z is a Markov time, for the equality $(\sigma_z \leq t) = (M(t) \geq z)$ implies that

$$(\sigma_z < t) = \bigcup_n (\sigma_z \leq t - n^{-1}) \in \mathbf{B}_t(M) \subseteq \mathbf{B}_t.$$

Taking $\sigma = \sigma_z$ in the previous result we form $\{X(t)\}$ as described in (2.72). Now put $\tau_z = \min\{s : X(s) = z\}$. Then $(\tau_z, \{X(t)\})$ and $(\sigma_z, \{B(t)\})$ have the same distribution, and therefore

$$P(\tau_z \leq t, X(t) < z - y) = P(\sigma_z \leq t, B(t) < z - y). \tag{2.77}$$

But it is immediate from the definition of $\{X(t)\}$ that $\sigma_z(\omega) = \tau_z(\omega)$ holds identically, and so we must have

$$(\tau_z \leq t, X(t) < z - y) = (\sigma_z \leq t, B(t) > z + y)$$

which, together with (2.77), implies

$$P(\sigma_z \leq t, B(t) > z + y) = P(\sigma_z \leq t, B(t) < z - y).$$

The left-hand side is equal to $P(M(t) \geq z, B(t) > z + y) = P(B(t) > z + y)$, the last equality following from the fact that $M(t) \geq z$ is a redundant condition in the presence of $B(t) > z + y$. By paraphasing the condition $\sigma_z \leq t$, the right-hand side of the above can be written as $P(M(t) \geq z, B(t) < z - y)$ and we see that (2.76) has been proved. □

Corollary. *The joint distribution of $(B(t), M(t))$ for $t > 0$ is absolutely continuous with respect to two-dimensional Lebesgue measure, and the density function $f(x, z)$ is given by*

$$f(x, z) = \begin{cases} 0, & x > z \text{ or } z < 0 \\ (2/\pi)^{1/2}(2z - x)t^{-3/2} \exp\left[\dfrac{-(2z - x)^2}{2t}\right], & z \geq 0, z \geq x. \end{cases} \tag{2.78}$$

PROOF. Set $z - y = x$ in (2.76) above and then take the second partial derivative with respect to x and z. Then (2.78) follows immediately. □

Proposition 2.9. *The following equalities hold for $z \geq 0$ and $t > 0$:*

$$P(M(t) \geq z) = 2P(B(t) \geq z) = P(|B(t)| \geq z). \tag{2.79}$$

PROOF. Put $y = 0$ in the equality (2.76) and we obtain

$$P(B(t) < z, M(t) \geq z) = P(B(t) > z).$$

We therefore have

$$P(M(t) \geq z) = P(B(t) < z, M(t) \geq z) + P(B(t) \geq z, M(t) \geq z)$$
$$= P(B(t) > z) + P(B(t) \geq z).$$

Now the first equality in (2.79) follows, for $P(B(t) = z) = 0$, and the second equality is a consequence of the symmetry of a zero-mean Gaussian distribution about zero. □

We turn now to another stochastic process $\{m(t): t \geq 0\}$ defined by

$$m(t, \omega) = \min_{0 \leq s \leq t} B(s, \omega), \qquad t \geq 0. \qquad (2.80)$$

Clearly $m(t)$ is $\mathbf{B}_t$-measurable. Since Brownian motion is symmetrically distributed and $m(t, \omega) = -\max_{0 \leq s \leq t}\{-B(t, \omega)\}$, we see that

$$\{-m(t): t \geq 0\} \sim \{M(t): t \geq 0\}, \qquad (2.81)$$

where $\sim$ means the two processes have the same distribution.

Using the Markov time technique we will now obtain Lévy's result concerning the joint distribution of $(m(t), M(t), B(t))$ [see P. Lévy (1948) Chapter VI, §42; also see L. Bachelier (1941)].

Let σ_x be the Markov time defined by $\sigma_x(\omega) = \min\{t: B(t, \omega) = x\}$ as before. Take constants a and b such that $a < 0 < b$, and define the events $A = (\sigma_a < \sigma_b)$, $C = (\sigma_b < \sigma_a)$, $A^* = (\sigma_a \leq t)$ and $C^* = (\sigma_b \leq t)$. Define the transformation γ_y by

$$\gamma_y(x) = 2y - x,$$

and extend it to a set transformation by writing

$$\gamma_y(H) = \{\gamma_y(h): h \in H\}, \qquad H \subseteq \mathbf{R}.$$

See Fig. 6.

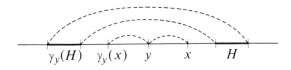

Figure 6

For a Borel subset $H \subseteq (-\infty, a]$ the relation $(B(t) \in H) \subseteq A^*$ holds, so that we also have

$$C \cap (B(t) \in H) \subseteq C^*.$$

Also the relation $\gamma_b(H) \subseteq [b, \infty)$ implies that

$$(B(t) \in \gamma_b(H)) \subseteq C^*.$$

See Fig. 7.

2.4 Markov Properties of Brownian Motion

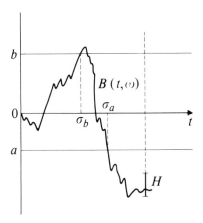

Figure 7

The reflection principle applied to σ_b gives

$$P(C \cap (B(t) \in H)) = P(C \cap \{B(t) \in \gamma_b(H)\}).$$

Since $C = \Omega \setminus A$, we have the following equality for $H \subseteq (-\infty, a]$:

$$P(C \cap \{B(t) \in H\}) = P(B(t) \in \gamma_b(H)) - P(A \cap (B(t) \in \gamma_b(H))). \quad (2.82)$$

Similarly, when $H \subset [b, \infty)$, we can prove that

$$P(A \cap (B(t) \in H)) = P(B(t) \in \gamma_a(H)) - P(C \cap \{B(t) \in \gamma_a(H)\}). \quad (2.83)$$

In order to state another interesting result, we introduce the following function, defined for $t > 0$, $a < 0 < b$ by:

$$k(x) = k(a, b, t; x)$$
$$= (2\pi t)^{-1/2} \sum_{n=-\infty}^{\infty} \left\{ \exp\left[\frac{-(x - 2nl)^2}{2t}\right] - \exp\left[\frac{-(x - 2a + 2nl)^2}{2t}\right] \right\},$$

where $l = b - a$.

Proposition 2.10 (P. Lévy). *For an Borel subset J of the interval $[a, b]$, we have*

$$P(a < m(t) \leq M(t) < b, B(t) \in J) = \int_J k(x)\, dx. \quad (2.84)$$

PROOF. The probability in question is

$$P(B(t) \in J) - P(A \cap A^* \cap \{B(t) \in J\}) - P(C \cap C^* \cap \{B(t) \in J\}). \quad (2.85)$$

We begin by computing the second term which we denote by p. By the reflection principle we may write $p = P(A \cap A^* \cap \{B(t) \in \gamma_a(J)\})$. Since $\gamma_a(J) \subset (-\infty, a]$, we have $\{B(t) \in \gamma_a(J)\} \subseteq A^*$ and so

$$p = P(A \cap \{B(t) \in \gamma_a(J)\}) = P(B(t) \in \gamma_a(J)) - P(C \cap \{B(t) \in \gamma_a(J)\}).$$

We are now in a position to use the relations (2.82) and (2.83) in turn to obtain

$$p = P(B(t) \in \gamma_a(J)) - P(B(t) \in \gamma_b(\gamma_a(J))) + P(B(t) \in \gamma_a(\gamma_b(\gamma_a(J)))) - \cdots.$$

Writing $\gamma_a(\gamma_b(x)) = (\gamma_a \gamma_b)(x)$ we find that $(\gamma_a \gamma_b)^n(x) = x - 2nl$ and $(\gamma_b \gamma_a)^n(x) = x + 2nl$ where $l = b - a$. We therefore have

$$P(B(t) \in (\gamma_b \gamma_a)^n(J)) = P(B(t) - 2nl \in J)$$
$$= \int_J (2\pi t)^{-1/2} \exp\left[\frac{-(x + 2nl)^2}{2t}\right] dx,$$
$$P(B(t) \in \gamma_a(J)) = \int_J (2\pi t)^{-1/2} \exp\left[\frac{-(x - 2a)^2}{2t}\right] dx.$$

With these formulae it is possible to obtain an explicit expression for p, and similarly for the third term in (2.85), finally obtaining the formula (2.84). □

There are many interesting properties of Brownian motion that are proved by the use of the strong Markov property, and these can be found in Chapter VI of the book of P. Lévy cited above. The reader will undoubtedly enjoy Lévy's clear and simple descriptions based upon a deep insight into and intuitive grasp of the subject. Some of these results would be rephrased and proved in the manner we have just indicated.

2.5 Applications of the Hille-Yosida Theorem

The beautiful theory due to K. Yosida and E. Hille concerning a one-parameter semigroup of linear operators acting on a Banach space is now known as the Hille-Yosida theory [see e.g. K. Yosida (1951) Chapter 12, (1965) Chapter IX], and it gives us an operator-theoretic method of analysing Markov processes in general and Brownian motion in particular.

Let $\{X(t): t \geq 0\}$ be a temporally homogeneous Markov process whose sample functions are right continuous, and let $\{P(t, x, A)\}$ be its system of transition probabilities [see (2.64) in the previous section]. We introduce the Banach space $\mathfrak{L}(\mathbf{R})$ consisting of all bounded Borel-measurable functions equipped with the norm $|f| = \sup_x |f(x)|$, $f \in \mathfrak{L}(\mathbf{R})$. Now define an operator P_t by

$$(P_t f)(x) = \int_{-\infty}^{\infty} f(y) P(t, x, dy) = E_x\{f(X(t))\}, \qquad t \geq 0, f \in \mathfrak{L}(\mathbf{R}). \quad (2.86)$$

2.5 Applications of the Hille-Yosida Theorem

Each P_t is a linear operator from $\mathfrak{L}(\mathbf{R})$ into itself and we denote the norm of P_t by $|P_t|$.

Theorem 2.11. *The collection $\{P_t: t \geq 0\}$ has the following properties:*

i. *Each P_t is a continuous linear operator on $\mathfrak{L}(\mathbf{R})$.*
ii. *P_t is positive: $(P_t f)(x) \geq 0$ if $f(x) \geq 0$.*
iii. *$\{P_t: t \geq 0\}$ is a one-parameter semigroup:*

$$P_t P_s = P_{t+s}, \qquad t, s \geq 0; \qquad (2.87)$$
$$P_0 = I \qquad \text{(the identity operator)}.$$

iv. *$|P_t| = 1$, $(P_t 1)(x) = 1(x)$ where $1(x)$ is the constant function taking the value 1.*
v. *For any $f \in \mathfrak{L}(\mathbf{R})$ the function*

$$u(x, t) = (P_t f)(x) \qquad (2.88)$$

is a measurable function of $t(> 0)$.
vi. *If $f \in \mathfrak{L}(\mathbf{R})$ is continuous at x_0, then*

$$\lim_{t \to 0+} (P_t f)(x_0) = f(x_0).$$

PROOF. i. Suppose that f_n converges in $\mathfrak{L}(\mathbf{R})$ to f, i.e. f is bounded and $\sup_x |f_n(x) - f(x)| \to 0$ as $n \to \infty$. Then

$$\sup_x |(P_t f_n)(x) - (P_t f)(x)| \leq \sup_x \int \sup_y |f_n(y) - f(y)| P(t, x, dy)$$
$$= \sup_y |f_n(y) - f(y)| \to 0$$

as $n \to \infty$. This shows that $|P_t f_n - P_t f| \to 0$ as $n \to \infty$.
ii. is obvious.
iii. Since $\{P(t, x, A)\}$ satisfies the Chapman-Kolmogorov equations (see (2.63) of the previous section)

$$P(t + s, x, A) = \int_{-\infty}^{\infty} P(t, x, dy) P(s, y, A) \qquad s, t \geq 0,$$

we have the relations

$$(P_t P_s f)(x) = \int \left[\int f(z) P(s, y, dz) \right] P(t, x, dy)$$
$$= \int f(z) \int P(s, y, dz) P(t, x, dy)$$
$$= \int f(z) P(t + s, x, dz) = (P_{t+s} f)(x).$$

The relation $P_0 = I$ is an immediate consequence of the equality $P(0, x, A) = \chi_A(x)$.

iv. is immediate from the definition of P_t.

v. is proved by using measurability of $u(t, x)$ in t and property (iii) above.

vi. Since f is supposed to be continuous at x_0, for any $\varepsilon > 0$ there exists a δ-neighbourhood $U_\delta(x_0)$ of x_0 such that when $y \in U_\delta(x_0)$ we have $|f(y) - f(x_0)| < \varepsilon$. With this choice of neighbourhood we use the relation

$$(P_t f)(x_0) - f(x_0) = \int_{U_\delta(x_0)} (f(y) - f(x_0))P(t, x_0, dy)$$
$$+ \int_{U_\delta(x_0)^c} (f(y) - f(x_0))P(t, x_0, dy)$$

to prove that

$$|(P_t f)(x_0) - f(x_0)| \leq \varepsilon P(t, x_0, U_\delta(x_0)) + 2|f|P(t, x_0, U_\delta(x_0)^c)$$
$$\leq \varepsilon + 2|f|P(t, x_0, U_\delta(x_0)^c).$$

To conclude the argument it now suffices to establish the following lemma.

Lemma 2.6. Let $U(a)$ be an open set involving a point a. Then we have

$$\lim_{t \to 0+} P(t, a, U(a)) = 1. \tag{2.89}$$

PROOF. Let t_n be a sequence decreasing to 0 and set $B_n = \{X(t_n) \in U(a)\}$. Since $X(t)$ is right-continuous in t we have

$$\{X(0) \in U(a)\} \subseteq \liminf_{n \to \infty} B_n$$

We therefore have

$$\liminf_{n \to \infty} P(t_n, a, U(a)) \geq P_a\left(\liminf_{n \to \infty} B_n\right) \geq P_a(X(0) \in U(a))$$
$$\geq P_a(X(0) = a) = 1.$$

[Here the notation P_a is that of (2.64)]. Thus we have proved (2.89). □

Definition 2.7. The system $\{P_t : t \geq 0\}$ in Theorem 2.11 is called the *semigroup* associated with the Markov process $\{X(t)\}$.

Given a semigroup $\{P_t : t \geq 0\}$ of operators on $\mathfrak{L}(\mathbf{R})$ satisfying the properties of Theorem 2.11 the system $\{P(t, x, A)\}$ defined by

$$P(t, x, A) = (P_t \chi_A)(x) \tag{2.90}$$

becomes the system of transition probabilities of a temporally homogeneous Markov process which, in addition, satisfies (2.89). Thus it is possible to

2.5 Applications of the Hille-Yosida Theorem

describe the properties of transition probabilities in terms of the semigroup $\{P_t\}$.

Once the semigroup $\{P_t\}$ associated with a Markov process is given, its Laplace transform introduces another system $\{G_\alpha: \alpha > 0\}$ of operators on $\mathfrak{L}(\mathbf{R})$ defined by

$$(G_\alpha f)(x) = \int_0^\infty e^{-\alpha t}(P_t f)(x)\, dt, \quad \alpha > 0, f \in \mathfrak{L}(\mathbf{R}). \tag{2.91}$$

By Theorem 2.11 $P_t f$ belongs to $\mathfrak{L}(\mathbf{R})$ and is bounded and continuous in t, and so the integral (2.91) can be expressed as the integral with respect to the measure $dP \times dt$. Then Fubini's theorem can be applied to obtain

$$(G_\alpha f)(x) = E_x\left[\int_0^\infty e^{-\alpha t} f(X(t))\, dt\right]. \tag{2.92}$$

Clearly G_α is a linear operator on $\mathfrak{L}(\mathbf{R})$ for $\alpha > 0$ and this operator is called the *Green operator* of order α. If in addition there is a measure $G(\alpha, x, dy)$ on $\mathbf{R}$ such that for all $f \in \mathfrak{L}(\mathbf{R})$, we can express $G_\alpha f$ in the form

$$(G_\alpha f)(x) = \int_{-\infty}^\infty f(y) G(\alpha, x, dy), \tag{2.93}$$

then $G(\alpha, x, dy)$ is called the *Green measure* of order α.

Theorem 2.12. *The system $\{G_\alpha: \alpha > 0\}$ of Green operators on $\mathfrak{L}(\mathbf{R})$ has the following properties:*

i. *Each G_α is a continuous linear operator on $\mathfrak{L}(\mathbf{R})$.*
ii. *G_α is positive.*
iii. *The resolvent equation, given below, is satisfied:*

$$G_\alpha - G_\beta + (\alpha - \beta) G_\beta G_\alpha = 0, \quad \alpha, \beta > 0. \tag{2.94}$$

iv. *$|G_\alpha| = \alpha^{-1}$, $(G_\alpha 1)(x) \equiv \alpha^{-1}$.*
v. *G_α can be extended to those complex α with $\operatorname{Re} \alpha > 0$, in such a way that $(G_\alpha f)(x)$ is an analytic function of α.*
vi. *If $f \in \mathfrak{L}(\mathbf{R})$ is continuous at x_0, then*

$$\lim_{\alpha \to \infty} \alpha (G_\alpha f)(x_0) = f(x_0).$$

PROOF. The assertions (i) to (vi) are all counterparts of those with the corresponding number of Theorem 2.11. Perhaps we need only prove (iii), (v) and (vi).

iii. The resolvent equation is proved as follows:

$$P_s(G_\alpha f)(x) = \int_{-\infty}^{\infty} P(s, x, dy) \int_0^{\infty} e^{-\alpha t}(P_t f)(y) \, dt$$

$$= \int_0^{\infty} e^{-\alpha t} \, dt \int_{-\infty}^{\infty} \int_{-\infty}^{\infty} f(z) P(s, x, dy) P(t, y, dz)$$

$$= \int_0^{\infty} e^{-\alpha t} \, dt \int_{-\infty}^{\infty} f(z) P(t + s, x, dz)$$

$$= e^{\alpha s} \int_s^{\infty} e^{-\alpha \tau}(P_\tau f)(x) \, d\tau.$$

Both sides are functions of s and their Laplace transforms are

$$(G_\beta G_\alpha f)(x) = \int_0^{\infty} e^{-\beta s} e^{\alpha s} \, ds \int_s^{\infty} e^{-\alpha \tau}(P_\tau f)(x) \, d\tau$$

$$= \int_0^{\infty} e^{-\alpha \tau}(P_\tau f)(x) \, d\tau \int_0^{\tau} e^{(\alpha - \beta)s} \, ds$$

$$= (\alpha - \beta)^{-1}\{(G_\beta f)(x) - (G_\alpha f)(x)\}$$

from which (2.94) follows.

v. It follows easily from the definition of G_α that when it is viewed as a function of α, it can be extended to complex $\alpha \in \mathbf{C}$ for which Re $\alpha > 0$. Moreover the resolvent equation guarantees the existence of the limit

$$\lim_{\beta \to \alpha} \frac{G_\alpha - G_\beta}{\beta - \alpha} f = \lim_{\beta \to \alpha} -G_\beta G_\alpha f = -G_\alpha^2 f.$$

In this way we see that G_α is an analytic function of α.

vi. follows from (vi) of Theorem 2.11 by using

$$\alpha(G_\alpha f)(x) = \int_0^{\infty} \alpha e^{-\alpha t}(P_t f)(x) \, dt = \int_0^{\infty} e^{-t}(P_{t/\alpha} f)(x) \, dt. \qquad \square$$

A detailed description of the infinitesimal generator $\mathfrak{g} = (d/dt)P_t|_{t=0}$ of a semigroup $\{P_t: t \geq 0\}$, together with a determination of its domain, can be obtained from family $\{G_\alpha: \alpha > 0\}$ of Green operators. Our discussion begins with the function space comprising the domain of $\mathfrak{g}$.

Setting $\mathfrak{R}_\alpha = \{G_\alpha f: f \in \mathfrak{L}(\mathbf{R})\}$ we see that $\mathfrak{R}_\alpha$ is a subspace of $\mathfrak{L}(\mathbf{R})$, for G_α is a continuous linear operator carrying $\mathfrak{L}(\mathbf{R})$ onto $\mathfrak{R}_\alpha$.

Proposition 2.11. *The space $\mathfrak{R}_\alpha$ is independent of α.*

PROOF. The resolvent equation implies that $G_\alpha G_\beta = G_\beta G_\alpha$ and further, that

$$G_\beta f = G_\alpha\{f + (\alpha - \beta)G_\beta f\}, \qquad f \in \mathfrak{L}(\mathbf{R}).$$

2.5 Applications of the Hille-Yosida Theorem

Now the function in the braces $\{\cdot\}$ belongs to $\mathfrak{L}(\mathbf{R})$, and so the above equation shows that $\mathfrak{R}_\beta \subseteq \mathfrak{R}_\alpha$. Interchanging the role of α and β, we prove that $\mathfrak{R}_\alpha \subseteq \mathfrak{R}_\beta$ and so deduce that $\mathfrak{R}_\alpha = \mathfrak{R}_\beta$, $\alpha, \beta > 0$. □

We now denote this range by $\mathfrak{R}$ instead of $\mathfrak{R}_\alpha$, and set $\mathfrak{N}_\alpha = \{f \in \mathfrak{L}(\mathbf{R}): G_\alpha f = 0\}$. The set $\mathfrak{N}_\alpha$ is a subspace of $\mathfrak{L}(\mathbf{R})$ which is also independent of α. For if $f \in \mathfrak{N}_\beta$, then the resolvent equation gives $G_\alpha f = G_\beta f - (\alpha - \beta) G_\alpha (G_\beta f) = 0$. Thus we may write $\mathfrak{N}$ instead of $\mathfrak{N}_\alpha$.

Proposition 2.12. *For $u \in \mathfrak{R}$ the function $\alpha u - G_\alpha^{-1} u$ is uniquely determined mod $\mathfrak{N}$, and is independent of α.*

PROOF. We begin by noting that the space $\mathfrak{N}$ is so defined that $G_\alpha^{-1} u$ is determined uniquely modulo $\mathfrak{N}$. Take a representative f of the class $\alpha u - G_\alpha^{-1} u$ and apply G_α. We get $G_\alpha f = \alpha G_\alpha u - u$, where this is a genuine equality (i.e. not mod $\mathfrak{N}$). Hence we have

$$G_\beta G_\alpha f = \alpha G_\beta G_\alpha u - G_\beta u,$$

to which we can apply the resolvent equation and obtain

$$(\beta - \alpha)^{-1}(G_\alpha - G_\beta)f = \alpha(\beta - \alpha)^{-1}(G_\alpha - G_\beta)u - G_\beta u,$$

and finally,

$$G_\beta f = G_\alpha f - \alpha G_\alpha u + \beta G_\beta u,$$
$$= -u + \beta G_\beta u.$$

Thus we have proved that $f = \beta u - G_\beta^{-1} u$, mod $\mathfrak{R}$ and so $\alpha u - G_\alpha^{-1} u$ is independent of α mod $\mathfrak{N}$. □

This proposition allows us to define an operator $\mathfrak{g}$ on $\mathfrak{R}$ mod $\mathfrak{N}$ by the formula

$$\mathfrak{g} u = \alpha u - G_\alpha^{-1} u, \qquad u \in \mathfrak{R}. \tag{2.95}$$

If $u = G_\alpha f$ for some $\alpha > 0$, then $\mathfrak{g} u = \alpha u - f$, mod $\mathfrak{N}$, i.e.

$$u = G_\alpha f \text{ is equivalent to } (\alpha - \mathfrak{g})u = f, \text{ mod } \mathfrak{R}. \tag{2.96}$$

The operator $\mathfrak{g}$ is called the *generator* of the semigroup $\{P_t\}$.

We now make a small digression to explain formally why $\mathfrak{g}$ may be regarded as a generator. Setting $(d/dt)P_t|_{t=0} = \mathfrak{g}$ we see that $(d/dt)P_t = P_t \mathfrak{g} = \mathfrak{g} P_t$. With this relation we may write $P_t = \exp(t\mathfrak{g})$ and so the Laplace transform becomes:

$$G_\alpha = \int_0^\infty \exp[-(\alpha I - \mathfrak{g})t]\, dt = (\alpha I - \mathfrak{g})^{-1}.$$

But this gives $(\alpha I - \mathfrak{g})G_\alpha = I$ or, equivalently, $\mathfrak{g} = \alpha I - G_\alpha^{-1}$, and hence we are led to the definition (2.95) of $\mathfrak{g}$. Of course what we have just done is a purely formal calculation, for since $\mathfrak{g}$ is unbounded more care is needed in using the exponential map.

Proposition 2.13. *The generator $\mathfrak{g}$ is a closed operator, and the domain $\mathfrak{R}$ of $\mathfrak{g}$ is dense in $\mathfrak{L}(\mathbf{R})$.*

PROOF. Let f_n be a sequence in $\mathfrak{L}(\mathbf{R})$ and set $u_n = G_\alpha f_n$. Assume that $u_n \to v$ and $\mathfrak{g} u_n = \alpha u_n - f_n \to \alpha v - g$ in $\mathfrak{L}(\mathbf{R})$ as $n \to \infty$. Since $f_n \to g$ and G_α is a bounded operator, we have $G_\alpha f_n \to G_\alpha g$, and hence $v = G_\alpha g$ follows from $u_n = G_\alpha f_n$. Thus $v \in \mathfrak{R}$ and $\mathfrak{g} v = \alpha v - g$ have been proved, completing the proof. □

The next problem is to show that given $\mathfrak{g}$ we can form a one-parameter semigroup $\{P_t\}$ with generator $\mathfrak{g}$. We repeat the warning that the exponential map is only defined formally by $\exp(t\mathfrak{g}) = \sum_0^\infty (t^n/n!)\mathfrak{g}^n$, since $\mathfrak{g}$ is not bounded. An exact definition of the exponential map is contained in the following theorem.

Theorem 2.12 (K. Yosida, E. Hille). *Let A be a closed operator with domain dense in a Banach space E, and suppose that for every $\alpha > 0$ the resolvent $G_\alpha = (\alpha I - A)^{-1}$ exists and satisfies*

$$|G_\alpha| \leq \alpha^{-1}. \tag{2.97}$$

Then there exists a unique semigroup $\{P_t\}$ with generator A such that $|P_t| \leq 1$. Furthermore P_t is given by

$$\lim_{\alpha \to \infty} \exp(t\alpha A G_\alpha) = P_t. \tag{2.98}$$

PROOF. We may write $AG_\alpha = \alpha G_\alpha - I$ and note that it is a bounded linear operator. Hence the exponential

$$\exp(t\alpha A G_\alpha) = P_t^{(\alpha)}, \qquad \alpha > 0, t \geq 0,$$

is well-defined, and since $G_\alpha I = I G_\alpha$ we have

$$\exp(t\alpha A G_\alpha) = \exp(t\alpha^2 G_\alpha)\exp(-t\alpha I).$$

Now $|\alpha G_\alpha| \leq 1$ holds by assumption, so that we have

$$|\exp(t\alpha A G_\alpha)| \leq \exp(t\alpha)\exp(-t\alpha) = 1. \tag{2.99}$$

For u in the domain $\mathfrak{D}(A)$ of A we have $G_\alpha Au(= -u + \alpha(\alpha I - A)^{-1}u) = AG_\alpha u$, and $G_\alpha A$ also commutes with $P_t^{(\beta)}$: $G_\alpha A P_t^{(\beta)} u = P_t^{(\beta)} G_\alpha Au$. In this case $dP_t^{(\alpha)}/dt$ exists and

$$\frac{d}{dt} P_t^{(\alpha)} u = \alpha A G_\alpha P_t^{(\alpha)} u = P_t^{(\alpha)}(\alpha A G_\alpha)u,$$

2.5 Applications of the Hille-Yosida Theorem

which proves that for $u \in \mathfrak{D}(A)$

$$(P_t^{(\alpha)} - P_t^{(\beta)})u = \int_0^t \frac{d}{ds}(P_{t-s}^{(\beta)} P_s^{(\alpha)} u)\, ds$$

$$= \int_0^t P_{t-s}^{(\beta)} P_s^{(\alpha)}(\alpha G_\alpha - \beta G_\beta) Au\, ds.$$

The relations $u = \alpha G_\alpha u - G_\alpha Au$ and $|G_\alpha Au| \le \alpha^{-1}|Au|$ imply that $\lim_{\alpha \to \infty} \alpha G_\alpha u = u$. By the use of the assumption $|\alpha G_\alpha| \le 1$ the relation $\lim_{\alpha \to \infty} \alpha G_\alpha f = f$ can be proved for any f in E, and with this result, (2.99) and the above computation, we can prove that

$$|(P_t^{(\alpha)} - P_t^{(\beta)})u| \to 0 \quad \text{as} \quad \alpha, \beta \to \infty$$

uniformly on any finite t-interval. Thus the existence of the limit (2.98) has been established and P_t is well-defined. By letting $\alpha \to \infty$ in (2.99) we can also prove that $P_t u$ is continuous in t and that $|P_t| \le 1$. Also the semigroup property

$$P_t P_s = P_s P_t = P_{t+s}, \qquad s, t \ge 0$$

can easily be proved by letting $\alpha \to \infty$ in the corresponding identity for $\{P_t^{(\alpha)}\}$.

We now show that the generator of the semigroup obtained above coincides with the original operator A. For $u \in \mathfrak{D}(A)$ we have

$$P_t^{(\alpha)} u - u = \int_0^t P_t^{(\alpha)} \alpha G_\alpha Au\, dt$$

since $P_t^{(\alpha)}$ is differentiable in t. Now let $\alpha \to \infty$ in this expression and obtain

$$P_t u - u = \int_0^t P_t Au\, dt,$$

which implies that $\lim_{t \to 0+} t^{-1}(P_t - I)u = Au$. If we denote the generator of $\{P_t\}$ by A' then we have proved that $\mathfrak{D}(A') \supseteq \mathfrak{D}(A)$, and also that $A'u = Au$ if $u \in \mathfrak{D}(A)$. Now $(\alpha I - A')^{-1}$ and $(\alpha I - A)^{-1}$ must be surjections from $\mathfrak{D}(A')$ and $\mathfrak{D}(A)$ respectively, onto E, and we therefore have $\mathfrak{D}(A') = \mathfrak{D}(A)$.

There only remains the uniqueness of $\{P_t\}$. Suppose that there was another semigroup $\{P_t'\}$ with the generator A. Then for any $u \in \mathfrak{D}(A)$

$$P_t' u - P_t^{(\alpha)} u = -\int_0^t \frac{d}{ds}(P_{t-s}' P_s^{(\alpha)})u\, ds$$

$$= \int_0^t P_{t-s}' P_s^{(\alpha)}(A - \alpha G_\alpha A)u\, ds \to 0 \quad \text{as} \quad \alpha \to \infty$$

and hence $P_t' = P_t$ follows. $\qquad\square$

We are now ready to apply this theorem to the semigroup, and its generator, of a temporally homogeneous Markov process. Take the basic Banach space E to be $\mathfrak{L}(\mathbf{R})$ and set $A = \mathfrak{g}$, where we suppose that $\mathfrak{g}$ satisfies

$$(\mathfrak{g}1)(x) \equiv 0. \tag{2.100}$$

This assumption is equivalent to $(\alpha G_\alpha 1)(x) \equiv 1$ and hence to

$$(P_t 1)(x) \equiv 1. \tag{2.101}$$

Now suppose that we have two Markov processes $\{X^{(i)}(t): t \in [0, \infty)\}$ $i = 1, 2$, and that the one-parameter semigroups $\{P_t^{(i)}\}$, $i = 1, 2$, derived from their transition probabilities satisfy the condition (2.101). If the generators of these semigroups coincide, and this includes their domains, then Theorem 2.13 asserts the equality $P_t^{(1)} = P_t^{(2)}$ for every t. In view of the definition (2.90), the transition probabilities for these two processes

$$P^{(i)}(t, x, A) = (P_t^{(i)} \chi_A)(x), \quad i = 1, 2, \qquad A \text{ a Borel subset of } \mathbf{R},$$

must also coincide. This means that provided an initial distribution is given, a generator determines a Markov process. Thus we have the following diagram:

$$\{X(t)\} \to \{P(t, x, A)\} \leftrightarrow \{P_t\} \to \{G_\alpha\} \to \mathfrak{g}.$$

In general a generator (together with its domain) is much easier to deal with, and even simpler in form, than a system of transition probabilities.

The general theory of Markov processes can now be applied to Brownian motion. We know that its transition probability has a density which can be written in terms of the Gauss kernel as $g(t; x - y)$, and for convenience we tacitly understand that $g(0; x - y)$ stands for the Dirac measure $\delta_0(x - y)$. The following lemma gives the elementary computations needed to obtain an explicit formula for the Green operator G_α, $\alpha > 0$.

Lemma 2.7. *The Laplace transform of the Gauss kernel $g(t; x - y)$, viewed as a function of t, exists and can be written in the form*

$$\int_0^\infty e^{-\alpha t} g(t; x - y) \, dt = (2\alpha)^{-1/2} \exp[-(2\alpha)^{1/2} |x - y|], \quad \alpha > 0. \tag{2.102}$$

OUTLINE OF THE PROOF. When $x > 0$

$$(2\pi)^{-1/2} \int_0^\infty t^{-1/2} \exp\left[-\alpha t - \frac{x^2}{2t}\right] dt$$

$$= \exp[-(2\alpha)^{1/2} x] (2\pi)^{-1/2} \int_0^\infty t^{-1/2} \exp\left\{-\frac{1}{2}[xt^{-1/2} - (2\alpha t)^{1/2}]^2\right\} dt$$

$$= \exp[-(2\alpha)^{1/2} x] (2\pi)^{-1/2} \int_{-\infty}^\infty \exp\left(-\frac{1}{2} u^2\right) \frac{du}{(2\alpha)^{1/2}}$$

$$= (2\alpha)^{-1/2} \exp[-(2\alpha)^{1/2} x].$$

2.5 Applications of the Hille-Yosida Theorem

In the last integral the variable u of integration is taken to be $xt^{-1/2} - (2\alpha t)^{1/2}$. The case $x < 0$ is handled similarly, with $u = xt^{-1/2} + (2\alpha t)^{1/2}$. □

Proposition 2.14. *For any fixed $\alpha > 0$ and $x \in \mathbf{R}$, $G_\alpha f(x)$ may be expressed in terms of the absolutely continuous measure*

$$G(\alpha, x, dy) \equiv (2\alpha)^{-1/2} \exp[-(2\alpha)^{1/2}|x - y|] \, dy$$

on $\mathbf{R}$ by

$$(G_\alpha f)(x) = \int_{-\infty}^{\infty} f(y) G(\alpha, x, dy)$$

$$= (2\alpha)^{-1/2} \int_{-\infty}^{\infty} f(y) \exp[-(2\alpha)^{1/2}|x - y|] \, dy. \quad (2.103)$$

PROOF. Fubini's theorem gives us

$$(G_\alpha f)(x) = \int_{-\infty}^{\infty} f(y) \, dy \int_0^{\infty} e^{-\alpha t} g(t; x - y) \, dt,$$

which, by the above lemma, turns out to be

$$\int_{-\infty}^{\infty} f(y)\{(2\alpha)^{-1/2} \exp[-(2\alpha)^{1/2}|x - y|]\} \, dy. \quad \square$$

By using formula (2.103) we can give explicit expressions for $\mathfrak{R}$, $\mathfrak{N}$, and $\mathfrak{g}$.

Theorem 2.14. i. *Let u be a function given by $u = G_\alpha f$ where $f \in \mathfrak{L}(\mathbf{R})$. Then $u'(x)$ exists, both $u(x)$ and $u'(x)$ are absolutely continuous, and the equation $u''(x) = 2\alpha u(x) - 2f(x)$ holds a.e., where u'' is the density function of u'.*
  ii. *The spaces $\mathfrak{R}$ and $\mathfrak{N}$ are given by*

$$\mathfrak{R} = \{u \in \mathfrak{L}(\mathbf{R}): u, u' \text{ absolutely continuous, } u'' \in \mathfrak{L}(\mathbf{R})\}$$

$$\mathfrak{N} = \{f \in \mathfrak{L}(\mathbf{R}): f(x) = 0, \text{ a.e.}\}.$$

  iii. $(\mathfrak{g}u)(x) = \frac{1}{2}u''(x)$, a.e., $u \in \mathfrak{R}$.

PROOF. From the formula (2.103) we have

$$u(x) = \int_{-\infty}^{x} (2\alpha)^{-1/2} \exp[-(2\alpha)^{1/2}(x - y)] f(y) \, dy$$

$$+ \int_{x}^{\infty} (2\alpha)^{-1/2} \exp[(2\alpha)^{1/2}(x - y)] f(y) \, dy,$$

from which we deduce the existence of $u'(x)$ and the formula

$$u'(x) = -\exp[-(2\alpha)^{1/2}x] \int_{-\infty}^{x} \exp[(2\alpha)^{1/2}y] f(y) \, dy$$

$$+ \exp[(2\alpha)^{1/2}x] \int_{x}^{\infty} \exp[-(2\alpha)^{1/2}y] f(y) \, dy.$$

With this expression we can prove that $u'(x)$ is absolutely continuous and that its density $u''(x)$ is expressible in the form

$$u''(x) = 2\alpha u(x) - 2f(x), \quad \text{a.e.}$$

Since $u, f \in \mathfrak{L}(\mathbf{R})$ it also follows that $u'' \in \mathfrak{L}(\mathbf{R})$.

ii. In the course of the proof of (i) we actually proved that $\mathfrak{R}$ is included in $\{u \in \mathfrak{L}(\mathbf{R}): u, u' \text{ are absolutely continuous}, u'' \in \mathfrak{L}(\mathbf{R})\}$ and so only the converse inclusion needs proving here. Take a function u of the type described and put $f = \alpha u - \frac{1}{2}u''$. Since $f \in \mathfrak{L}(\mathbf{R})$ we can define $v = G_\alpha f$ and by (i) we must have

$$v''(x) = 2\alpha v(x) - 2f(x), \quad \text{a.e.}$$

Thus the difference $w = u - v$ must satisfy

$$w''(x) = 2\alpha w(x), \quad \text{a.e.,}$$

and, since u and v are continuous, so also is w, which implies that the last equation is valid everywhere. Any function with this property can be written in the form $c_1 \exp[(2\alpha)^{1/2} x] + c_2 \exp[-(2\alpha)^{1/2} x]$, and the further property of boundedness implies that $c_1 = c_2 = 0$. Thus we have $w = 0$ so that $u = G_\alpha f$ and assertion (ii) is proved.

The result concerning $\mathfrak{N}$ is easier to establish. Indeed if $f \in \mathfrak{N}$ then $u(x) = (G_\alpha f)(x) = 0$, and with this and the equation $u''(x) = 2\alpha u(x) - 2f(x)$ a.e., we conclude that $f(x) = 0$, a.e. The converse is obvious.

iii. Now suppose that $u = G_\alpha f$. By definition this means that $\mathfrak{g}u = \alpha u - f$, mod $\mathfrak{N}$ is true, but we know from (i) that $\frac{1}{2}u'' = \alpha u - f$, mod $\mathfrak{N}$. Thus

$$\mathfrak{g}u = \frac{1}{2} u'', \text{ mod } \mathfrak{N},$$

and assertion (iii) has been proved. □

Following the diagram given earlier, we start with the system of transition probabilities which determines the distribution of Brownian motion, then the semigroup $\{P_t\}$ and the system $\{G_\alpha\}$ of Green operators are obtained, and finally we arrive at the generator $\mathfrak{g}$ above. Together with the associated subspaces $\mathfrak{R}$ and $\mathfrak{N}$, $\mathfrak{g}$ provides an adequate description of Brownian motion as a Markov process. We pause now to introduce the Dynkin formula which gives a probabilistic interpretation to the generator $\mathfrak{g}$. It is based on the following theorem due to E. B. Dynkin.

Theorem 2.15. *Let σ be a Markov time. If $u(x)$ is given by $u = G_\alpha f$ with $f \in \mathfrak{L}(\mathbf{R})$, then it can be expressed in the form*

$$u(x) = E_x\left[\int_0^\sigma e^{-\alpha t} f(B(t)) \, dt\right] + E_x[e^{-\alpha\sigma} u(B(\sigma))]. \tag{2.104}$$

2.5 Applications of the Hille-Yosida Theorem

PROOF. We may assume that $\sigma(\omega) < \infty$, a.e. (P). From formula (2.92) we have

$$u(x) = E_x\left[\int_0^\infty e^{-\alpha t} f(B(t))\, dt\right]$$

$$= E_x\left[\int_0^\sigma e^{-\alpha t} f(B(t))\, dt\right] + E_x\left[\int_\sigma^\infty e^{-\alpha t} f(B(t))\, dt\right].$$

We now examine the second term.

$$E_x\left[e^{-\alpha\sigma}\int_0^\infty e^{-\alpha t} f(B(t+\sigma))\, dt\right]$$

$$= \int_0^\infty e^{-\alpha t} E_x[e^{-\alpha\sigma} f(B(t+\sigma))]\, dt$$

$$= \int_0^\infty e^{-\alpha t} E_x[e^{-\alpha\sigma} E_{B(\sigma)}\{f(B(t))\}]\, dt \quad \text{by the strong Markov property,}$$

$$= E_x\left[e^{-\alpha\sigma}\int_0^\infty E_{B(\sigma)}\{e^{-\alpha t} f(B(t))\}\, dt\right]$$

$$= E_x[e^{-\alpha\sigma} u(B(\sigma))].$$

The formula (2.104) follows from this last expression. $\square$

This theorem is valid for a general Markov process as long as it possesses the strong Markov property.

The *Dynkin formula* stated below is an immediate consequence of Theorem 2.15.

Corollary. *If σ is a Markov time for Brownian motion such that $E_x(\sigma) < \infty$ and if $u \in \Re$, then*

$$E_x\left[\int_0^\sigma (\mathfrak{g}u)(B(t))\, dt\right] = E_x[u(B(\sigma))] - u(x). \tag{2.105}$$

PROOF. Since u can be written $u = G_\alpha f$ the formula from Theorem 2.15:

$$u(x) = E_x\left[\int_0^\sigma e^{-\alpha t} f(B(t))\, dt\right] + E_x[e^{-\alpha\sigma} u(B(\sigma))]$$

and the equation $f(x) = \alpha u(x) - \mathfrak{g}u(x)$, a.e. imply that

$$u(x) = E_x\left[\int_0^\sigma e^{-\alpha t}\{\alpha u(B(t)) - (\mathfrak{g}u)(B(t))\}\, dt\right] + E_x[e^{-\alpha\sigma} u(B(\sigma))].$$

Letting $\alpha \to \infty$, and noting that $P_x(\sigma < \infty) = 1$ follows from the assumption that $E_x(\sigma) < \infty$, we obtain Dynkin's formula (2.105). $\square$

EXAMPLE. Let σ_ε be the first time at which a Brownian motion starting from x escapes from the interval $[x - \varepsilon, x + \varepsilon]$, $\varepsilon > 0$. As in Example 1 of the last section, we can prove that σ_ε is a Markov time. From the symmetry of Brownian motion the equalities $P_x(B(\sigma_\varepsilon) = x + \varepsilon) = P_x(B(\sigma_\varepsilon) = x - \varepsilon) = \frac{1}{2}$ are obvious. Further the fact that the distribution of Brownian motion is invariant under the space-time transformation [see Proposition 2.1(B)] we can see that $E(\sigma_\varepsilon) = C\varepsilon^2$ where C is a positive constant. [In fact C can be proved to equal 1, see Dynkin and Yushkevich (1969) Chapter 2]. We therefore have

$$E_x\left[\int_0^{\sigma_\varepsilon} (\mathfrak{g}u)(B(t))\, dt\right] = \frac{1}{2}[u(x + \varepsilon) + u(x - \varepsilon)] - u(x). \qquad (2.106)$$

Now we know that $\sigma_\varepsilon \to 0$ a.e. as $\varepsilon \to 0$ and so we see that whenever $\mathfrak{g}u$ is continuous at x, the left-hand side of (2.106) equals $\sigma_\varepsilon((\mathfrak{g}u)(x) + o(1))$ as $\varepsilon \to 0$. Dividing both sides of (2.106) by $E(\sigma_\varepsilon) = \varepsilon^2$ and letting $\varepsilon \to 0$ we obtain $(\mathfrak{g}u)(x) = \frac{1}{2}u''(x)$.

Using the fact that $\mathfrak{g} = \frac{1}{2}d^2/dx^2$ it is interesting to see how Theorem 2.13 applies in giving the operator P_t, and we do this next. Firstly consider the differential equation

$$\alpha u - \frac{d^2}{dx^2} u = f.$$

Since the Green function for this equation is the well-known kernel $g_\alpha(x - y) = (2\alpha)^{-1/2} \exp[-(2\alpha)^{1/2}|x - y|]$, we can easily see that the resolvent $G_\alpha = (\alpha - \mathfrak{g})^{-1}$ is given by (2.103). Using this formula we can compute the exponential $\exp(t\alpha g G_\alpha)$ explicitly. Take the formal Taylor series expansion

$$\exp(t\alpha \mathfrak{g} G_\alpha)f = \sum_0^\infty \frac{(t\alpha)^n}{n!} \left(\frac{1}{2}\frac{d^2}{dx^2}\right)^n G_\alpha^n f, \qquad f \in \mathfrak{L}(\mathbf{R}). \qquad (2.107)$$

On the right-hand side G_α^n denotes the integral operator with kernel g_α^{n*}, the n-fold convolution of g_α. Since the Fourier transform of g_α is $2(2\pi)^{-1/2} \times (\lambda^2 + 2\alpha)^{-1}$, that of g_α^{n*} must be $2^n(2\pi)^{-1/2}(\lambda^2 + 2\alpha)^{-n}$ and its inverse Fourier transform defines the kernel of G_α^n. What is required now is to apply G_α^n to f and take the $2n$-th derivative. This operation is equivalent to an integral operator such that the kernel is obtained by taking the Fourier inverse transform of the product of $(-\lambda^2)^n$ and the Fourier transform of g_α^{n*}. Thus the right-hand side of (2.107) may be expressed in the form

$$\sum_0^\infty \frac{(t\alpha)^n}{n!} \mathscr{F}^{-1}\left[\left(-\frac{1}{2}\lambda^2\right)^n (2\pi)^{-1/2} 2^n (\lambda^2 + 2\alpha)^{-n}\right] \cdot f$$

$$= (2\pi)^{-1/2} \mathscr{F}^{-1}\left[\sum_0^\infty \frac{1}{n!}\left(-\frac{1}{2}t\lambda^2\right)^n (2\alpha)^n (\lambda^2 + 2\alpha)^{-n}\right] \cdot f$$

$$\to \int_{-\infty}^\infty g(t; x - y)f(y)\, dy \quad \text{as} \quad \alpha \to \infty.$$

In the above $\mathscr{F}^{-1}$ denotes the inverse Fourier transform, and we have proved that the density of the transition probability is $g(t; x - y)$.

One final remark should be made. So far we have taken the basic Banach space to be $\mathfrak{L}(\mathbf{R})$ but it is known that for certain kinds of Markov processes, the corresponding P_t becomes a continuous linear operator on $\mathscr{C}_0$ (see §1.3; this is the Banach space consisting of all continuous functions vanishing at $\pm \infty$, equipped with the topology of uniform convergence). Brownian motion is, of course, one such Markov process, as is easily seen from the properties of the transition probability density $g(t; x - y)$. The Hille-Yosida theorem can certainly be applied in this case, the space $\mathfrak{N}$ turning out to be the trivial space consisting of only the identically zero function. The exponential map also becomes much simpler.

However, for a Markov process less special than Brownian motion, further discussion is needed in order to find conditions under which the associated P_t are operators on $\mathscr{C}_0$. We will also be discussing Markov processes different from, but related to, Brownian motion, and for this reason we developed a general theory of Markov processes using the space $\mathfrak{L}(\mathbf{R})$ in the present section.

2.6 Processes Related to Brownian Motion

We will now discuss some processes which are obtained from Brownian motion by applying transformations which preserve the Markov property.

(i) Brownian Motion with a Reflecting Barrier

Take a Brownian motion $\{B(t, \omega) + a: t \geq 0\}$ starting from a point $a(> 0)$, and as soon as it reaches the origin 0, reflect it, whilst as long as it remains within the open interval $(0, \infty)$ let it behave exactly as a standard Brownian motion. A process which satisfies these intuitive requirements may be constructed by forming

$$X(t, \omega) = |B(t, \omega) + a|, \qquad t \geq 0. \tag{2.108}$$

Denoting $\mathbf{B}_t(X)$ by $\mathbf{B}_t^+$, we see that $\mathbf{B}_t^+ \subseteq \mathbf{B}_t(= \mathbf{B}_t(B))$.

Proposition 2.15. *The process $\{X(t, \omega): t \geq 0\}$ given by (2.108) is a Markov process.*

PROOF. We note that for any Borel subset A of $\mathbf{R}^+ = [0, \infty)$, the event $(X(t, \omega) \in A)$ may be expressed $(B(t, \omega) + a \in A \cup (-A))$, where $-A = \{-x : x \in A\}$. Let $t > s = s_0 > s_1 > \cdots > s_n$ be given, and take Borel subsets $A_1, A_2, \ldots, A_n$ and A of $\mathbf{R}^+$. Furthermore set

$$C = (X(s_n) \in A_n, \ldots, X(s_1) \in A_1, X(s) \in A_0).$$

The set C belongs to $\mathbf{B}_s^+$ and so we have

$$\int_C P(X(t) \in A | \mathbf{B}_s^+) \, dP = \int_C \chi_{(X(t) \in A)} \, dP$$
$$= P(B(s_n) + a \in A_n \cup (-A_n), \ldots, B(s)$$
$$+ a \in A_0 \cup (-A_0), B(t) + a \in A \cup (-A)).$$

The remainder of the proof is the same as that of (2.65). By using the transition probability density of Brownian motion we see that the above probability is

$$= \int_C P(X(t) \in A | X(s)) \, dP$$

which proves the assertion. □

Since the sample functions of Brownian motion are continuous, so also are those of the process $X(t, \omega)$. The distribution of $\{X(t)\}$ is therefore a probability measure on the space $\mathscr{C}^+ = \mathscr{C}^+([0, \infty)) = \{x \in \mathscr{C} : x(t) \geq 0\}$. If we define a mapping γ of $\mathscr{C} = \mathscr{C}([0, \infty))$ onto $\mathscr{C}^+$ by

$$(\gamma x)(t) = |x(t)|, \qquad x \in \mathscr{C}, \tag{2.109}$$

and let $\mathfrak{B}, \mathfrak{B}^+$ be the σ-fields of subsets of $\mathscr{C}, \mathscr{C}^+$ respectively, generated by cylinder sets, then γ is a measurable mapping of $(\mathscr{C}, \mathfrak{B})$ onto $(\mathscr{C}^+, \mathfrak{B}^+)$. Therefore the conditional Wiener measure μ_a^W on $(\mathscr{C}, \mathfrak{B})$ induces a probability measure ν_a on $(\mathscr{C}^+, \mathfrak{B}^+)$ through γ by putting $\nu_a = \gamma^{-1} \circ \mu_a^W$, i.e.

$$\nu_a(B) = \mu_a^W(\gamma^{-1}(B)), \qquad B \in \mathfrak{B}^+. \tag{2.110}$$

The probability space $(\mathscr{C}^+, \mathfrak{B}^+, \nu_a)$ so obtained is the distribution of $\{X(t)\}$. An explicit expression for this distribution will now be given in terms of the system $P^+(s, x; t, A)$ of transition probabilities.

As before we let $g(t; x)$ denote the Gauss kernel, and set

$$\int_A g(t; x - y) \, dy = G(t, x, A).$$

It is clear that $P(X(t) \in A) = G(t, a, A \cup (-A))$.

Proposition 2.16. *The Markov process $\{X(t)\}$ is temporally homogeneous, with transition probability given by*

$$P^+(s, x, t, A) = G(t - s, x, A \cup (-A)), \qquad t > s, \tag{2.111}$$

and transition probability density $g(t - s; x - y) + g(t - s; x + y)$.

2.6 Processes Related to Brownian Motion

PROOF. Let A and B be Borel subsets of $\mathbf{R}^+$. Then we have

$$\int_{(X(s) \in B)} P(X(t) \in A \mid X(s)) \, dP$$

$$= P(X(s) \in B, X(t) \in A)$$

$$= \int_{B \cup (-B)} \int_{A \cup (-A)} g(s; a - y) g(t - s; y - z) \, dy \, dz$$

$$= \int_{B \cup (-B)} g(s; a - y) \{ G(t - s, y, A) + G(t - s, y, -A) \} \, dy$$

$$= \int_B G(t - s, y, A \cup (-A)) P(X(s) \in dy),$$

which implies (2.111). The assertion concerning the density function is obvious. □

As a result of this proposition we are now able to use the notation $P^+(t - s, x, A)$ instead of $P^+(s, x, t, A)$, and the Chapman-Kolmogorov equation can then be expressed in the form

$$\int_0^\infty P^+(s, x, dy) P^+(t, y, A) = P^+(s + t, x, A). \tag{2.112}$$

We turn now to the semigroup associated with $\{X(t): t \geq 0\}$. Let $\mathfrak{L}(\mathbf{R}^+)$ be the Banach space consisting of all bounded measurable functions defined on $\mathbf{R}^+$ with the sup norm as in $\mathfrak{L}(\mathbf{R})$. Define the linear operator P_t^+ on $\mathfrak{L}(\mathbf{R}^+)$ by

$$(P_t^+ f)(x) = \int_0^\infty f(y) P^+(t, x, dy), \qquad f \in \mathfrak{L}(\mathbf{R}^+). \tag{2.113}$$

As before the system $\{P_t^+ : t \geq 0\}$ forms a one-parameter semigroup and it is of interest to examine its relationship to $\{P_t\}$. Firstly we extend any $f \in \mathfrak{L}(\mathbf{R}^+)$ to a symmetric function $\hat{f}$ by writing

$$\hat{f}(y) = \begin{cases} f(y) & y \geq 0, \\ f(-y) & y < 0. \end{cases}$$

Clearly $\hat{f} \in \mathfrak{L}(\mathbf{R})$ and the operator P_t associated with Brownian motion may be applied to it.

Proposition 2.17. *For any $f \in \mathfrak{L}(\mathbf{R}^+)$ we have*

$$(P_t^+ f)(x) = (P_t \hat{f})(x), \qquad x \geq 0. \tag{2.114}$$

PROOF. The assertion follows from the simple calculation

$$(P_t^+ f)(x) = \int_0^\infty f(y)\{g(t; x-y) + g(t; x+y)\} \, dy$$

$$= \int_0^\infty f(y)g(t; x-y) \, dy + \int_{-\infty}^0 f(-y)g(t; x-y) \, dy$$

$$= (P_t \hat{f})(x), \quad \text{if } x \geq 0. \qquad \square$$

Remark. P_t carries symmetric functions into symmetric functions, and in particular

$$(P_t \hat{f})(-x) = (P_t \hat{f})(x).$$

Many of the properties of $\{P_t^+\}$ may be deduced from those of $\{P_t\}$ (see Theorem 2.11) through the relation (2.114), and so will not be reproduced here. We denote the Green operator for $\{X(t)\}$ by the symbol G_α corresponding to the case of Brownian motion, but with a superscript $+$:

$$(G_\alpha^+ f)(x) = \int_0^\infty e^{-\alpha t}(P_t^+ f)(x) \, dt.$$

It follows immediately from Proposition 2.17 that

$$(G_\alpha^+ \hat{f})(x) = (G_\alpha \hat{f})(x), \qquad x \geq 0. \tag{2.115}$$

and G_α^+ satisfies the resolvent equation. This proves that $G_\alpha^+(\mathfrak{L}(\mathbf{R}^+))$ is independent of α and so can be denoted by $\mathfrak{R}^+$. If f is in $\mathfrak{L}(\mathbf{R}^+)$, then $\hat{f}$ belongs to $\mathfrak{L}(\mathbf{R})$, so that $u = G_\alpha \hat{f}$ is in $\mathfrak{R}$ and thus u' exists. By the remark above, $(G_\alpha \hat{f})(x)$ is again a symmetric function: $u(-x) = u(x)$, and since u' is continuous, $u'(0)$ must vanish. Restricting u to $[0, \infty)$ i.e. viewing u as $G_\alpha^+ f$, we deduce that $u'(0+) = 0$. This must hold for every member of $\mathfrak{R}^+$.

Conversely, given $u \in \mathfrak{R}$ such that $u'(0+) = 0$, there exists a symmetric $f \in \mathfrak{L}(\mathbf{R})$ such that $\hat{u} = G_\alpha f$. If we restrict f to $[0, \infty)$, then u is expressed in the form $u = G_\alpha^+ f$ and belongs to $\mathfrak{R}^+$. The foregoing observations give us the expressions for $\mathfrak{R}^+$, $\mathfrak{N}^+$ and the generator $\mathfrak{g}^+$ explained below.

$$\mathfrak{R}^+ = G_\alpha^+ \mathfrak{L}(\mathbf{R}^+)$$
$$= \{u \in \mathfrak{L}(\mathbf{R}^+): u, u' \text{ are absolutely continuous}, u'' \in \mathfrak{L}(\mathbf{R}^+), u'(0+) = 0\}. \tag{2.116}$$

The kernel $\mathfrak{N}^+$ of G_α^+ is given by

$$\mathfrak{N}^+ = \{f \in \mathfrak{L}(\mathbf{R}^+): f(x) = 0, \quad \text{a.e.}\} \tag{2.117}$$

as in the case discussed in §2.5.

The generator $\mathfrak{g}^+$ is obtained in a manner similar to that in (2.95) and is an operator such that for $u \in \mathfrak{R}^+$

$$\mathfrak{g}^+ u = \alpha u - (G_\alpha^+)^{-1} u, \quad \text{mod } \mathfrak{N}^+. \tag{2.118}$$

2.6 Processes Related to Brownian Motion

We can then see that if we extend f and u on $\mathbf{R}^+$ to symmetric functions on $\mathbf{R}$, $\mathfrak{g}^+$ has the same form as the generator $\mathfrak{g}$ of Brownian motion, so that discussion can be reduced to that concerning Brownian motion. Summing up, the above discussion has yielded:

Proposition 2.18. i. *The space* $\mathfrak{R}^+ = G_\alpha^+ \mathfrak{L}(\mathbf{R}^+)$ *and the kernel* $\mathfrak{N}^+$ *of* G_α^+ *are independent of* α *and given by* (2.116), (2.117) *respectively*.
ii. *The generator* $\mathfrak{g}^+$ *acts in such a way that*

$$(\mathfrak{g}^+ u)(x) = \tfrac{1}{2} u''(x), \quad \text{a.e.} \quad x \geq 0.$$

As we saw in the Example of the last section, the generator of a *diffusion process* (a strong Markov process with continuous sample paths) is known to be locally determined in space. The process $X(t, \omega)$ discussed in this subsection is certainly a diffusion process, and as long as it remains within the open interval $(0, \infty)$, the local behaviour of its sample path is exactly the same as that of a Brownian motion. Therefore it is quite natural that $\mathfrak{g}^+$ coincides with the generator of Brownian motion.

(ii) The Kac Formula and Some Applications

In this subsection we establish a formula concerning Brownian functionals due to M. Kac, and use it to discuss further properties of Brownian motion and its transformations. To simplify notation we will omit ω in what follows.

Let $V(x)$ be a bounded non-negative measurable function on $\mathbf{R}$ and let E_x denote the expectation operator corresponding to a Brownian motion starting from x, for example $E_x\{f(B(t))\} = E\{f(B(t) + x)\}$. Now suppose that $\alpha > 0$ and $f \in \mathfrak{L}(\mathbf{R})$ and set

$$v(x) = E_x\left\{\int_0^\infty e^{-\alpha t} f(B(t)) \exp\left[-\int_0^t V(B(s))\, ds\right] dt\right\}. \quad (2.119)$$

Since the integral inside the braces $\{\cdot\}$ exists and is uniformly bounded in ω, $v(x)$ belongs to $\mathfrak{L}(\mathbf{R})$. Keeping the notation $\mathfrak{R}$ and $\mathfrak{g}$ introduced in the last section for Brownian motion we have

Theorem 2.16 (M. Kac). *The function* $v(x)$ *given by* (2.119) *belongs to* $\mathfrak{R}$ *and satisfies*

$$(\alpha - \mathfrak{g} + V(x)) v(x) = f(x) \quad \text{a.e.} \quad (2.120)$$

PROOF. Still continuing with our earlier notation we set

$$u(x) = (G_\alpha f)(x) = E_x\left[\int_0^\infty e^{-\alpha t} f(B(t))\, dt\right].$$

Integration by parts gives us the equality

$$-\int_0^t V(B(s)) \exp\left[-\int_s^t V(B(\tau))\, d\tau\right] ds = \exp\left[-\int_0^t V(B(s))\, ds\right] - 1$$

which implies that

$$v(x) - u(x) = -E_x\left\{\int_0^\infty e^{-\alpha t} f(B(t)) \int_0^t V(B(s)) \exp\left[-\int_s^t V(B(\tau))\, d\tau\right] ds\, dt\right\}.$$

Here we note that

$$\int_0^\infty \left\{\int_0^t \left|e^{-\alpha t} f(B(t)) V(B(s)) \exp\left[-\int_s^t V(B(\tau))\, d\tau\right]\right| ds\right\} dt$$

$$\leq \int_0^\infty |f||V| t e^{-\alpha t}\, dt < \infty,$$

and so we can use Fubini's theorem. Using this, $v(x) - u(x)$ turns out to be

$$-E_x\left\{\int_0^\infty V(B(s))\, ds \int_s^\infty e^{-\alpha t} f(B(t)) \exp\left[-\int_s^t V(B(\tau))\, d\tau\right] dt\right\}$$

$$= -E_x\left\{\int_0^\infty V(B(s))\, ds \right.$$
$$\left. \cdot \int_0^\infty e^{-\alpha(t+s)} f(B(t+s)) \exp\left[-\int_s^{t+s} V(B(\tau))\, d\tau\right] dt\right\}$$

$$= -\int_0^\infty e^{-\alpha s}\, ds E_x\left\{V(B(s)) \right.$$
$$\left. \cdot \int_0^\infty e^{-\alpha t} f(B(t+s)) \exp\left[-\int_0^t V(B(\tau+s))\, d\tau\right] dt\right\}$$

$$= -\int_0^\infty e^{-\alpha s}\, ds E_x\left\{V(B(s)) \right.$$
$$\left. \cdot E_x\left(\int_0^\infty e^{-\alpha t} f(B(t+s)) \exp\left[-\int_0^t V(B(\tau+s))\, d\tau\right] dt \,\middle|\, \mathbf{B}_s\right)\right\}$$

$$= -\int_0^\infty e^{-\alpha s}\, ds E_x\left\{V(B(s)) \right.$$
$$\left. \cdot E_{B(s)}\left(\int_0^\infty e^{-\alpha t} f(B(t)) \exp\left[-\int_0^t V(B(\tau))\, d\tau\right]\right)\right\}$$

$$= -\int_0^\infty e^{-\alpha s}\, ds E_x\{V(B(s)) v(B(s))\}$$

$$= -G_\alpha(V \cdot v)(x),$$

2.6 Processes Related to Brownian Motion

where $(V \cdot v)(x) = V(x)v(x)$. Therefore $u - v \in \mathfrak{R}$ and hence $v \in \mathfrak{R}$, and in addition the above calculation gives us

$$G_\alpha^{-1}(v - u) = -V \cdot v, \mod \mathfrak{N},$$

$$(\alpha - \mathfrak{g})v - (\alpha - \mathfrak{g})u = -V \cdot v, \mod \mathfrak{N}.$$

We now use the relation $(\alpha - \mathfrak{g})u = f$ to conclude that

$$(\alpha - \mathfrak{g})v - f = -V \cdot v, \quad \text{a.e.,}$$

which is the same as (2.120). □

Remark. Assume that f is continuous and let $\mathfrak{g}$ be $\frac{1}{2}d^2/dx^2$ in the differential equation (2.120). Then the solution is given by (2.119), and indeed this solution has been constructed probabilistically.

Setting $\mathfrak{g}^V = \mathfrak{g} - V$ we can rewrite (2.120) as

$$(\alpha - \mathfrak{g}^V)v = f. \tag{2.121}$$

This equation can then be compared with (2.96): $(\alpha - \mathfrak{g})u = f$, and the question arises as to the existence of a Markov process with generator $\mathfrak{g}^V$ for which a discussion analogous to that concerning Brownian motion might take place. More precisely, we might ask whether there exists a one-parameter semigroup $\{P_t^V\}$ associated with a system $\{P^V(t, x, A)\}$ of transition probabilities for a Markov process through the relation $(P_t^V f)(x) = \int f(y) P^V(t, x, dy)$, and which has generator $\mathfrak{g}^V = (d/dt)P_t^V|_{t=0}$. The answer is indeed yes, but to examine this question fully we would need much more background and further discussion concerning Markov processes than it is appropriate to give here. Thus we will omit a complete discussion of this topic [referring to K. Itô and H. P. McKean Jr. (1965), and W. Feller (1966)], but we do present particular cases which can be discussed without further background, and which provide a glimpse of the general theory.

Let us consider the case $V(x) \equiv \lambda$, a positive constant. We know that in the case of Brownian motion the exponential map $\exp(t\mathfrak{g}) = P_t$ is well-defined, and this, coupled with the obvious fact that the generator $\mathfrak{g}$ commutes with λ ($= \lambda I$, I the identity), allows us to give the correct interpretation to

$$\exp[t(\mathfrak{g} - \lambda)] = e^{-\lambda t} \exp(t\mathfrak{g}) = e^{-\lambda t} P_t.$$

If we denote this expression by P_t^V and look for the system $\{P^V(t, x, A)\}$, it is not hard to see that it is obtained from the transition probabilities $\{P(t, x, A)\}$ of Brownian motion by

$$P^V(t, x, A) = e^{-\lambda t} P(t, x, A), \quad t \geq 0.$$

Clearly the P^V satisfy all the properties appropriate to a system of Markov transition probabilities, and hence there exists a temporally homogeneous Markov process, unique if the initial distribution is also specified, having P^V

as its transition probability. A crucial difference between this system and that of Brownian motion lies in the fact that total mass is $P^V(t, x, \mathbf{R}) = e^{-\lambda t} P(t, x, \mathbf{R}) = e^{-\lambda t}$. Here "particles disappear" at an exponential rate λ with a proportion $1 - e^{-\lambda t}$ "disappearing" during the time interval $[0, t]$. For a more general function $V(x)$ depending on x, the "proportion of particles disappearing" depends upon "the location of the particles", the parameter in this case being replaced by the function $V(x)$, and in this case P^V no longer has the above simple form.

Another simple example may be found where there is a *drift* associated with a Brownian motion. Consider a diffusion process with generator $\mathfrak{g} + b(d/dx)$, b a constant. The term $b(d/dx)$ corresponds to the constant drift and since it commutes with $\mathfrak{g} = \frac{1}{2} d^2/dx^2$, the one-parameter semigroup $\{\tilde{P}_t\}$ associated with the generator $\mathfrak{g} + b(d/dx)$ is given by

$$\tilde{P}_t = \exp\left[t\left(\mathfrak{g} + b\frac{d}{dx}\right)\right] = \exp\left(bt\frac{d}{dx}\right)\exp(t\mathfrak{g})$$

$$= S_{-bt} \exp(t\mathfrak{g}),$$

where S_a is the shift operator defined by $(S_a f)(x) = f(x - a)$. This shift will be re-introduced and further discussed below. The operator $\tilde{P}_t$ acts in such a way that $(\tilde{P}_t f)(x) = (P_t f)(x + bt)$, and is an integral operator with the transition probability density $g(t; x - y + bt)$ as kernel.

Returning now to Kac's formula, we present two important examples due to M. Kac (1951).

EXAMPLE 1. The function $V(x)$ in Theorem 2.16 is taken to be

$$V(x) = \begin{cases} \beta & x > 0, (\beta > 0 \text{ constant}); \\ 0 & x \leq 0. \end{cases} \quad (2.122)$$

Using the same constant β we define a function $\Phi(t)$ by

$$\Phi(t) = \beta^{-1} \int_0^t V(B(s))\, ds. \quad (2.123)$$

Denoting the Lebesgue measure of a set $B \subseteq \mathbf{R}$ by $|B|$, we see that $\Phi(t)$ coincides with $|\{s \leq t : B(s) > 0\}|$. Setting $f \equiv 1$ in (2.119) we obtain

$$v(x) = E\left\{\int_0^\infty e^{-\alpha t} \exp\left[-\int_0^t V(B(s))\, ds\right] dt\right\},$$

and this function satisfies

$$\begin{cases} \left(\alpha + \beta - \frac{1}{2}\frac{d^2}{dx^2}\right)v(x) = 1 & x > 0, \\ \left(\alpha - \frac{1}{2}\frac{d^2}{dx^2}\right)v(x) = 1 & x < 0, \end{cases} \quad (2.124)$$

2.6 Processes Related to Brownian Motion

where we have substituted $\mathfrak{g} = \frac{1}{2} d^2/dx^2$. A general solution $v(x)$ to this equation may be expressed in the form

$$v(x) = \begin{cases} (\alpha + \beta)^{-1} + A_1 \exp[-2^{1/2}(\alpha + \beta)^{1/2}x] + A_2 \exp[2^{1/2}(\alpha + \beta)^{1/2}x], \\ \hspace{20em} x > 0 \\ \alpha^{-1} + B_1 \exp[-(2\alpha)^{1/2}x] + B_2 \exp[(2\alpha)^{1/2}x], \hspace{3em} x < 0, \end{cases}$$

for suitable constants A_1, A_2, B_1 and B_2. In addition $v(x)$ is a continuous function and $v'(x)$ is continuous at $x = 0$, since $v \in \mathfrak{R}$, so that we must have

$$v(0) = \{\alpha(\alpha + \beta)\}^{-1/2}.$$

By definition we have the relation

$$v(0) = E_0\left[\int_0^\infty e^{-\alpha t} \exp[-\beta\Phi(t)] \, dt\right]$$

and the following identity can also be established:

$$\{\alpha(\alpha + \beta)\}^{-1/2} = \int_0^\infty e^{-\alpha t} \left\{\int_0^t e^{-\beta s}[\pi(t-s)(\pi s)]^{-1/2} \, ds\right\} dt.$$

From the uniqueness theorem for Laplace transforms we can deduce the equality

$$E_0\{\exp[-\beta\Phi(t)]\} = \pi^{-1} \int_0^t e^{-\beta s}\{s(t-s)\}^{-1/2} \, ds.$$

Since the left-hand side is the Laplace transform of $\Phi(t)$, this equality proves that the distribution function $F_t(s)$ say, of $\Phi(t)$, is absolutely continuous with density

$$\frac{dF_t(s)}{ds} = \pi^{-1}\{s(t-s)\}^{-1/2} = \frac{d}{ds}\left[2\pi^{-1} \arcsin\left(\frac{s}{t}\right)^{1/2}\right], \quad 0 \le s \le t. \quad (2.125)$$

Thus we have derived the arc-sine law for the distribution of the sojourn time on the positive half-line.

EXAMPLE 2. We now use the method of Example 1 to give an alternative derivation of the distribution of $M(t)$. For any constant $a > 0$ we set

$$V_a(x) = \begin{cases} 0 & x < a, \\ 1 & x \ge a. \end{cases} \quad (2.126)$$

It is easy to see that

$$\lim_{\lambda \to \infty} E_0\left\{\exp\left[-\lambda \int_0^t V_a(B(s)) \, ds\right]\right\} = P_0\left(\max_{0 \le s \le t} B(s) < a\right) \quad (2.127)$$

whilst we also know that the function v given by

$$v(x) = E_x\left\{\int_0^\infty e^{-\alpha t} \exp\left[-\lambda \int_0^t V_a(B(s)) \, ds\right]\right\}$$

satisfies the differential equation $(\mathfrak{g} - \alpha - \lambda V_a)v = 1$. That is

$$\begin{cases} \dfrac{1}{2} v''(x) - \alpha v(x) = 1 & x < a, \\ \dfrac{1}{2} v''(x) - (\alpha + \lambda)v(x) = 1 & x > a. \end{cases}$$

As in Example 1 we can solve this equation and obtain

$$v(0) = \alpha^{-1}\left(1 - \exp[-a(2\alpha)^{1/2}]\left\{1 - \left[\frac{\alpha}{(\alpha + \lambda)}\right]^{1/2}\right\}\right).$$

Letting $\lambda \to \infty$ and taking note of (2.127) we obtain a formula for the distribution of $M(t) = \max_{0 \leq s \leq t} B(s)$:

$$\int_0^\infty e^{-\alpha t} P(M(t) < a)\, dt = \alpha^{-1}\{1 - \exp[-a(2\alpha)^{1/2}]\},$$

and therefore

$$P(M(t) < a) = \left(\frac{2}{\pi}\right)^{1/2} \int_0^{at^{-1/2}} \exp\left[-\frac{u^2}{2}\right] du$$

$$= 2(2\pi t)^{-1/2} \int_0^a \exp\left[-\frac{x^2}{2t}\right] dx,$$

which agrees with Proposition 2.9.

This example can be generalised in the following manner. Set

$$m(t) = \min_{0 \leq s \leq t} B(s) \qquad (2.128)$$

and take the function V to be

$$V_{ab}(x) = \begin{cases} 0 & -b < x < a \quad (a > 0, b > 0), \\ 1 & x \leq -b \text{ or } x \geq a. \end{cases} \qquad (2.129)$$

Then we could find the joint distribution $P(-b \leq m(t) \leq M(t) < a)$, and it is good exercise to show that the result coincides with the correct particular case of Proposition 2.10.

(iii) The Brownian Bridge

Take a Brownian motion which begins at the point x, such motion being realised by $B(t, \omega) + x$. We wish to "deform" it so that it passes through a fixed point y at a fixed instant t_0 of time. In other words we are interested in the modified motion of a Brownian particle constrained at the extreme points of the time interval $[0, t_0]$. There are several ways to construct such a process, and typical methods include the following. (a) Given the process

2.6 Processes Related to Brownian Motion

$\{B(t) + x: t \in [0, t_0]\}$ and the fixed y, we can introduce the conditional distribution given the condition $\{B(t_0) + x = y\}$. This will uniquely determine the desired process. (b) Use the method of §2.3 (i) to construct Brownian motion over the interval $[0, t_0]$ but taking the $X_1(\omega)$ defining $B(t, \omega)$ to be the constant $y - x$, rather than what it is there. Although the time parameter extends over $[0, t_0]$, rather than the unit interval of §2.3, the argument is unchanged. (c) The conditional Wiener measure μ_x^W is restricted to $\mathscr{C}(t_0) = \mathscr{C}([0, t_0])$ and a conditional probability measure $\mu_{(0, x)}^{(t_0, y)}$ is taken such that we have

$$\mu_x^W(\cdot) = \int \mu_{(0, x)}^{(t_0, y)}(\cdot)(2\pi t_0)^{-1/2} \exp\left[-\frac{(y - x)^2}{2t_0}\right] dy.$$

The measure $\mu_{(0, x)}^{(t_0, y)}$ is then the distribution of the process under discussion. In what follows we adopt approach (b) for this makes use of the important property of Brownian motion that it can be obtained by successively interpolating independent Gaussian random variables.

As indicated in (b) above, we assume that $\{B(t, \omega): t \in [0, t_0]\}$ has been formed by the method of §2.3(i) where $X_1(\omega)$ is set to be $y - x$. We then add x and obtain

$$X(t, \omega) = B(t, \omega) + x - \frac{t}{t_0}\{B(t_0, \omega) - y + x\}$$

$$= B(t, \omega) - \frac{t}{t_0} B(t_0, \omega) + \frac{t_0 - t}{t_0} x + \frac{t}{t_0} y, \quad 0 \leq t \leq t_0. \tag{2.130}$$

Clearly $\{X(t, \omega)\}$ is a Gaussian process satisfying the constraints $X(0, \omega) = x$, $X(t_0, \omega) = y$ and so is, in a sense, a kind of conditional Brownian motion. Since it is Gaussian, it is uniquely determined by its mean and covariance function (Theorem 1.10, §1.6), and these are

$$m(t) = E\{X(t)\} = \frac{t_0 - t}{t_0} x + \frac{t}{t_0} y,$$

$$\Gamma(t, s) = E[\{X(t) - E(X(t))\}\{X(s) - E(X(s))\}] = s\frac{t_0 - t}{t_0}, \quad s < t. \tag{2.131}$$

Definition 2.8. A *Brownian bridge* or *tied down Brownian motion* is a Gaussian process whose mean $m(t)$ and covariance function $\Gamma(t, s)$ are of the form given in (2.131).

The process $\{X(t)\}$ of (2.130) is understood to be a realisation of a Brownian bridge, and as far as distributional properties of the Brownian bridge are concerned, it gives us enough information.

A Brownian bridge depends on the choice $[0, t_0]$ of the time interval, the starting point x and the terminal point y, and when we wish to emphasise its dependence on these aspects, we will use the notation $\{X_{(0, x)}^{(t_0, y)}(t): t \in [0, t_0]\}$. Otherwise either or both the subscript and superscript will be omitted.

Proposition 2.19. *The two Gaussian processes* $\{X_{(0, x)}^{(t_0, y)}(t): t \in [0, t_0]\}$ *and* $\{X_{(0, y)}^{(t_0, x)}(t_0 - t): t \in [0, t_0]\}$ *have the same distribution.*

PROOF. The expectation of $X_{(0, y)}^{(t_0, x)}(t_0 - t)$ is $[t_0 - (t_0 - t)]y/t_0 + (t_0 - t)x/t_0$ which agrees with that of $X_{(0, x)}^{(t_0, y)}(t)$, and the covariance function of the process $\{X_{(0, y)}^{(t_0, x)}(t_0 - t)\}$ is independent of x and y, and is the same as that of $\{B(t_0 - t) - [(t_0 - t)/t_0]B(t_0)\}$. A simple computation proves that the latter coincides with $\Gamma(t, s)$ in (2.131), as was to be proved. □

Proposition 2.20. *A Brownian bridge is a Markov process.*

PROOF. Take a pair s, t with $0 < s < t < t_0$ and write $X(t) = X_{(0, x)}^{(t_0, y)}(t)$ in the form

$$X(t) = \frac{t_0 - t}{t_0 - s} X(s) + \{X(t) - \frac{t_0 - t}{t_0 - s} X(s)\}. \quad (2.132)$$

The covariance between the random variable inside the braces $\{\cdot\}$ and $X(\tau)$ with $\tau \leq s$ is $\tau(t_0 - t)/t_0 - [(t_0 - t)/(t_0 - s)]\tau(t_0 - s)/t_0 = 0$, so that those two random variables are independent. Thus the bracketted random variable is independent of $\mathbf{B}_s(X)$. Now it is clear that $[(t_0 - t)/(t_0 - s)]X(s)$ is $\mathbf{B}_s(X)$-measurable, so that we have

$$E\{X(t) | \mathbf{B}_s(X)\} = \frac{t_0 - t}{t_0 - s} X(s).$$

This means that $\{(t_0 - t)^{-1}X(t)\}$ is an additive process and hence that $\{X(t)\}$ is a Markov process. □

Now let us normalise the process $X(t)$ to have zero mean and variance unity. Since the variance of $X(t)$ is $t(t_0 - t)/t_0$, by (2.131), the normalised process $Y(t)$ is given by

$$Y(t, \omega) = t_0^{1/2}\{t(t_0 - t)\}^{-1/2}[X(t, \omega) - \left|\frac{t_0 - t}{t_0}\right|x - \frac{t}{t_0}y], \quad (2.133)$$

$$0 < t < t_0.$$

The process $\{Y(t): 0 < t < t_0\}$ depends on neither x nor y and is a Gaussian process with (zero mean and) covariance function

$$\Gamma(t, s) = E\{Y(t)Y(s)\} = \left|\frac{s(t_0 - t)}{t(t_0 - s)}\right|^{1/2}, \quad 0 < s < t < t_0. \quad (2.134)$$

Looking carefully at the expression inside the braces $\{\cdot\}$ of this covariance function, we recognize it as the *anharmonic ratio* of the four numbers $(0, s, t, t_0)$. From this fact we can easily derive the *projective invariance* of Brownian motion, a result due to P. Lévy which is stated in the next proposition.

2.6 Processes Related to Brownian Motion

Proposition 2.21 (P. Lévy). *Let $p(t)$, $0 < t < t_0$, be a projective transformation of the interval $(0, t_0)$ onto itself. Then the two processes $\{Y(t): 0 < t < t_0\}$ and $\{Y(p(t)): 0 < t < t_0\}$ coincide.*

This important property will be discussed again (§5.4) in the context of white noise.

Remark 1. With a suitable change in notation this result holds even for an infinite time interval. In particular, for the interval $(0, \infty)$, we arrive at the conclusion stated in §2.1 Proposition 2.1.

Remark 2. The process $\{Y(t): 0 < t < t_0\}$ under discussion is a Gaussian process having a canonical representation (see §4.4 below for related topics). More fully, there is a Brownian motion $\{\tilde{B}(t): 0 \le t \le t_0\}$ such that for all t, $\mathbf{B}_t(Y) = \mathbf{B}_t(\tilde{B})$, and $Y(t)$ has a representation

$$Y(t) = \{t_0(t_0 - t)\}^{1/2} t^{-1/2} \int_0^t (t_0 - u)^{-1} d\tilde{B}(u), \text{ a Wiener integral, see §4.2.}$$

From this expression we can easily derive the stochastic differential equation satisfied by the Brownian bridge [see also K. Itô (1951a), H. P. McKean Jr. (1969) and the brief review in Chapter 4 below]. It is

$$dX(t) = d\tilde{B}(t) - \{(t_0 - t)^{-1} X(t) + t_0^{-1}(x - y)\} dt$$

with

$$X(0) = x, \text{ the initial state.}$$

We now return to the Brownian bridge $X(t) = X_{(0, x)}^{(t_0, y)}(t)$ introduced earlier. Since it can be realised as in (2.130), almost all of its sample functions are continuous, and so the distribution of $\{X(t): t \in [0, t_0]\}$ may then be defined on $\mathscr{C}(t_0) = \mathscr{C}([0, t_0])$. As usual the σ-field $\mathfrak{B}(t_0)$ of subsets of $\mathscr{C}(t_0)$ is taken to be that generated by the cylinder sets, and we denote the distribution on $(\mathscr{C}(t_0), \mathfrak{B}(t_0))$ of $\{X(t)\}$ by $\mu_{(0, x)}^{(t_0, y)}$. We will continue to use previously established notation such as $g(t; x)$ for the Gauss kernel, and μ_x for the conditional Wiener measure.

Proposition 2.22. *For any $A \in \mathfrak{B}(t_0)$ we have the following equality*

$$\int_{A(t_0)} \mu_{(0, x)}^{(t_0, y)}(A) g(t_0; x - y) \, dy = \mu_x(A), \tag{2.135}$$

where $A(t_0) = \{\xi(t_0): \xi \in A\}$.

PROOF. If we recall the construction of $X(t)$ then the proof of this assertion is almost obvious, intuitively at least. A formal proof proceeds as follows. It clearly suffices to prove (2.135) when A is a cylinder subset of $\mathscr{C}(t_0)$, and so let us set

$$A = \{\xi \in \mathscr{C}(t_0): \xi(s_1) \in B_1, \ldots, \xi(s_n) \in B_n, \xi(t_0) \in B_0\},$$

where $s_1 < s_2 < \cdots < s_n < t_0$ are time points and $B_0, B_1, \ldots, B_n$ are Borel subsets of $\mathbf{R}$. Then we have

$$\mu_x(A) = P(B(s_1) + x \in B_1, \ldots, B(s_n) + x \in B_n, B(t_0) + x \in B_0)$$

$$= \int_{(B(t_0) + x \in B_0)} P(B(s_1) + x \in B_1, \ldots, B(s_n) + x \in B_n | B(t_0)) \, dP.$$

Noting that the density of the distribution of $B(t_0)$ is just $g(t_0; x)$, we can express the above integral in terms of distributions and we immediately obtain the integral on the left-hand side of (2.135). □

By using this proposition, $v(x)$ given by (2.119) can be expressed in the form

$$v(x) = E_x \left\{ \int_0^\infty e^{-\alpha t} f(B(t)) \exp\left[-\int_0^t V(B(s)) \, ds \right] dt \right\}$$

$$= \int_0^\infty e^{-\alpha t} E_x \left\{ f(B(t)) \exp\left[-\int_0^t V(B(s)) \, ds \right] \right\} dt$$

$$= \int_0^\infty e^{-\alpha t} \, dt \int_{C(t)} f(\xi(t)) \exp\left[-\int_0^t V(\xi(s)) \right] d\mu_x(\xi)$$

$$= \int_0^\infty e^{-\alpha t} \, dt \int_{-\infty}^\infty f(y) g(t; y - x) \, dy$$

$$\cdot \left\{ \int_{C(t)} \exp\left[-\int_0^t V(\xi(s)) \, ds \right] d\mu_{(0, x)}^{(t, y)}(\xi) \right\}.$$

Our interest in this expression focusses on the value of the expression inside the braces $\{\cdot\}$ in the last formula. Setting

$$\psi(t, x, y) = \int_{C(t)} \exp\left[-\int_0^t V(\xi(s)) \, ds \right] d\mu_{(0, x)}^{(t, y)}(\xi), \quad (2.136)$$

we note that ψ does not involve the f of the previous expressions. When V is a bounded function a standard approach [see M. Kac (1951)] is to expand the exponent of (2.136) in a power series, thereby obtaining an explicit formula for $\psi(t, x, y)$. In the unbounded case where, say, $V(x) \to \infty$ as $|x| \to \infty$, we need conditions on the growth of V to enable us to make use of the eigenfunction expansions developed by E. C. Titchmarsh (1946). For example $V(x) = |x|^\alpha$ for some $\alpha > 1$ is a sufficient condition. Let $\{\lambda_n\}$ and

$\{\psi_n\}$ be the system of eigenvalues and associated normalised eigenfunctions, respectively, of the differential operator $-\frac{1}{2}d^2/dx^2 + V(x)$:

$$\frac{1}{2}\frac{d^2}{dx^2}\psi_n - V(x)\psi_n = -\lambda_n\psi_n.$$

Then it is known that the ψ of (2.136) may be expressed in terms of $\{\lambda_n\}, \{\psi_n\}$ and the Gauss kernel g as

$$g(t; x - y)\psi(t, x, y) = \sum_n e^{-\lambda_n t}\psi_n(x)\psi_n(y), \qquad (2.137)$$

[see the paper of M. Kac cited above, and K. Itô and H. P. McKean Jr. (1965)]. The particular case (i) in which $V(x) = x^2$ can be dealt with in a quite different manner as we will see in §7.6 (iii). Thus alternative intepretations may be provided for $v(x)$.

If we denote the left-hand side of (2.137) by $\varphi(0, x; t, y)$, then we may replace the time interval $[0, t]$ with $[t, u]$, and a function $\varphi(t, y; u, z)$ can then be defined in a natural way using the distribution $\mu_{(t, y)}^{(u, z)}$ of the Brownian bridge on the interval $[t, u]$. For $t_0 < t_1 < t_2$ we can prove the relation

$$\varphi(t_0, x; t_2, z) = \int_{-\infty}^{\infty} \varphi(t_0, x; t_1, y)\varphi(t_1, y; t_2, z)\, dy$$

and the fundamental solution of the differential equation

$$\frac{\partial}{\partial t}\varphi(x, t) = \frac{1}{2}\frac{\partial^2}{\partial x^2}\varphi(x, t) - V(x)\varphi(x, t)$$

is given by $\varphi(0, x_0; t, x)$. The reader interested in this field is recommended to consult I. M. Gelfand and A. M. Yaglom (1960), where a description of the so-called Feynman path integral can be found.

As we have seen, albeit briefly, several significant theories arise out of Brownian motion, in areas extending from analysis to quantum mechanics. In each case excellent books exist which cover these theories, and so we do not stop to develop them here, but rather hurry on to our next topic.

3 Generalised Stochastic Processes and Their Distributions

As was announced in §1.3, (ii), this chapter is devoted to a study of the distributions of generalised stochastic processes. In the finite-dimensional case Bochner's theorem establishes a one-to-one correspondence between distributions and characteristic functions, and this result has a counterpart in the present context, where the Bochner-Minlos theorem (§3.2) links probability measures on the space of generalised functions with characteristic functionals on the space of test functions. A secondary aim of this chapter is to provide a discussion of the generalised stochastic process white noise from several points of view (§§3.3, 3.4), this serving as background to much of what will be considered in the next chapter.

3.1 Characteristic Functionals

When considering the distribution of a generalised stochastic process it is helpful to have in mind the time-derivative $(d/dt)B(t, \omega)$ of a Brownian motion as a typical example. The distribution of such a process is a probability measure on a space of generalised functions, and, strictly speaking, the space should be specified for each process. However we may generally assume that the time parameter runs over some interval T (finite or infinite; occasionally the circle is considered), and we can then use the Hilbert space $H = L^2(T)$ as follows. Take a triple

$$E \subseteq H \subseteq E^*, \tag{3.1}$$

where the injections are both continuous, and E is a countably Hilbert nuclear space (see §A.3). The dual space E^* of E is the basic space on which the distribution of a generalised stochastic process is located.

3.1 Characteristic Functionals

Some preliminaries are needed before we can state the Bochner-Minlos theorem. The canonical bilinear form that links E and E^* is denoted by

$$\langle x, \xi \rangle, \qquad x \in E^*, \xi \in E. \tag{3.2}$$

In particular, if $x \in H$, then $\langle x, \xi \rangle$ coincides with the inner product on H.

We now recall the definition of a generalised stochastic process. It is a system of random variables with parameter space E

$$\{X(\xi, \omega): \xi \in E\},$$

defined on a probability space $(\Omega, \mathbf{B}, P)$ in such a way that $X(\xi, \omega)$ is continuous and linear in ξ. The joint distribution of $(X(\xi_1), X(\xi_2), \ldots, X(\xi_n))$ is uniquely determined by the characteristic function

$$\int \exp\left[i \sum_j z_j X(\xi_j, \omega)\right] dP(\omega), \qquad z_j \in \mathbf{R},$$

which, because of the linearity of X in ξ, turns out to be the characteristic function of $X(\sum_j z_j \xi_j)$. Since $\sum_j z_j \xi_j \in E$, we may conclude that the distribution of the generalised stochastic process under consideration is completely determined by

$$C_X(\xi) = \int \exp[iX(\xi, \omega)] \, dP(\omega). \tag{3.3}$$

The functional $C_X(\xi)$, $\xi \in E$, is called the *characteristic functional* of the generalised stochastic process X, and satisfies the following properties:

1. C_X is continuous in $\xi \in E$;
2. C_X is positive definite, i.e. for any n, complex numbers $\alpha_1, \alpha_2, \ldots, \alpha_n$, and $\xi_1, \xi_2, \ldots, \xi_n \in E$, we have

$$\sum_{j,k} \alpha_j \bar{\alpha}_k C_X(\xi_j - \xi_k) \geq 0; \tag{3.4}$$

3. $C_X(0) = 1$.

These conditions are identical in form to those which apply in the finite-dimensional case (cf. Chapter 1, §1.3). Following §1.3, (i) we consider the problem of finding a probability measure μ such that

$$C_X(\xi) = \int_{E^*} e^{i\langle x, \xi \rangle} \, d\mu(x), \tag{3.5}$$

where C_X is a given functional satisfying (3.4).

As has been pointed out above [§1.3, (iii)], measures on infinite-dimensional spaces are in some ways different from those on finite-dimensional spaces. The Bochner theorem, which provides a one-to-one correspondence between distributions and characteristic functions, needs to be modified and investigated more carefully in the infinite-dimensional case.

We do this in several steps, beginning with an example which illustrates the difficulties. Whereas on $\mathbf{R}^n$ a continuous positive definite function determines a probability measure on the same space $\mathbf{R}^n$, this is not true in the case of a Hilbert space.

EXAMPLE 1. Let $E = H = L^2(\mathbf{R})$ (whence $E^* = H$ also) and let us take

$$C(\xi) = \exp\left[-\frac{1}{2}\|\xi\|^2\right], \qquad \xi \in E, \qquad \|\cdot\| \text{ the } H\text{-norm.} \qquad (3.6)$$

It is easy to check that $C(\xi)$ satisfies the three conditions of (3.4). Suppose that a probability measure μ can be defined on H which relates to $C(\xi)$ in the manner of (3.5). Then for any complete orthonormal system $\{\xi_n\}$ in H the definition (3.6) gives

$$\int_H e^{iz\langle x, \xi_n\rangle} d\mu(x) = e^{-z^2/2}, \qquad z \in \mathbf{R}. \qquad (3.7)$$

On the other hand, for any $x \in H$ we must have

$$\lim_{n\to\infty} \langle x, \xi_n\rangle = 0.$$

But this is incompatible with the identity (3.7) and so we have a contradiction.

The equation (3.7) shows that $\langle x, \xi_n\rangle$ is a standard Gaussian variable, and in addition the relation

$$C\left(\sum_1^n z_k \xi_k\right) = \exp\left[-\frac{1}{2}\sum_1^n z_k^2\right], \qquad z_k \in \mathbf{R},$$

implies that $\{\langle x, \xi_n\rangle : n \geq 1\}$ is an independent system of random variables. The law of large numbers then implies that

$$\lim_{N\to\infty} N^{-1} \sum_1^N \langle x, \xi_n\rangle^2 = 1 \qquad \text{a.e. }(\mu).$$

Recalling that $\{\langle x, \xi_n\rangle\}$ are the coordinates of x we see that, roughly speaking, the support of μ is much wider than H, and can begin to understand why the above contradiction occurs.

The probability measure determined by the characteristic functional (3.6) is the white noise that will be introduced later (§3.3) and indeed is the main subject of this chapter.

3.2 The Bochner-Minlos Theorem

In this section we describe a method of constructing a measure (distribution) μ on the space E^* when given a characteristic functional $C(\xi)$ satisfying (3.4). Proceeding in steps, we first describe the measurable space on which μ is to

3.2 The Bochner-Minlos Theorem

be defined. Collections of generalised functions, that is, subsets of E^*, that can be specified by a finite number of test functions should be considered. We therefore require that subsets of E^* of the form

$$A_{\xi_1, \ldots, \xi_n, B} = \{x \in E^* : (\langle x, \xi_1 \rangle, \ldots \langle x, \xi_n \rangle) \in B\} \tag{3.8}$$

where $n \geq 1$, $\xi_1, \ldots, \xi_n \in E$ and B is a Borel subset of $\mathbf{R}^n$, all belong to the σ-field $\mathfrak{B}$ which is to be the domain of μ. A subset of E^* of the form (3.8) is called a *cylinder set*. If the ξ_i in (3.8) are all taken from a finite-dimensional subspace F of E, then the cylinder set is said to be based on F, and the collection of all cylinder sets based on a fixed such F forms a σ-field $\mathfrak{A}_F$. The union

$$\mathfrak{A} = \bigcup_{F \subset E} \mathfrak{A}_F$$

taken over all finite-dimensional subspaces F is only a field of subsets of E^*, and so we take the smallest σ-field $\mathfrak{B}$ containing $\mathfrak{A}$ to define our measurable space $(E^*, \mathfrak{B})$.

We are now ready to define the probability measure (distribution) determined by the characteristic functional of the given stochastic process. Starting from the given $C(\xi)$ we

1. construct a measure space $(E^*, \mathfrak{A}_F, m_F)$ for each fixed finite-dimensional subspace $F \subset E$; and then
2. let F vary to obtain a system of measure spaces which determine a finitely-additive measure space $(E^*, \mathfrak{A}, m)$; and finally
3. extend $(E^*, \mathfrak{A}, m)$ to a countably-additive measure space $(E^*, \mathfrak{B}, \mu)$.

It is this last step which presents difficulties unique to the infinite-dimensional situation. Details of the three steps now follow.

1. Suppose that F is a fixed n-dimensional subspace of E. The annihilator F^a of F is the subspace of E^* defined by

$$F^a = \{x \in E^* : \langle x, \xi \rangle = 0 \quad \text{for all} \quad \xi \in F\},$$

and the quotient space E^*/F^a is isomorphic to F^* ($\cong F$) and, in particular, n-dimensional. The bilinear form $\langle \cdot, \cdot \rangle_F$ that connects E^*/F^a with F is defined in such a way that if x is a representative of $\bar{x} \in E^*/F^a$,

$$\langle \bar{x}, \xi \rangle_F = \langle x, \xi \rangle, \qquad \xi \in F.$$

Indeed the right-hand side is independent of the choice of the representative x and $\langle \cdot, \cdot \rangle_F$ may be regarded as the inner produce in an n-dimensional Euclidean space.

On the other hand the restriction $C_F(\xi)$ of $C(\xi)$ to F may be viewed as a characteristic function on $\mathbf{R}^n$ by (3.4), and so the Bochner theorem (Theorem 1.1) guarantees the existence of a unique probability measure m_F on F^*, i.e. on E^*/F^a, such that

$$C_F(\xi) = \int_{E^*/F^a} e^{i \langle \bar{x}, \xi \rangle} \, dm_F(\bar{x}).$$

Thus we have a probability space $(E^*/F^a, \mathfrak{B}_F, m_F)$, where $\mathfrak{B}_F$ is the Borel field of subsets of E^*/F^a.

2. Let ρ_F be the canonical projection from E^* to E^*/F^a, i.e.

$$\rho_F \colon x \to \bar{x} = x + F^a \in E^*/F^a, \qquad x \in E^*.$$

With this notation we have

$$\rho_F^{-1}(\mathfrak{B}_F) = \mathfrak{A}_F.$$

If $B \in \mathfrak{B}_F$ then $A \equiv \rho_F^{-1}(B) \in \mathfrak{A}_F$ and we can then define $\tilde{m}_F$ by

$$\tilde{m}_F(A) = m_F(B),$$

obtaining a measure space $(E^*, \mathfrak{A}_F, \tilde{m}_F)$.

Now let us suppose that F and G are finite-dimensional subspaces of E with $F \subset G$. Then the canonical projection

$$T \colon E^*/G^a \to E^*/F^a$$

is well-defined, and for $B \in \mathfrak{B}_F$ we have

$$m_F(B) = m_G(T^{-1}B).$$

Indeed this equality follows from Bochner's theorem, since $C_F(\xi)$ coincides with the restriction of $C_G(\xi)$ to F. Thus the probability space $(E^*, \mathfrak{A}_G, \tilde{m}_G)$ is an extension of $(E^*, \mathfrak{A}_F, \tilde{m}_F)$.

We are now in a position to define m on $\mathfrak{A}$. For any $A \in \mathfrak{A}$ there is a suitable finite-dimensional subspace F of E such that $A \in \mathfrak{A}_F$, and we define $m(A)$ by

$$m(A) = \tilde{m}_F(A). \tag{3.9}$$

By the preceding remarks m is well-defined. We now show that m is finitely-additive.

Let $A_1, A_2, \ldots, A_n$ be mutually exclusive elements of $\mathfrak{A}$ based upon subspaces $F_1, F_2, \ldots, F_n$, and suppose that these subspaces span the subspace F of E. Clearly F is also finite-dimensional, $A_i \in \mathfrak{A}_F$, and

$$\tilde{m}_{F_i}(A_i) = \tilde{m}_F(A_i), \qquad i = 1, 2, \ldots, n.$$

Since $\tilde{m}_F$ is additive on $\mathfrak{A}_F$ and $\bigcup_1^n A_i \in \mathfrak{A}_F$, we have

$$m\left(\bigcup_1^n A_i\right) = \tilde{m}_F\left(\bigcup_1^n A_i\right) = \sum_1^n \tilde{m}_F(A_i)$$

$$= \sum_1^n \tilde{m}_{F_i}(A_i) = \sum_1^n m(A_i).$$

Thus we have proved that $(E^*, \mathfrak{A}, m)$ is a finitely-additive probability space, i.e. a probability space in the weak sense (cf. Example 4, §1.2).

3. Our aim now is to extend m to measure on the space $(E^*, \mathfrak{B})$. Such an extension is possible if and only if m is σ-additive on $\mathfrak{A}$ (Lemma 2.1, §2.1),

3.2 The Bochner-Minlos Theorem

and under the assumption (3.1) made at the beginning of this chapter, σ-additivity of m on $\mathfrak{A}$ follows from the conditions (1), (2), (3) of (3.4) on $C(\xi)$. The following Lemma [R. A. Minlos (1959)] is the key to the proof of this fact.

Lemma 3.1. *Let μ be a probability measure on $\mathbf{R}^n$ and denote by $\mathscr{E}$ the ellipsoid*

$$\left\{ z = (z_1, z_2, \ldots, z_n) : \sum_1^n a_i^2 z_i^2 \leq \gamma^2 \right\}.$$

If the characteristic function $\varphi(z)$, $z \in \mathbf{R}^n$, of μ satisfies

$$|\varphi(z) - 1| < \varepsilon \quad \text{for } z \in \mathscr{E}, \tag{3.10}$$

then for the ball $S(t)$ in $\mathbf{R}^n$ of radius t we have the inequality

$$\mu(S(t)^c) < \beta^2 \left(\varepsilon + \frac{2}{\gamma^2 t^2} \sum_1^n a_i^2 \right), \tag{3.11}$$

where $S(t)^c = \mathbf{R}^n \setminus S(t)$ and β is a positive constant independent of n and t.

PROOF. We begin with an easy estimate of an integral:

$$I \equiv \int_{\mathbf{R}^n} \left[1 - \exp\left(-\frac{1}{2t^2} \sum_1^n x_i^2 \right) \right] d\mu(x)$$

$$\geq \int_{S(t)^c} \left[1 - \exp\left(-\frac{1}{2t^2} \sum_1^n x_i^2 \right) \right] d\mu(x)$$

$$\geq [1 - \exp(-\tfrac{1}{2})]\mu(S(t)^c).$$

Now the integral I can be written in the form

$$I = 1 - \int_{\mathbf{R}^n} \exp\left(-\frac{1}{2t^2} \sum_1^n x_i^2 \right) d\mu(x)$$

and the integrand $\exp(-(1/2t^2)\sum_1^n x_i^2)$ may be regarded as the characteristic function of a Gaussian distribution. We therefore have

$$I = 1 - (t^2/2\pi)^{n/2} \int_{\mathbf{R}^n} \varphi(z) \exp\left(-\frac{1}{2} t^2 \sum_1^n z_i^2 \right) dz$$

$$= \left(\frac{t^2}{2\pi} \right)^{n/2} \int_{\mathbf{R}^n} [1 - \varphi(z)] \exp\left(-\frac{1}{2} t^2 \sum_1^n z_i^2 \right) dz$$

$$\leq \left(\frac{t^2}{2\pi} \right)^{n/2} \left| \int_{\mathscr{E}} + \int_{\mathscr{E}^c} \right|$$

$$< \varepsilon + \left(\frac{t^2}{2\pi} \right)^{n/2} \frac{2}{\gamma^2} \int_{\mathscr{E}^c} \left(\sum_1^n a_i^2 z_i^2 \right) \exp\left(-\frac{1}{2} t^2 \sum_1^n z_i^2 \right) dz$$

$$< \varepsilon + \frac{2}{\gamma^2 t^2} \sum_1^n a_i^2.$$

These two evaluations from above and below give

$$\mu(S(t)^c) < \left[1 - \exp\left(-\frac{1}{2}\right)\right]^{-1}\left(\varepsilon + \frac{2}{\gamma^2 t^2}\sum_1^n a_i^2\right). \qquad \square$$

A further lemma gives a condition on the space E^* for a measure μ to be extendable. We shall use in what follows the notations $\| \quad \|_p$ the p-th norm of E and E_p the corresponding Hilbert space, which are illustrated in §A.3.

Lemma 3.2. *A necessary and sufficient condition for a finitely additive measure m to be extendable to a σ-additive measure on $(E^*, \mathfrak{B})$ is: for any $\varepsilon > 0$ there exists a natural number n and a ball $S_n = \{x \in E^*: \|x\|_{-n} \leq \gamma_n\}$, such that for any $A \in \mathfrak{A}$ disjoint from S_n we have*

$$\mu(A) < \varepsilon.$$

PROOF. Necessity. Suppose that m has an extension μ. Choose a sequence S_n of balls with increasing radii γ_n such that $\gamma_n \to \infty$. Then $\bigcup_n S_n = E^*$ and so, as μ is countably additive,

$$\mu(S_n^c) < \varepsilon$$

must hold for sufficiently large n. The required inequality for $\mu(A)$ now follows.

Sufficiency is proved by reductio ad absurdum. Suppose that $\{A_n\}$ is a sequence of pairwise disjoint elements of $\mathfrak{A}$ such that $\sum_n A_n = E^*$. Since m is finitely additive, $m(\sum_1^n A_k) = \sum_1^n m(A_k) \leq 1$ and so

$$\sum_1^\infty m(A_n) \leq 1.$$

Suppose that the above inequality is strict. Then there exists $\varepsilon > 0$ such that

$$\sum_1^\infty m(A_n) = 1 - 3\varepsilon < 1.$$

For each A_n we can find an open cylinder set A_n' [i.e. the set B in (3.8) is open] such that $A_n' \supset A_n$ and

$$m(A_n' \setminus A_n) < \frac{\varepsilon}{2^n}.$$

Clearly $\bigcup_j A_j' \supseteq S_n$ and, since S_n is weakly compact, we can choose a finite number $A_1', A_2', \ldots, A_k'$ of the A_j' which cover S_n. Setting $A' = \bigcup_1^k A_j'$ we have $A' \in \mathfrak{A}$ and

$$1 = m(A' + A'^c) = m(A') + m(A'^c),$$

$$m(A') \leq \sum_1^k m(A_j') + \varepsilon,$$

$$m(A'^c) < \varepsilon.$$

The last inequality comes from the hypothesis in the lemma, and the three inequalities combine to give

$$1 \le \sum_{1}^{k} m(A_j) + \varepsilon + \varepsilon \le (1 - 3\varepsilon) + 2\varepsilon = 1 - \varepsilon,$$

a contradiction. □

Theorem 3.1. *Let $C(\xi)$, $\xi \in E$, be a functional which is*

1. *continuous in the norm $\|\cdot\|_p$ for some p;*
2. *positive definite; and such that*
3. *$C(0) = 1$;*

i.e. $C(\xi)$ is a characteristic functional.

If for some $n(>p)$ the injection $T_p^n: E_n \to E_p$ is of Hilbert-Schmidt type, then there exists a unique countably additive extension μ of m to $(E^*, \mathfrak{B})$ and μ is supported by E_n^*.

PROOF. It follows from our assumptions that for any $\varepsilon > 0$ there exists a ball U in E_p with radius γ (a neighbourhood of 0) such that

$$|C(\xi) - 1| < \frac{\varepsilon}{2\beta^2}, \qquad \xi \in U,$$

where β is the constant in Lemma 3.1. Also the assumptions imply that there is a neighbourhood V of 0 in E_n such that

$$T_p^n V \subseteq U.$$

We can then show that the ball S_n in E_n^* with radius $t = 2\beta \|T_p^n\|_2/(\gamma^2 \varepsilon)^{1/2}$ corresponds to the ball in Lemma 3.2. Indeed let A be a cylinder set based upon the finite-dimensional subspace F and disjoint from S_n. Then there exists an n-dimensional Borel set B such that $A = \rho_F^{-1}(B)$ and

$$B \cap \rho_F(S_n) = \varnothing.$$

Passing now to finite-dimensional measures we begin by noting that since $V \cap F$ is a finite-dimensional ellipsoid in the norm $\|\cdot\|_n$, it can be expressed in cartesian co-ordinates in the form $\sum_i a_i^2 z_i^2 \le \gamma^2$. We also note that $\sum_i a_i^2 \le \|T_p^n\|_2^2$. Using Lemma 3.1 we have

$$m_F(\rho_F(S_n)^c) < \beta^2 \left(\frac{\varepsilon}{2\beta^2} + \frac{2}{\gamma^2 t^2} \sum_{1}^{k} a_i^2 \right)$$

$$< \frac{1}{2}\varepsilon + \frac{2\beta^2}{\gamma^2 t^2} \|T_p^n\|_2^2 = \varepsilon.$$

This guarantees the existence of an extension of m.

The uniqueness comes from general measure theory as we are dealing with finite (in fact, probability) measures. □

With minor modifications Theorem 3.1 implies the following theorem.

Theorem 3.2. Let $C(\xi)$ be a characteristic functional on E, i.e. a functional satisfying (1), (2) and (3) of (3.4). Then there exists a unique probability measure μ on $(E^*, \mathfrak{B})$ such that

$$C(\xi) = \int_{E^*} e^{i\langle x, \xi \rangle} \, d\mu(x). \tag{3.12}$$

Definition 3.1. Let $C(\xi)$ be the characteristic functional of a generalised stochastic process $\{X(\xi): \xi \in E\}$. Then the measure μ on $(E^*, \mathfrak{B})$ determined by Theorem 3.2 is called the *distribution* of $\{X(\xi)\}$. The given functional is often referred to as the *characteristic functional* of this distribution.

Having established the one-to-one correspondence between characteristic functionals on E and distributions on E^*, we could turn to considering convergence of distributions, cf. the Lévy convergence theorem of §1.3 (i). However we will not do this, but rather consider examples of generalised stochastic processes and their distributions.

3.3 Examples of Generalised Stochastic Processes and Their Distributions

EXAMPLE 1. The first example is our paradigm of a generalised stochastic process. Here the nuclear space is taken to be $\mathscr{S}$ (it could be D_0) and the characteristic functional is given by

$$C_{\sigma^2}(\xi) = \exp\left(-\frac{1}{2}\sigma^2 \|\xi\|^2\right), \qquad \xi \in \mathscr{S}. \tag{3.13}$$

By using Theorem 3.2 we immediately see that the probability measure μ_σ on $(\mathscr{S}^*, \mathfrak{B})$ exists and is unique.

Definition 3.2. The measure space $(\mathscr{S}^*, \mathfrak{B}, \mu_\sigma)$ determined by $C_{\sigma^2}(\xi)$ given by (3.13) is called the *white noise* with variance σ^2.

Let $\mathscr{S}_n$ be the Hilbert space obtained by completing $\mathscr{S}$ with respect to the n-th norm $\|\cdot\|_n$, as explained in Example 4 of §A.3 of the Appendix. The functional C_{σ^2} given by (3.13) is continuous in $L^2(\mathbf{R})$ and the injection

$$\mathscr{S}_1 \to L^2(\mathbf{R})$$

is of Hilbert-Schmidt type as is seen in §A.3. Thus by Theorem 3.1 μ_σ is supported by $\mathscr{S}_1^*$, i.e.

$$\mu_\sigma(\mathscr{S}_1^*) = 1.$$

3.3 Examples of Generalised Stochastic Processes and Their Distributions

We now turn our attention to marginal distributions of $(\mathscr{S}^*, \mathfrak{B}, \mu_\sigma)$, i.e. the restrictions of μ_σ to subspaces of $\mathscr{S}^*$. For simplicity in what follows we suppose $\sigma^2 = 1$ and write $C(\xi)$, μ for $C_1(\xi)$, μ_1 respectively. Let F be a finite-dimensional subspace of $\mathscr{S}$, and let ρ_F denote the canonical projection of $\mathscr{S}^*$ onto $\mathscr{S}^*/F^a$ as in the last section. Then the product $\rho_F \circ \mu$ is nothing but m_F introduced in §3.2, and the restriction of $C(\xi)$ to F may be viewed as the characteristic function of m_F. Introducing cartesian co-ordinates for F, we obtain the expression

$$C(\xi) = \exp\left(-\frac{1}{2}\sum_1^n z_j^2\right), \quad \xi = \sum_1^n z_j \xi_j, \quad n = \dim F.$$

Thus m_F is the n-dimensional Gaussian distribution invariant under the rotations (i.e. an $SO(n)$-invariant measure). The co-ordinates $\langle x, \xi_1 \rangle$, $\langle x, \xi_2 \rangle$, ..., $\langle x, \xi_n \rangle$ may be thought of as random variables on $(\mathscr{S}^*, \mathfrak{B}, \mu)$, and as such they are mutually independent and subject to the standard Gaussian distribution. Their joint distribution determines the marginal distribution corresponding to F.

Another case of interest is the marginal distribution of μ when the co-dimension of F is finite. In this case F is a nuclear space, for we can take a suitable Hilbert subspace H' of $L^2(\mathbf{R})$ of finite co-dimension, and a subspace $F^* \cong \mathscr{S}^*/F^a$ of $\mathscr{S}^*$, to give a triple

$$F \subset H' \subset F^*.$$

Then we can proceed as in (3.1). The probability μ_F on F^* is obtained in such a way that the characteristic functional of μ_F is the restriction, $C_F(\xi)$, of $C(\xi)$ to the subspace F. Thus defined, μ_F is also a marginal distribution of μ, and the measure space (F^*, μ_F) is, of course, isomorphic to $(\mathscr{S}^*, \mu)$.

EXAMPLE 1'. Let us take the second-order derivative $B''(t)$, $-\infty < t < \infty$, of a Brownian motion, this derivative being in the sense of generalised functions. The characteristic functional of the generalised stochastic process $\{B''(t)\}$ is given by

$$C'(\xi) = \exp\left(-\frac{1}{2}\|\xi'\|^2\right), \quad \xi \in \mathscr{S}, \quad \|\cdot\| \text{ the } L^2(\mathbf{R})\text{-norm}.$$

In the notation of §A.3 we can assert that the measure μ' determined by $C'(\xi)$ is supported by $\mathscr{S}_2^*$ (see Theorem 3.1). Thus almost all sample functions are thought of as generalized functions belonging to $\mathscr{S}_2^*$. Indeed $\{B''(t)\}$ is a generalized stochastic process.

The measures μ' and μ of Example 1 are both probability measures on the same space $(\mathscr{S}^*, \mathfrak{B})$, and are mutually singular, as we will show below. Let $\xi_n(u)$ be given by

$$\xi_n(u) = (2^n n! \sqrt{\pi})^{-1/2} H_n(u) \exp\left(-\frac{1}{2}u^2\right), \quad n \geq 0.$$

Then $\{\xi_n: n \geq 0\}$ forms a complete orthonormal system in $L^2(\mathbf{R})$, and hence on $(\mathscr{S}^*, \mathfrak{B}, \mu)$ the system

$$\{\langle x, \xi_{3n}\rangle: n \geq 1\} \tag{3.14}$$

forms a sequence of mutually independent random variables, all subject to the standard Gaussian distribution. Further, the formula

$$\xi'_n(u) = \left(\frac{1}{2}n\right)^{1/2}\xi_{n-1}(u) - \left(\frac{1}{2}(n+1)\right)^{1/2}\xi_{n+1}(u)$$

[see (A.26) of §A.5] tells us that $\langle x, \xi_{3n}\rangle$ is a Gaussian random variable on $(\mathscr{S}^*, \mathfrak{B}, \mu')$ with mean 0 and variance $\|\xi'_{3n}\|^2 = 3n + \frac{1}{2}$. Also the expression for $C'(\xi)$ proves that (3.14) forms an independent sequence on this space. Now consider the series of independent random variables

$$\sum_1^\infty n^{-1}\langle x, \xi_{3n}\rangle. \tag{3.15}$$

It converges for almost all x with respect to μ but diverges for almost all x under μ', for in the latter case the variance of the n-th term is of order n^{-1}. Letting X denote the collection of all x for which the series (3.15) converges, we see that $X \in \mathfrak{B}$, $\mu(X) = 1$ and $\mu'(X) = 0$, i.e. μ and μ' are mutually singular.

EXAMPLE 2. Let $E = \mathscr{D}(\pi)$ (notation as in §A.3) and let the characteristic functional be of the same form as in Example 1:

$$C(\xi) = \exp\left(-\frac{1}{2}\|\xi\|^2\right), \quad \xi \in \mathscr{D}(\pi).$$

Then a probability measure $\bar{\mu}$ is uniquely determined on the space $\mathscr{D}(\pi)^*$ of generalised functions on the interval $[0, 2\pi]$. Denoting by $\mathfrak{B}(\pi)$ the σ-field generated by all cylinder subsets of the space $\mathscr{D}(\pi)^*$, the measure space $(\mathscr{D}(\pi)^*, \mathfrak{B}(\pi), \bar{\mu})$ is also referred to as *white noise* with time interval $[0, 2\pi]$.

We can introduce one more white noise as follows. The basic nuclear space is now taken to be $\hat{\mathscr{D}}(\pi)$, the subspace of C^∞ consisting of all functions with period 2π. (This space may be viewed as the collection of all C^∞-functions on the unit circle.) Once again we let the characteristic functional be given by (3.13) with $\sigma = 1$. A probability measure $\hat{\mu}$ is then obtained and, as in Example 1, $\hat{\mu}$ may be viewed as a marginal distribution of $\bar{\mu}$. The σ-field $\hat{\mathfrak{B}}(\pi)$ can then be defined, and we obtain a measure space $(\hat{\mathscr{D}}(\pi), \hat{\mathfrak{B}}(\pi), \hat{\mu})$ called *periodic white noise*.

EXAMPLE 3. Here the characteristic functional of the generalised stochastic process is expressed in the somewhat more general form

$$C(\xi) = \exp\left[im(\xi) - \frac{1}{2}K(\xi, \xi)\right], \quad \xi \in E, \tag{3.16}$$

3.3 Examples of Generalised Stochastic Processes and Their Distributions

where m is a continuous linear functional on E, and $K(\xi, \eta)$ is a continuous positive definite bilinear form on $E \times E$. In such cases the generalised stochastic process is said to be *Gaussian* with m and K being called the mean and covariance functional, respectively. It is easy to see that the examples discussed so far are all Gaussian.

We will not be giving details of the general theory of Gaussian generalised stochastic processes save for one basic result concerning any process with characteristic functional (3.16). This is the fact that the functional in question determines a measure space $(E^*, \mathfrak{B}, \mu)$ on which there is a Gaussian system $\{\langle x, \xi \rangle : \xi \in E\}$ of random variables with mean $m(\xi)$ and covariance

$$\int [\langle x, \xi \rangle - m(\xi)][\langle x, \eta \rangle - m(\eta)] \, d\mu(x) = K(\xi, \eta).$$

Firstly we show that the system is Gaussian. Any finite linear combination

$$\sum_j a_j \langle x, \xi_j \rangle, \qquad a_j \in \mathbf{R}, \qquad j = 1, 2, \ldots, n,$$

of the $\langle x, \xi \rangle$ is a random variable on $(E^*, \mathfrak{B}, \mu)$ and has characteristic function

$$\int \exp\left[it \sum_j a_j \langle x, \xi_j \rangle\right] d\mu(x) = \int \exp\left[i\left\langle x, \sum_j t a_j \xi_j \right\rangle\right] d\mu(x)$$

$$= \exp\left[im\left(\sum_j t a_j \xi_j\right) - \frac{1}{2} K\left(\sum_j t a_j \xi_j, \sum_j t a_j \xi_j\right)\right]$$

$$= \exp\left[itm\left(\sum_j a_j \xi_j\right) - \frac{1}{2} t^2 K\left(\sum_j a_j \xi_j, \sum_j a_j \xi_j\right)\right].$$

This proves that $\sum_j a_j \langle x, \xi_j \rangle$ is a Gaussian random variable with mean $m(\sum_j a_j \xi_j)$ and variance $K(\sum_j a_j \xi_j, \sum_j a_j \xi_j)$. Setting $t = 1$ in the above and regarding the a_j as variables, we see that the mean of $\langle x, \xi_j \rangle$ is $m(\xi_j)$ because $m(\sum_j a_j \xi_j) = \sum_j a_j m(\xi_j)$, and that the covariance of $\langle x, \xi_j \rangle$ with $\langle x, \xi_k \rangle$ is $K(\xi_j, \xi_k)$, since $K(\sum_j a_j \xi_j, \sum_j a_j \xi_j) = \sum_{j,k} a_j a_k K(\xi_j, \xi_k)$. In view of this, the measure μ determined by the characteristic functional $C(\xi)$ of (3.16) is said to be a Gaussian measure.

The distribution of a Brownian motion $\{B(t); t \geq 0\}$ (Wiener measure) can be given as the measure on E^* which has characteristic functional

$$C(\xi) = \exp\left(-\frac{1}{2} \int_0^\infty |\hat{\xi}(t)|^2 \, dt\right), \qquad \xi \in E \, [\subseteq L^2([0, \infty))], \tag{3.17}$$

$$\hat{\xi}(t) = \int_t^\infty \xi(u) \, du, \qquad t \geq 0.$$

Indeed the random variable

$$\int_0^\infty B(t)\xi(t)\,dt, \qquad \xi \in E$$

is Gaussian, with mean 0 and variance

$$\int_0^\infty \int_0^\infty (t \wedge s)\xi(t)\xi(s)\,dt\,ds = \int_0^\infty |\hat{\xi}(t)|^2\,dt.$$

These observations lead us to view $\{B(t)\}$ as a generalised stochastic process with characteristic functional (3.17).

EXAMPLE 4. We take the nuclear space E to be a class of $\mathscr{C}^\infty$-functions on $(0, \infty)$ which is dense in $L^2([0, \infty))$. For $1 < \alpha < 2$ the expression

$$C_\alpha(\xi) = \exp\left[-\int_0^\infty |\xi(t)|^\alpha\,dt\right], \qquad \xi \in E, \tag{3.18}$$

is a characteristic functional, and defines a measure space $(E^*, \mathfrak{B}, \mu^{(\alpha)})$. This will be the distribution of the derived process $\{X'_\alpha(t): t > 0\}$ of the symmetric stable process $\{X_\alpha(t): t > 0\}$ with characteristic exponent α.

For a more general case we can take

$$C(\xi) = \exp\left\{\int_0^\infty dt \int [e^{i\xi(t)u} - 1 - i\xi(t)u]\,d\mathfrak{N}(u)\right\} \tag{3.19}$$

where $d\mathfrak{N}(u)$ is a Lévy measure satisfying

$$\int \frac{u^2}{1+u^2}\,d\mathfrak{N}(u) < \infty.$$

The functional $C(\xi)$ given by (3.19) is a characteristic functional, and determines a probability measure on E^* which coincides with the distribution of the derivative of some Lévy process. In particular if the Lévy measure $d\mathfrak{N}(u) = c_\alpha|u|^{-(\alpha+1)}\,du$ for some constant $c_\alpha > 0$, then the $C(\xi)$ of (3.19) coincides with $C_\alpha(\xi)$ of (3.18).

There is an important property satisfied by the characteristic functionals of white noise in Example 1 and 1', as well as by (3.19), namely:

$$C(\xi_1 + \xi_2) = C(\xi_1) \times C(\xi_2), \qquad \text{whenever} \quad \xi_1(t)\xi_2(t) \equiv 0. \tag{3.20}$$

Definition 3.3. A generalised stochastic process whose characteristic functional satisfies the functional equation (3.20) is said to have *independent values at every moment.*

This definition is based upon the fact that (3.20) implies the result:

$$\langle x, \xi_1 \rangle \quad \text{and} \quad \langle x, \xi_2 \rangle \quad \text{are independent if} \quad \xi_1(t)\xi_2(t) \equiv 0, \tag{3.21}$$

which is valid on the probability space determined by $C(\xi)$. This last assertion is a consequence of the Kac Theorem asserting that the factorization of characteristic functions implies that the sum is of independent random variables.

Generalised stochastic processes with independent values at every moment, of which white noise is one example, are the continuous analogues of independent sequences of random variables. Such generalised processes exhibit the most basic form of dependence, and consequently arise as the "innovation" in prediction theory and the mathematical theory of communication. The problem of forming the innovation process of a given stochastic process, and the expressing of the process in terms of its innovation process [see P. Lévy (1937) §§39, 64; N. Wiener (1958) Lectures 5, 6] plays an important role in the investigation of stochastic processes. The contents of the next chapter are along these lines.

3.4 White Noise

Recall that the measure μ_σ on E^* is determined by the characteristic functional $C(\xi) = \exp(-\frac{1}{2}\sigma^2 \|\xi\|^2)$ on E, that the measure space $(E^*, \mathfrak{B}, \mu_\sigma)$ is called white noise with variance σ^2 (see Example 1 of the last section), and that this measure μ_σ is a Gaussian measure. In what follows only the measure $\mu_1 = \mu$ with $\sigma^2 = 1$ will be termed white noise, and in this case $\langle x, \xi \rangle$ is a standard Gaussian measure when $\|\xi\| = 1$. In this discussion the basic nuclear space E could be $\mathscr{S}$, $\mathscr{D}$ or D_0, but we will only remark upon this when the time interval is not the entire real line **R**.

Now let us return to the characteristic functional $C(\xi)$ given at the beginning of this section, viewing $\sigma > 0$ as a parameter. That is, we consider the system $\{\mu_\sigma : \sigma > 0\}$ of measures on E^*.

Proposition 3.1. *Two measures μ_{σ_1} and μ_{σ_2} with different variances $(\sigma_1 \neq \sigma_2)$ are mutually singular.*

PROOF. We first note that if ξ and η are orthogonal in $L^2(\mathbf{R})$, then $\langle x, \xi \rangle$ and $\langle x, \eta \rangle$ are independent random variables on the probability space $(E^*, \mathfrak{B}, \mu_\sigma)$. For, in the case of μ_σ, the bilinear functional $K(\xi, \eta)$ of Example 3 in §3.3 above corresponds to $(\xi, \eta)\sigma^2$, where $(\cdot, \cdot)$ is the inner product in $L^2(\mathbf{R})$.

Let $\{\xi_n\} \subset E$ be a complete orthonormal system in $L^2(\mathbf{R})$. The collection $\{\langle x, \xi_n \rangle\}$ is, with respect to both μ_{σ_1} and μ_{σ_2}, an independent sequence of zero-mean Gaussian random variables. The variance with respect to μ_{σ_i} of $\langle x, \xi_n \rangle$ is σ_i^2, $i = 1, 2$, and so the strong law of large numbers gives

$$\lim_{n \to \infty} N^{-1} \sum_{1}^{N} \langle x, \xi_n \rangle^2 = \sigma_i^2, \quad \text{a.e. } (\mu_{\sigma_i}), \quad i = 1, 2. \tag{3.22}$$

Now consider the event $A(\sigma_1) = \{x: \lim_N N^{-1} \sum_1^N \langle x, \xi_n \rangle^2 = \sigma_1^2\}$. We have seen that $\mu_{\sigma_1}(A(\sigma_1)) = 1$, whilst $\mu_{\sigma_2}(A(\sigma_1)) = 0$ since $\sigma_1 \neq \sigma_2$, and so μ_{σ_1} and μ_{σ_2} are mutually singular. □

We have thus introduced uncountably many mutually singular Gaussian measures μ_σ, $\sigma > 0$ on the space E^*. Now let us change our point of view and consider the relation of white noise to Gaussian measures on $\mathbf{R}^\infty$. Using the complete orthonormal system $\{\xi_n\}$ in $L^2(\mathbf{R})$, we define the mapping

$$T: x \to x_n = \langle x, \xi_n \rangle, \quad n = 1, 2, \ldots; \quad x \in E^*, \qquad (3.23)$$

from E^* to $\mathbf{R}^\infty = \{\tilde{x} = (x_1, x_2, \ldots): x_n \in \mathbf{R}\}$. Clearly T is surjective. The σ-field generated by cylinder subsets of $\mathbf{R}^\infty$ will be denoted by $\tilde{\mathfrak{B}}$. Then T is seen to be a measurable transformation from $(E^*, \mathfrak{B})$ to $(\mathbf{R}^\infty, \tilde{\mathfrak{B}})$, for the inverse image $T^{-1}A$ of any cylinder subset A of $\mathbf{R}^\infty$ is a cylinder subset of E^*. Thus we may introduce a probability measure $\tilde{\mu}$ on $(\mathbf{R}^\infty, \tilde{\mathfrak{B}})$ by writing

$$\tilde{\mu}(B) = \mu(T^{-1}B), \quad B \in \tilde{\mathfrak{B}}. \qquad (3.24)$$

In other words, $\tilde{\mu} = T^{-1} \circ \mu$. Now recall that the $\langle x, \xi_n \rangle$ are mutually independent standard Gaussian random variables. This fact, and the relation (3.24) prove that $\{X_n(\tilde{x})\}$, where $X_n(\tilde{x}) = x_n$, also forms a system of mutually independent standard Gaussian random variables on $(\mathbf{R}^\infty, \tilde{\mathfrak{B}}, \tilde{\mu})$. Thus $\tilde{\mu}$ on $\mathbf{R}^\infty$ may be regarded as the direct product of countably many standard Gaussian distributions. The measure space $(\mathbf{R}^\infty, \tilde{\mathfrak{B}}, \tilde{\mu})$ is called *discrete parameter white noise*, and arises naturally within our framework. If $X_n(\tilde{x})$ has a variance σ^2 different from 1, we call it white noise with variance σ^2.

In order to define such a measure $\tilde{\mu}$ on $\mathbf{R}^\infty$, we can appeal to the Kolmogorov extension theorem by starting with a consistent family of finite-dimensional distributions. To this end we take an element of the subset $(\mathbf{R}^\infty)_0$ consisting of all sequences $a = (a_1, a_2, \ldots) \in \mathbf{R}^\infty$ which are zero from some point onwards, and give the distribution or characteristic function of $\sum_j a_j X_j(\tilde{x})$. The latter is actually given by

$$\int_{\mathbf{R}^\infty} \exp\left[i \sum_j a_j X_j(\tilde{x})\right] d\tilde{\mu}(\tilde{x}) = \int_{E^*} \exp\left[i \sum_j a_j \langle x, \xi_j \rangle\right] d\mu(x)$$

$$= \exp\left[-\frac{1}{2} \sum_j a_j^2 \|\xi_j\|^2\right] = \exp\left[-\frac{1}{2} \sum_j a_j^2\right].$$

We now examine the support of the measure $\tilde{\mu}$. The computations are simple, but they are helpful in understanding Theorem 3.1.

Proposition 3.2. *The subset $l^2 = \{\tilde{x} = (x_1, x_2, \ldots): \sum x_n^2 < \infty\}$ of $\mathbf{R}^\infty$ is $\tilde{\mathfrak{B}}$-measurable, and $\tilde{\mu}(l^2) = 0$.*

3.4 White Noise

PROOF. A necessary and sufficient condition for $\tilde{x}$ to belong to l^2 is that for any $\varepsilon > 0$ there exists N such that for any $n > m > N$, $\sum_m^n x_k^2 < \varepsilon$. With this in mind we write l^2 in the form

$$l^2 = \bigcap_p \bigcup_N \bigcap_{n > m \geq N} \left\{ \tilde{x} : \sum_m^n x_k^2 < \frac{1}{p} \right\}. \tag{3.25}$$

Since the set in braces is a cylinder set, we have proved that $l^2 \in \tilde{\mathfrak{B}}$.

Now the argument in the proof of Proposition 3.1 gives

$$\lim_{N \to \infty} N^{-1} \sum_1^N x_n^2 = \lim_{N \to \infty} N^{-1} \sum_1^N X_n(\tilde{x})^2 = 1 \quad \text{a.e. } (\tilde{\mu}),$$

and so we have $\sum_n x_n^2 = \infty$, except for an $\tilde{x}$-set of $\tilde{\mu}$-measure zero. But the set l^2 is included in the exceptional null set and so $\tilde{\mu}(l^2) = 0$. □

It is of some interest to consider the convergence of weighted sums from the sequence $\{x_n\}$. Let $\{\lambda_n\}$ be a sequence of real numbers satisfying

$$\sum_n \lambda_n^2 < \infty, \quad \lambda_n \neq 0 \quad \text{for all} \quad n, \tag{3.26}$$

and define a space $\tilde{E}$ by

$$\tilde{E} = \left\{ \tilde{a} = (a_1, a_2, \ldots) : \sum_n \frac{a_n^2}{\lambda_n^2} = \|\tilde{a}\|_1^2 < \infty \right\}.$$

With the norm $\| \cdot \|_1$ the space $\tilde{E}$ becomes a Hilbert space. Furthermore

$$\sum_n \lambda_n^2 x_n^2 = \sum_n \lambda_n^2 X_n(\tilde{x})^2 < \infty, \quad \text{a.e. } (\tilde{\mu}), \tag{3.27}$$

for the expectation is $\sum_n \lambda_n^2 < \infty$, and the sum $\sum_n \lambda_n^2 (X_n(\tilde{x})^2 - 1)$ converges in $L^2(\mathbf{R}^\infty, \tilde{\mathfrak{B}}, \tilde{\mu})$ since $\sum_n \lambda_n^4 < \infty$. Now introduce the norm $\|\tilde{x}\|_{-1} = (\sum_n \lambda_n^2 x_n^2)^{1/2}$. For $\tilde{a} \in \tilde{E}$ the series $\sum_n a_n x_n$ is convergent for almost all x and

$$\left| \sum_n a_n x_n \right| \leq \left(\sum_n \frac{a_n^2}{\lambda_n^2} \sum_n \lambda_n^2 x_n^2 \right)^{1/2} = \|\tilde{a}\|_1 \|\tilde{x}\|_{-1}. \tag{3.28}$$

We now set

$$\tilde{E}^* = \{ \tilde{x} \in \mathbf{R}^\infty : \|\tilde{x}\|_{-1} < \infty \}.$$

This is a Hilbert space with Hilbertian norm $\| \cdot \|_{-1}$ and is the dual space of $\tilde{E}$ via the canonical bilinear form $\sum_n a_n x_n$. The inclusions amongst $\tilde{E}$, l^2 and $\tilde{E}^*$ are

$$\tilde{E} \subset l^2 \subset \tilde{E}^*, \tag{3.29}$$

where the injections from left to right are easily seen from (3.26) to be of Hilbert–Schmidt type.

Furthermore we can prove the $\tilde{\mathfrak{B}}$-measurability of $\tilde{E}^*$ as we did in (3.25), and, with this and (3.29), we obtain the following

Proposition 3.3. $\tilde{E}^*$ is $\mathfrak{B}$-measurable and $\tilde{\mu}(\tilde{E}^*) = 1$.

We can now speak of the connection mentioned in the Preface, as well as at the beginning of this chapter, between the white noise of this section and the time derivative (in the sense of generalised functions) of Brownian motion. Firstly (a) we can illustrate it in terms of characteristic functionals, for the characteristic functional of the time derivative of a Brownian motion is given by replacing ξ in (3.17) of Example 3 in §3.3 by $-\xi'$. We therefore have

$$\exp\left[-\frac{1}{2}\int_0^\infty |-\hat{\xi}'(t)|^2 \, dt\right] = \exp\left[-\frac{1}{2}\|\xi\|^2\right],$$

where $\|\cdot\|$ is the $L^2([0, \infty))$-norm, which agrees with the characteristic functional of white noise. Secondly (b) we can offer a naive formal approach as follows. Set

$$B(t, x) = \langle x, \chi_{[t \wedge 0, t \vee 0]}\rangle, \quad -\infty < t < \infty, \tag{3.30}$$

on the measure space $(E^*, \mathfrak{B}, \mu)$ of white noise. The indicator function in (3.30) does not belong to E but we can approximate it by a sequence $\{\xi_n\}$ in E for which the random variables $\langle x, \xi_n\rangle$ are mean-square convergent. Thus we regard the right-hand side of (3.30) as such a limit, leaving fuller details on this matter until Chapter 4. The random variable $B(t, x)$ is Gaussian with mean zero and variance $\|\chi\|^2 = |t|$. In addition we see that $\{B(t, x): -\infty < t < \infty\}$ is a Brownian motion with time parameter space $\mathbf{R}$. As a formal relation we have $(d/dt)B(t, x) = x(t), t > 0$, and a member x of E^* equipped with the measure μ may be viewed as the derivative of a sample function of Brownian motion. However, as we foreshadowed in §2.2 (i), we have to be careful in this matter.

Let $\{B(t): -\infty < t < \infty\}$ be a Brownian motion and form its increment process $\Delta B(t) = B(t + b) - B(t + a)$, $\Delta = (a, b]$. By definition $\{\Delta B(t): -\infty < t < \infty\}$ is a stationary Gaussian process, and this stationarity is inherited by white noise. That is, μ is invariant under the simultaneous translations, by a constant, of the variable of all sample (generalised) functions. We can formalise this intuitive notion of stationarity quite readily in the framework of this chapter. The shift of the variable of $x \in E^*$, called simply the shift, can be defined via the shift of ξ (see §4.4), and the invariance of μ can be defined via that of its characteristic functional $C(\xi)$.

Define *the shift S_t* by $(S_t \xi)(u) = \xi(u - t)$.

Definition 3.4. If S_t is an automorphism of E and we also have

$$C(S_t \xi) = C(\xi), \tag{3.31}$$

$t \in \mathbf{R}$, then $(E^*, \mathfrak{B}, \mu)$ is called a *stationary generalised stochastic process* or a *stationary random distribution*.

3.4 White Noise

Remark. We usually require continuity in t of the shift S_t. When E is one of the spaces $\mathscr{S}$, $\mathscr{D}$ or D_0, this is automatically satisfied.

The following proposition is easily proved.

Proposition 3.4. *White noise is a stationary generalised stochastic process.*

The final aim of this chapter is to give a plausible explanation of the terminology "white" in white noise. In general any mean-square continuous stationary process $\{X(t): -\infty < t < \infty\}$ with $E\{X(t)\} \equiv 0$ has a spectral decomposition

$$X(t) = \int_{-\infty}^{\infty} e^{i\lambda t}\, dM(\lambda),$$

where $dM(\lambda)$ is a suitable random measure [see J. L. Doob (1953) Chapter XI]. Corresponding to this there is a spectral decomposition of a stationary generalised stochastic process, expressed in the form

$$\langle x, \xi \rangle = \int \hat{\xi}(\lambda)\, dM(\lambda), \tag{3.32}$$

where $\hat{\xi}$ denotes the Fourier transform of ξ. In the case of white noise the above decomposition involves a "flat" spectrum, $E|dM(\lambda)|^2 = d\lambda$. That is, the spectrum is distributed uniformly over the entire real line, hence the term *white*. Such an x may be thought of as an ideal form of the thermal noise arising in an electrical network, and these considerations have led us to adopt the term white noise. Alongside Brownian motion, white noise undoubtedly occupies an honoured place in the theory of probability, communication theory, and other areas in which random fluctuations arise.

Remark. Other stationary generalised stochastic processes with flat spectra are also called white noise, even though they are not necessarily Gaussian. Amongst these we find processes of the Poisson type, which also arise in applications.

4 Functionals of Brownian Motion

We are going to discuss those functionals, in general non-linear, of Brownian motion, which can be expressed in the form:

$$\Phi(B(t, \omega): t \in T). \tag{4.1}$$

The reasons why we are interested in such functionals will become clear as the present and following chapters proceed. Functionals of the form (4.1), or systems of such functionals, often arise as the output of a non-linear device with Brownian motion as input, or as the solution to certain stochastic differential equations. It is therefore an important practical problem to analyse such functionals, and apply the theory developed to actual problems such as non-linear prediction. Since the functional Φ in (4.1) may be viewed as a function of the infinite-dimensional argument $\{B(t): t \in T\}$, the analysis of Φ is therefore a typical example of infinite-dimensional analysis. Thus we adopt the viewpoint of functional analysis, and we find applications of our results in physics, specifically mechanics, as well as in electrical engineering.

4.1 Basic Functionals

Since Brownian motion has independent increments, a discrete parameter analogue would be a sum of independent random variables. In order to analyse functions of such sums, we usually express them in terms of the independent random variables directly, and by analogy we write (4.1) in the form

$$\varphi(B'(t, \omega): t \in T), \qquad B'(t) = \frac{d}{dt} B(t) \quad (= \dot{B}(t)) \tag{4.2}$$

4.1 Basic Functionals

Recall that when $T = (-\infty, \infty)$, the distribution of $\{B'(t): t \in T\}$ is given by the white noise $(\mathscr{S}^*, \mathfrak{B}, \mu)$ introduced in Example 1, §3.3. Assuming that the random variables in question have finite variances, our problem now is to analyse the functionals

$$\varphi(x) \in L^2(\mathscr{S}^*, \mu), \qquad (4.3)$$

where $L^2(\mathscr{S}^*, \mu)$ is the Hilbert space of all complex-valued functions of $x \in \mathscr{S}^*$ which are square integrable with respect to μ. This space appears frequently, and will be denoted simply by (L^2).

Having thus clarified our aim, we turn now to the most basic classes of functionals of x. Unless otherwise stated, the basic nuclear space in this chapter is always taken to be $\mathscr{S}$.

i. Let $P(t_1, t_2, \ldots, t_n)$ be an ordinary polynomial in real variables $t_1, t_2, \ldots, t_n$ with complex coefficients. With this P we may define

$$\varphi(x) = P(\langle x, \xi_1 \rangle, \langle x, \xi_2 \rangle, \ldots, \langle x, \xi_n \rangle). \qquad (4.4)$$

Such a functional is called a *polynomial* in $x \in \mathscr{S}^*$, and the degree of φ is defined by that of P. If P is of degree d, then φ is said to be of degree at most d; if a polynomial in x is of degree at most d but not at most $d-1$, then the degree of the polynomial is exactly d. A representation of $\varphi(x)$ in the form (4.4) is, of course, not unique, but if the $\xi_1, \ldots, \xi_n$ are linearly independent, then the degree of $\varphi(x)$ as just defined coincides with the degree of P. We will omit definitions of terms such as monomial and homogeneous polynomial in x, for these are used with just the same meanings as in the finite-dimensional case.

Proposition 4.1. *All polynomials in* $x \in \mathscr{S}^*$ *belong to* (L^2).

PROOF. Recall that for fixed ξ, $\langle x, \xi \rangle$ is a Gaussian random variable on the probability space $(\mathscr{S}^*, \mathfrak{B}, \mu)$ with mean 0 and variance $\|\xi\|^2$ ($\|\cdot\|$ the $L^2(\mathbf{R})$-norm). We therefore have the result

$$\int |\langle x, \xi \rangle|^{2n} d\mu(x) = \frac{(2n)!}{2^n n!} \|\xi\|^{2n}, \qquad n = 1, 2, \ldots,$$

and this implies that $\langle x, \xi \rangle^n \in (L^2)$. The Schwarz inequality can be used to prove that a monomial of the form $\langle x, \xi_1 \rangle^{k_1} \langle x, \xi_2 \rangle^{k_2} \cdots \langle x, \xi_n \rangle^{k_n}$ belongs to (L^2), and hence also any polynomial. □

ii. *Exponential functions.* A functional of the form

$$\varphi(x) = \exp[\alpha \langle x, \xi \rangle], \qquad \alpha \in \mathbf{C}, \qquad (4.5)$$

is said to be an *exponential function* on $\mathscr{S}^*$.

Proposition 4.2. *All exponential functions on* $\mathscr{S}^*$ *belong to* (L^2).

PROOF. Set $\alpha = a + ib$, $a, b \in \mathbf{R}$, in the expression (4.5) for $\varphi(x)$. Then we have

$$\int |\varphi(x)|^2 \, d\mu(x) = (2\pi)^{-1/2} \|\xi\|^{-1} \int_{-\infty}^{\infty} \exp\left[2at - \frac{t^2}{2} \|\xi\|^{-2}\right] dt < \infty,$$

which proves the assertion. □

As we have just seen, polynomials and exponential functions on $\mathscr{S}^*$ are quite easily defined. This is of particular interest because, as will be proved in the next section, the subspace spanned by the polynomials or by the exponential functions is dense in (L^2). Because of this, operations or transformations on (L^2) will be first defined on these basic functionals, and then extended to the entire space (L^2). In this way both of these classes of basic functionals play an important role in analysis on (L^2).

4.2 The Wiener-Itô Decomposition of (L^2)

The complex vector space spanned by the exponential functions of the form

$$\exp[i\langle x, \xi\rangle], \qquad \xi \in \mathscr{S}, \tag{4.6}$$

forms an algebra which we denote by **A**.

Theorem 4.1. *The algebra* **A** *is dense in* (L^2).

PROOF. It suffices to prove that if any $\varphi(x) \in (L^2)$ is orthogonal to all functionals of the form

$$\prod_{1}^{n} \exp[it_k \langle x, \xi_k \rangle] = \exp\left[i\langle x, \sum_k t_k \xi_k \rangle\right], \qquad t_k \in \mathbf{R}, \quad \xi_k \in \mathscr{S},$$

then $\varphi(x) = 0$, a.e. (μ). Let us fix $\xi_1, \ldots, \xi_n$ and suppose that for every t_k, $1 \le k \le n$, we have

$$\int_{\mathscr{S}^*} \prod_{1}^{n} \exp[it_k \langle x, \xi_k \rangle] \overline{\varphi(x)} \, d\mu(x) = 0.$$

If $\mathfrak{B}_n \subseteq \mathfrak{B}$ is the σ-field generated by $\langle x, \xi_1 \rangle, \ldots, \langle x, \xi_n \rangle$, then the above relation can be rewritten

$$\int_{\mathscr{S}^*} \exp\left[i \sum_{1}^{n} \langle x, \xi_k \rangle t_k\right] E\{\overline{\varphi(x)} | \mathfrak{B}_n\} \, d\mu(x) = 0.$$

This equality holds for any $t_k \in \mathbf{R}$, $1 \le k \le n$, and as it is an integral with respect to the joint distribution of the $\langle x, \xi_k \rangle$, a property of Fourier transforms can be used to imply that

$$E\{\overline{\varphi(x)} | \mathfrak{B}_n\} = 0, \quad \text{a.e.} \quad (\mu).$$

Now let us add ξ_k, $k \geq n+1$ so that $\{\xi_n\}$ becomes a complete system in $L^2(\mathbf{R})$. Then $\mathfrak{B}_n$ increases monotonically to $\mathfrak{B}$ and so we deduce that

$$E\{\varphi(x)|\mathfrak{B}\} = \overline{\varphi(x)} = 0, \quad \text{a.e.} \quad (\mu)$$

which was to be proved. □

The collection **P** of all polynomials in x also forms an algebra.

Corollary 1. *The algebra* **P** *is dense in* (L^2).

PROOF. The Taylor expansion

$$\exp[i\langle x, \xi\rangle] = \sum_0^\infty \frac{[i\langle x, \xi\rangle]^n}{n!}$$

is almost surely and (L^2)-convergent and so any member of **A** can be approximated arbitrarily closely by a polynomial. Thus **P** is also dense in (L^2). □

Take a complete orthonormal system $\{\xi_n\} \subseteq L^2(\mathbf{R})$ such that for all n, $\xi_n \in \mathscr{S}$. Following on from the expansion of $\xi \in \mathscr{S}$ as

$$\xi = \sum_0^\infty a_n \xi_n,$$

we have the mean-square limit expressing $\langle x, \xi\rangle$ as

$$\langle x, \xi\rangle = \underset{N\to\infty}{\text{l.i.m.}} \sum_1^N a_n \langle x, \xi_n\rangle.$$

Moreover, any power of $\langle x, \xi\rangle$ can be approximated by a sequence of polynomials in the $\langle x, \xi_n\rangle$, $n \geq 1$, and so we have the following:

Corollary 2. *Let* $\{\xi_n\} \subseteq \mathscr{S}$ *be a complete orthonormal system for* $L^2(\mathbf{R})$. *Then the collection of all finite products of the form*

$$\prod_j \langle x, \xi_j\rangle^{n_j}, \quad n_j \geq 0, \quad j = 1, 2, \ldots, \tag{4.7}$$

spans a dense subspace of (L^2).

Let $H_n(u)$, $n \geq 0$, be the sequence of Hermite polynomials in $u \in \mathbf{R}$, for details and formulae see §A.5 (i), and let $\{\xi_n\}$ be as above.

Definition 4.1. A *Fourier-Hermite* polynomial based on $\{\xi_n\}$ is a polynomial in x expressible as a finite product in the form

$$\varphi(x) = c \prod_j H_{n_j}\left(\frac{\langle x, \xi_j\rangle}{\sqrt{2}}\right), \quad c \text{ a constant.} \tag{4.8}$$

We now compute the (L^2)-norm of such a polynomial. Since the $\langle x, \xi_j \rangle$, $j \geq 1$, are mutually independent, we have the following:

$$\int_{\mathscr{S}^*} |\varphi(x)|^2 \, d\mu(x) = |c|^2 \prod_j \int_{\mathscr{S}^*} H_{n_j}\left(\frac{\langle x, \xi_j \rangle}{\sqrt{2}}\right)^2 d\mu(x)$$

$$= |c|^2 \prod_j (2\pi)^{-1/2} \int_{-\infty}^{\infty} H_{n_j}\left(\frac{t}{\sqrt{2}}\right)^2 \exp\left[-\frac{1}{2}t^2\right] dt \quad (4.9)$$

$$= |c|^2 \prod_j n_j! \, 2^{n_j},$$

where we refer to formula (A.24') of §A.5 for the last integral. Let us take $c = (\prod_j n_j 2^{n_j})^{-1/2}$ and arrange the Fourier-Hermite polynomials in a sequence $\varphi_1, \varphi_2, \ldots, \varphi_n, \ldots$.

Theorem 4.2. *The collection $\{\varphi_n : n \geq 1\}$ of normalised Fourier-Hermite polynomials just defined forms a complete orthonormal system in (L^2).*

Remark. This theorem just paraphrases a result of Cameron and Martin (1947, b).

PROOF OF THEOREM 4.2. First we show that the $\{\varphi_n\}$ are orthogonal. Just as in the computation (4.9) we have

$$\int_{\mathscr{S}^*} \prod_j H_{n_j}\left(\frac{\langle x, \xi_j \rangle}{\sqrt{2}}\right) \cdot \prod_j H_{m_j}\left(\frac{\langle x, \xi_j \rangle}{\sqrt{2}}\right) d\mu(x)$$

$$= \prod_j \int_{\mathscr{S}^*} H_{n_j}\left(\frac{\langle x, \xi_j \rangle}{\sqrt{2}}\right) H_{m_j}\left(\frac{\langle x, \xi_j \rangle}{\sqrt{2}}\right) d\mu(x)$$

$$= \prod_j n_j! \, 2^{n_j} \, \delta_{n_j, m_j}.$$

Thus if $n_j \neq m_j$ for some j, i.e. if the φ_n are distinct, then they are orthogonal.

The fact that any monomial of the form (4.7) can be expressed as a sum of Fourier-Hermite polynomials of degree $\leq \sum_j n_j$, together with Corollary 2 above, shows that $\{\varphi_n\}$ spans (L^2), and the theorem is proved. □

Denote by $\mathscr{H}_n$ the subspace of (L^2) spanned by all the Fourier-Hermite polynomials of degree n.

Definition 4.2. The space $\mathscr{H}_n$ is referred to as the *multiple Wiener integral* of degree n.

Remark. The meaning of the term "integral" in the above will be clarified in the next section when we give an integral representation of the elements of $\mathscr{H}_n$.

Theorem 4.3 [N. Wiener (1938), K. Itô (1951b)]. *The Hilbert space (L^2) admits a direct-sum decomposition:*

$$(L^2) = \sum_{n=0}^{\infty} \oplus \mathcal{H}_n. \tag{4.10}$$

The proof is an immediate consequence of Theorem 4.2 and the definition of $\mathcal{H}_n$.

Observations. Let S^N denote the N-dimensional sphere in $\mathbf{R}^{N+1}$ and let σ_N be the uniform (i.e. rotation invariant) measure on S^N. As is well known we have the following direct-sum decomposition

$$L^2(S^N, \sigma_N) = \sum_{n=0}^{\infty} \oplus \mathcal{H}_n \tag{4.11}$$

where $\mathcal{H}_n$ in this decomposition is the vector space spanned by the spherical harmonics of degree n. When (4.10) and (4.11) are compared, we see a great similarity, but since the former space is built over an infinite-dimensional space (unlike the finite-dimensional S^N), we find certain differences as well. Certain kinds of differences are of importance in any discussion of infinite-dimensional analysis, but it is also of interest to let the differences suggest methods of finite-dimensional approximation. Incidentally we might recall the observations made in §3.1, which led us to think intuitively of the support of the measure μ of white noise as an infinite-dimensional sphere.

4.3 Representations of Multiple Wiener Integrals

We first introduce a transformation $\mathcal{T}$ which carries members of (L^2) into functionals on $\mathcal{S}$:

$$(\mathcal{T}\varphi)(\xi) = \int_{\mathcal{S}^*} \exp[i\langle x, \xi \rangle] \varphi(x)\, d\mu(x), \qquad \varphi \in (L^2). \tag{4.12}$$

Remark. Although the integral (4.12) is similar in form to the finite-dimensional Fourier transform, the transformation $\mathcal{T}$ does not have all the properties one would expect of a Fourier transform acting on (L^2). For example, $\mathcal{T}$ maps (L^2) into another function space (a space of functionals on $\mathcal{S}$, not $\mathcal{S}^*$) different from (L^2) itself.

Le **F** be the image of (L^2) under the transformation $\mathcal{T}$. Since $\mathcal{T}$ is linear by definition, **F** is a vector space. Indeed we have

Proposition 4.3. *The transformation $\mathcal{T}$ gives an isomorphism between the vector spaces (L^2) and* **F**.

PROOF. By Theorem 4.1

$$(\mathcal{T}\varphi)(\xi) \equiv 0 \quad \text{implies that} \quad \varphi(x) = 0, \quad \text{a.e.} \quad (\mu),$$

and so $\mathcal{T}$ is a one-to-one mapping. Since $\mathcal{T}$ is linear the two vector spaces are isomorphic. □

We are also able to introduce an inner product into **F** which topologises it, and makes $\mathcal{T}$ a Hilbert space isomorphism. Firstly, we take a functional φ in the algebra **A** expressible as a finite sum

$$\varphi(x) = \sum_j a_j \exp[i\langle x, \eta_j\rangle].$$

Then

$$\begin{aligned}(\mathcal{T}\varphi)(\xi) &= \sum_j a_j \int_{\mathcal{S}^*} \exp[i\langle x, \xi + \eta_i\rangle]\, d\mu(x) \\ &= \sum_j a_j C(\xi + \eta_j).\end{aligned} \quad (4.13)$$

Take another functional $\psi(x) = \exp[-i\langle x, \eta\rangle]$ and obtain the result

$$(\mathcal{T}\psi)(\xi) = C(\xi - \eta).$$

Then we have the relation

$$(\varphi, \psi)_{(L^2)} = \int_{\mathcal{S}^*} \varphi(x)\overline{\psi(x)}\, d\mu(x) = \sum_j a_j C(\eta + \eta_j).$$

Thus we see that if $\mathcal{T}$ is to give a Hilbert space isomorphism of (L^2) onto **F**, the inner product $(\cdot, \cdot)_\mathbf{F}$ must be introduced into **F** in such a way that

$$(\mathcal{T}\varphi, \mathcal{T}\psi)_\mathbf{F} = (\varphi, \psi)_{(L^2)}.$$

But this means that the left-hand side must coincide with $\sum_j a_j C(\eta + \eta_j)$, which is just $\mathcal{T}\varphi$ evaluated at η:

$$(\mathcal{T}\varphi)(\eta) = \sum_j a_j C(\eta + \eta_j).$$

Now the element φ above is the general form of an element of **A**, and **A** is dense in (L^2) (by Theorem 4.1), and so we see that it is necessary to have

$$(f(\cdot), C(\cdot - \eta))_\mathbf{F} = f(\eta) \quad \text{for any} \quad f(\cdot) \in \mathbf{F}. \quad (4.14)$$

If we now vary the η in this equation and take linear combinations of the $C(\cdot - \eta_j)$, the inner product being introduced into $\mathcal{T}(\mathbf{A}) \subseteq \mathbf{F}$ will be completely determined. Further the completion of $\mathcal{T}(\mathbf{A})$ with respect to the induced norm will coincide with **F** (now isomorphic to (L^2)).

4.3 Representations of Multiple Wiener Integrals

We now sum up what has been obtained in the following statement.

1. The kernel $C(\xi - \eta)$, $(\xi, \eta) \in \mathscr{S} \times \mathscr{S}$, is (strictly) positive definite:
$$\sum_{j,k} a_j \bar{a}_k C(\xi_j - \xi_k) \geq 0, \qquad a_j \in \mathbf{C}, \qquad \xi_j \in \mathscr{S},$$
and $= 0$ if and only if $a_j = 0$ for all j.
2. $(f(\cdot), C(\cdot - \eta)) = f(\eta)$ for any $f \in \mathbf{F}$.
3. $\mathbf{F}$ is spanned by the $C(\cdot - \eta)$, $\eta \in \mathscr{S}$.

A Hilbert space such as $\mathbf{F}$ is referred to as a *reproducing kernel Hilbert space* and $C(\xi - \eta)$ is called the reproducing kernel of $\mathbf{F}$.

Theorem 4.4. *The Hilbert space* (L^2) *is isomorphic under the transformation* $\mathscr{T}$ *to the reproducing kernel Hilbert space* $\mathbf{F}$ *with reproducing kernel* $C(\xi - \eta)$, $(\xi, \eta) \in \mathscr{S} \times \mathscr{S}$.

As soon as $\mathscr{T}$ is restricted to the subspace $\mathscr{H}_n$, an explicit form of the transformation can be given. To illustrate this fact we shall start with an exponential function $\varphi(x)$ given by the generating function of the Hermite polynomials:

$$\varphi(x) = \exp\left[\frac{2t\langle x, \eta\rangle}{\sqrt{2}} - t^2\right] = \sum_{0}^{\infty} \frac{t^k}{k!} H_k\left(\frac{\langle x, \eta\rangle}{\sqrt{2}}\right), \qquad (4.15)$$

where $\|\eta\| = 1$. Since $\varphi \in (L^2)$ the transformation $\mathscr{T}$ can be applied, and we get

$$\begin{aligned}(\mathscr{T}\varphi)(\xi) &= \exp(-t^2) \int_{\mathscr{S}^*} \exp[\sqrt{2}\,t\langle x, \eta\rangle + i\langle x, \xi\rangle]\, d\mu(x) \\ &= \exp\left[-t^2 - \frac{1}{2}\|\xi\|^2 + i\sqrt{2}\,t(\eta, \xi) + t^2\right] \\ &= C(\xi) \sum_{k=0}^{\infty} \frac{[(i\sqrt{2}\,t)^k]}{k!}(\eta, \xi)^k,\end{aligned} \qquad (4.16)$$

where (η, ξ) is the inner product in $L^2(\mathbf{R})$. Comparing this with (4.15) and still assuming that $\|\eta\| = 1$, we obtain

$$\left(\mathscr{T}H_k\left(\frac{\langle\cdot, \eta\rangle}{\sqrt{2}}\right)\right)(\xi) = C(\xi)(i\sqrt{2})^k(\eta, \xi)^k. \qquad (4.17)$$

The relation (4.17) can be generalised to Fourier-Hermite polynomials based on a fixed complete orthonormal system $\{\eta_n\}$ say, where $\{\eta_n\} \subset \mathscr{S}$. Suppose that $\varphi(x)$ is of the form

$$\varphi(x) = \prod_j H_{k_j}\left(\frac{\langle x, \eta_j\rangle}{\sqrt{2}}\right). \qquad (4.18)$$

Then we can write $\xi = \sum_j t_j \eta_j + \xi'$ where ξ' is orthogonal to the η_j in the first term, and these are taken to be those involved in (4.18). With this we have

$$(\mathcal{T}\varphi)(\xi) = \int_{\mathcal{S}^*} \exp\left[i\sum_j t_j \langle x, \eta_j\rangle + i\langle x, \xi'\rangle\right] \prod_j H_{k_j}\left(\frac{\langle x, \eta_j\rangle}{\sqrt{2}}\right) d\mu(x)$$

$$= \int_{\mathcal{S}^*} \exp[i\langle x, \xi'\rangle] \, d\mu(x)$$

$$\cdot \prod_j \int_{\mathcal{S}^*} \exp[it_j\langle x, \eta_j\rangle] H_{k_j}\left(\frac{\langle x, \eta_j\rangle}{\sqrt{2}}\right) d\mu(x)$$

$$= \exp\left[-\frac{1}{2}\|\xi'\|^2\right] \prod_j \exp\left[-\frac{1}{2}t_j^2\right](\sqrt{2}\,i)^{k_j}(\eta_j, t_j\eta_j)^{k_j}$$

by (4.17). Noting that $t_j = (\eta_j, \xi)$ and that $\|\xi\|^2 = \sum_j t_j^2 + \|\xi'\|^2$, we can express this last formula in the form

$$(\sqrt{2}\,i)^n C(\xi) \int_{\mathbf{R}^n} \cdots \int [\eta_1(t_1) \cdots \eta_1(t_{k_1})\eta_2(t_{k_1+1}) \cdots \eta_2(t_{k_1+k_2}) \cdots]$$
$$\times \xi(t_1) \cdots \xi(t_n) \, dt_1 \cdots dt_n,$$

where $n = \sum_j k_j$. We will denote by $F(t_1, t_2, \ldots, t_n)$ the expression inside the bracket $[\,\cdot\,]$ multiplied by $2^{n/2}$, and we see that the value of the integral is unchanged if F is replaced by its symmetrisation $\hat{F}$

$$\hat{F}(t_1, t_2, \ldots, t_n) = \frac{1}{n!} \sum_\pi F(t_{\pi(1)}, t_{\pi(2)}, \ldots, t_{\pi(n)}),$$

where π denotes a permutation of $\{1, 2, \ldots, n\}$ and the sum $\sum_\pi$ extends over all permutations. Both F and $\hat{F}$ are functions in $L^2(\mathbf{R}^n)$ and have norms

$$\|F\|_{L^2(\mathbf{R}^n)} = 2^{n/2},$$

$$\|\hat{F}\|_{L^2(\mathbf{R}^n)} = \left(\prod_j k_j!\right)^{1/2}(n!)^{-1/2}2^{n/2}.$$

Furthermore, we may use (4.9) to prove that the (L^2)-norm of φ in (4.18) is

$$\|\varphi\|_{(L^2)} = \left(\prod_j k_j\right)^{1/2} 2^{n/2}.$$

Thus a symmetric $L^2(\mathbf{R}^n)$-function $\hat{F}$ can be chosen in such a way that $\mathcal{T}\varphi$ may be written

$$(\mathcal{T}\varphi)(\xi) = i^n C(\xi) \int_{\mathbf{R}^n} \cdots \int \hat{F}(t_1, t_2, \ldots, t_n)\xi(t_1)\xi(t_2) \cdots \xi(t_n) \, dt_1 \, dt_2 \cdots dt_n,$$

(4.20)

4.3 Representations of Multiple Wiener Integrals

where $n = \sum_j k_j$, and also

$$\|\varphi\|_{(L^2)} = (n!)^{1/2} \|\hat{F}\|_{L^2(\mathbf{R}^n)}. \tag{4.21}$$

Suppose that we have a complete orthonormal system $\{\varphi_m\}$ in $\mathcal{H}_n$ consisting of Fourier-Hermite polynomials. Associated with each φ_m is a symmetric $L^2(\mathbf{R}^n)$-function $\hat{F}_m$ such that

$$\|\hat{F}_m\|_{L^2(\mathbf{R}^n)} = (n!)^{-1/2},$$

see the relation (4.21). Another important property of this representation is the fact that the collection $\{\hat{F}_m\}$ associated with $\{\varphi_m\}$ forms an orthogonal system in $L^2(\mathbf{R}^n)$. With any linear combination of the $\{\varphi_m\}$ we can associate the corresponding linear combination of the $\{F_m\}$, so that the relations (4.20) and (4.21) can be extended to ones which apply for any $\varphi(x)$ in $\mathcal{H}_n$. Thus we are led to define the subspaces

$$\hat{L}^2(\mathbf{R}^n) = \{F \in L^2(\mathbf{R}^n): F \text{ is symmetric}\} \subseteq L^2(\mathbf{R}^n),$$

and state the following theorem, where the $\hat{F}$ above is now denoted simply by F.

Theorem 4.5. i. *For all $\varphi \in \mathcal{H}_n$ there is a function F in $\hat{L}^2(\mathbf{R}^n)$ such that*

$$(\mathcal{T}\varphi)(\xi) = i^n C(\xi) \int_{\mathbf{R}^n} \cdots \int F(t_1, t_2, \ldots, t_n) \xi(t_1) \xi(t_2) \cdots \xi(t_n) \, dt_1 dt_2 \cdots dt_n,$$

and the correspondence

$$\varphi \to F \in \hat{L}^2(\mathbf{R}^n), \qquad \varphi \in \mathcal{H}_n,$$

is bijective.

ii. *In the above correspondence we also have the relation*

$$\|\varphi\|_{(L^2)} = (n!)^{1/2} \|F\|_{L^2(\mathbf{R}^n)}.$$

Through this theorem the space $\hat{L}^2(\mathbf{R}^n)$ can be regarded as a representation or realisation of $\mathcal{H}_n$, often called the *integral representation*, with the $\hat{L}^2(\mathbf{R}^n)$-function F being called the *kernel* of the integral representation.

Remark. Our representation leads to the symbolic statement that the transformation $\mathcal{T}$ gives an isomorphism

$$(L^2) \cong \sum_{n=0}^{\infty} \oplus (n!)^{1/2} \hat{L}^2(\mathbf{R}^n) \qquad \text{(Fock space).} \tag{4.22}$$

Since a general element $\varphi(x)$ of (L^2) is a non-linear function of the generalised function variable x, easy analytic expressions will in general not be available. Thus we expand $\varphi(x)$ as

$$\varphi(x) = \sum_{n=0}^{\infty} \varphi_n(x), \qquad \varphi_n(x) \in \mathcal{H}_n, \tag{4.23}$$

and then we can apply Theorem 4.5 to each φ_n in turn, obtaining an integral representation for it involving a kernel $F_n \in \hat{L}^2(\mathbf{R}^n)$. In other words, we have the following representation for φ:

$$\varphi \leftrightarrow F = \{F_n: n \geq 0\} \tag{4.24}$$

where F_0 is tacitly understood to be the constant term, which is in fact the inner product of φ with 1, i.e. the expectation of φ.

The great advantage of this representation is the fact that many operators of interest on $\mathcal{H}_n$ are transformed via the transformation $\mathcal{T}$ to operators on $\hat{L}^2(\mathbf{R}^n)$, and then standard tools from finite-dimensional analysis can be efficiently utilised. This will become clear in the sections which follow. For further details concerning the integral representation, we refer the reader to T. Hida (1970b), T. Hida and N. Ikeda (1967) and G. Kallianpur (1970).

4.4 Stochastic Processes

Our original aim was to analyse stochastic processes $\{X(t): t \in T\}$ expressed as functionals of Brownian motion and, as explained at the beginning of this chapter, this has stimulated considerable research. The discussion in this section continues along these lines.

Let a "flow" (i.e., a one-parameter group of measure-preserving transformations) $\{T_t: -\infty < t < \infty\}$ be given on the measure space $(\mathcal{S}^*, \mathfrak{B}, \mu)$ of white noise, and for $\varphi(x) \in (L^2)$ set

$$X(t, x) = \varphi(T_t x), \quad -\infty < t < \infty. \tag{4.25}$$

Then $\{X(t, x): -\infty < t < \infty\}$ is a stochastic process on $(\mathcal{S}^*, \mathfrak{B}, \mu)$. Now for any Borel subset B_n of $\mathbf{R}^n$ and any $t_1, t_2, \ldots, t_n$ we have the relation:

$$\{x: (X(t_1 + h, x), \ldots, X(t_n + h, x)) \in B_n\}$$
$$= T_{-h}\{x: (\varphi(T_{t_1} x), \ldots, \varphi(T_{t_n} x)) \in B_n\},$$

which, since T_{-h} is a measure-preserving transformation, gives the equality

$$\mu((X(t_1 + h), \ldots, X(t_n + h)) \in B_n) = \mu((X(t_1), \ldots, X(t_n)) \in B_n).$$

But this states that any finite-dimensional distribution of the process $\{X(t, x): -\infty < t < \infty\}$ is invariant under the time-shift, i.e. that $\{X(t)\}$ is a stationary process.

All "flows" discussed below will be assumed to satisfy the condition that the mapping

$$(t, x) \to T_t x \in \mathcal{S}^*, \quad (t, x) \in \mathbf{R} \times \mathcal{S}^*, \tag{4.26}$$

is $\mathfrak{B}(\mathbf{R}) \times \mathfrak{B}$-measurable (cf. the shift of §3.4).

4.4 Stochastic Processes

Now the unitary operators $\{U_t: -\infty < t < \infty\}$ defined on (L^2) by

$$(U_t\varphi)(x) = \varphi(T_t x), \qquad \varphi \in (L^2), \tag{4.27}$$

form a one-parameter group:

$$U_t U_s = U_s U_t = U_{t+s}, \qquad -\infty < t, s < \infty,$$
$$U_0 = I \quad \text{(the identity)}.$$

And with the measurability assumption just explained, the unitary group $\{U_t\}$ becomes strongly continuous in t, i.e.

$$\operatorname*{s-lim}_{h \to t} U_h = U_t.$$

Here s-lim means the strong limit of operators. This implies that the stationary process $\{X(t, x): -\infty < t < \infty\}$ given by (4.25) is continuous and weakly stationary [see §1.3, (ii)].

Let P_n denote the orthogonal projection from (L^2) onto the subspace $\mathcal{H}_n$. For t fixed we may view $X(t, x)$ as a member of (L^2) and define $X_n(t, x)$ by

$$X_n(t, x) = P_n X(t, x),$$

giving a stochastic process in $\mathcal{H}_n$. The original process $X(t)$ may itself be expressed as a sum

$$X(t, x) = \sum_{n=0}^{\infty} X_n(t, x), \tag{4.28}$$

where the sum is a limit in the sense of mean square convergence.

Returning to the U_t we can prove that $U_t \mathcal{H}_n = \mathcal{H}_n$, i.e. that

$$P_n U_t = U_t P_n, \qquad n \geq 0, \tag{4.29}$$

for each U_t carries a Fourier–Hermite polynomial into another Fourier–Hermite polynomial of the same degree. From this relation we obtain the equality

$$X_n(t, x) = P_n U_t X(0, x) = U_t P_n X(0, x) = U_t X_n(0, x), \tag{4.30}$$

and the family $\{X_n(t, x): -\infty < t < \infty\}$, $n = 0, 1, 2, \ldots$, is proved to be a system of mutually orthogonal stationary processes. We use (4.28) once again to show that $\{X(t, x)\}$ admits an orthogonal decomposition into stationary processes in $\mathcal{H}_n$, $n \geq 0$. Thus we might anticipate that an effective approach to the investigation of $\{X(t)\}$ will be to decompose it in the form (4.28), and then to use the integral representation for the studying the part $\{X_n(t)\}$ living in $\mathcal{H}_n$. We shall be able to present a theory in which this idea is most satisfactorily realised.

Before we come to our main topic we have to introduce the flow that comes from the shift $\{S_t\}$ introduced in §3.4. The basic nuclear space on which S_t acts is specified to be $\mathscr{S}$, so that many detailed properties of the shift, or the flow derived from it, can be explicitly obtained.

The transformation S_t on $\mathscr{S}$ is defined by

$$S_t: \xi(u) \to (S_t\xi)(u) = \xi(u-t), \qquad -\infty < t < \infty, \qquad (4.31)$$

and satisfies the following properties:

1. it is a linear isomorphism;
2. the equality $\|S_t\xi\| = \|\xi\|$ holds for all $\xi \in \mathscr{S}$;
3. $\{S_t: -\infty < t < \infty\}$ forms a one-parameter group:

$$S_t S_s = S_s S_t = S_{s+t}, \qquad -\infty < t, s < \infty,$$

$$S_0 = I,$$

$$S_t \to I \quad \text{as} \quad t \to 0.$$

The adjoint operator $S_t^* = T_t$ may be defined by using (1) above and the canonical bilinear form $\langle x, \xi \rangle$, $x \in \mathscr{S}^*$, $\xi \in \mathscr{S}$, linking $\mathscr{S}$ and $\mathscr{S}^*$:

$$\langle T_t x, \xi \rangle = \langle x, S_t \xi \rangle. \qquad (4.32)$$

Property (2) above proves that the characteristic functional $C(\xi)$ of white noise is S_t-invariant, that is:

$$C(S_t \xi) = \exp\left[-\frac{1}{2}\|S_t\xi\|^2\right] = C(\xi),$$

so that $C(S_t \xi)$ and $C(\xi)$ define the same probability measure. Indeed

$$C(S_t\xi) = \int_{\mathscr{S}^*} \exp[i\langle x, S_t\xi \rangle]\, d\mu(x) = \int_{\mathscr{S}^*} \exp[i\langle x, \xi \rangle]\, d\mu(T_t^{-1}x)$$

implies that μ is T_t-invariant, i.e.

$$T_t \circ \mu = \mu. \qquad (4.33)$$

Moreover (3) proves that $\{T_t: -\infty < t < \infty\}$ forms a one-parameter group, i.e. a flow on $(\mathscr{S}^*, \mathfrak{B}, \mu)$.

Returning to S_t we note that the measurability of the mapping (4.26) is easily proved, and so we can give

Definition 4.3. The flow $\{T_t: -\infty < t < \infty\}$ obtained from the shift as above is called the *flow of Brownian motion*.

For the remainder of this section the term flow always means the flow of Brownian motion, and accordingly the operator U_t will always be the one defined by (4.26) using T_t.

Remark. Let us take Brownian motion $\{\langle x, \chi_{[0, t]} \rangle: t > 0\}$ as it was introduced in §3.4, and apply the operator U_t to its increments. Then we have, for example,

$$U_t\{\langle x, \chi_{[0,b]}\rangle - \langle x, \chi_{[0,a]}\rangle\} = U_t\{\langle x, \chi_{(a,b]}\rangle\} = \langle T_t x, \chi_{(a,b]}\rangle$$
$$= \langle x, \chi_{[0,b+t]}\rangle - \langle x, \chi_{[0,a+t]}\rangle,$$

where $b > a > 0$, and so U_t defines the time shift. Definition 4.3 above may be viewed as a formal response to this observation.

Further background needed concerns a family $\{\mathfrak{B}_t: -\infty < t < \infty\}$ of sub-σ-fields of $\mathfrak{B}$. Let us write

$$\mathfrak{B}_t = \mathfrak{B}\{\langle x, \xi\rangle: \operatorname{supp}(\xi) \subseteq (-\infty, t]\},$$

that is, the σ-field generated by the random variables in the braces $\{\cdot\}$. Clearly

$$\mathfrak{B}_t \supseteq \mathfrak{B}_s \quad \text{when} \quad t > s, \tag{4.34}$$

$$\bigvee_t \mathfrak{B}_t = \mathfrak{B}, \tag{4.35}$$

where the left-hand side denotes the σ-field generated by $\bigcup_t \mathfrak{B}_t$, and

$$T_t \mathfrak{B}_s = \mathfrak{B}_{t+s}, \qquad -\infty < t, s < \infty. \tag{4.36}$$

Moreover we have the following:

Proposition 4.4. *The family* $\{\mathfrak{B}_t: -\infty < t < \infty\}$ *satisfies*

$$\bigcap_t \mathfrak{B}_t = \mathbf{2} \ (\operatorname{mod} \mu), \tag{4.37}$$

where $\mathbf{2}$ *denotes the trivial σ-field consisting of* $\varnothing$ *and* $\mathscr{S}^*$.

PROOF. For any ξ we can find a sequence of $\{\xi_n\}$ with compact support such that $\langle x, \xi_n\rangle$ converges almost surely with respect to μ to $\langle x, \xi\rangle$. Take measurable set $F \in \bigcap_t \mathfrak{B}_t$, and note that it follows from the definition of $\mathfrak{B}_t$ that $\chi_F(x)$ is independent of $\langle x, \xi\rangle$ for any ξ with compact support (see §3.4). Hence $\mathfrak{B}_c = \mathfrak{B}\{\langle x, \xi\rangle: \operatorname{supp}(\xi) \text{ is compact}\}$ and F are independent. But it was proved earlier that $\mathfrak{B}_c$ and $\mathfrak{B}$ coincide, whilst $\mathfrak{B}$ contains F, and so F is independent of itself, i.e. $\mu(F) = 0$ or 1. Thus $F = \varnothing$ or $F = \mathscr{S}^*$ (mod μ). □

We pause here to give a general definition.

Definition 4.4. A flow $\{T_t\}$ on a probability space $(\Omega, \mathbf{B}, P)$ is called a *Kolmogorov flow* if there exists a sub-σ-field $\mathbf{B}_0$ of $\mathbf{B}$ such that $T_t \mathbf{B}_0 = \mathbf{B}_t$,

i. $\mathbf{B}_t \supseteq \mathbf{B}_s$ if $t > s$;
ii. $\bigvee_t \mathbf{B}_t = \mathbf{B}$;
iii. $\bigcap_t \mathbf{B}_t = \mathbf{2}$.

For the flow of Brownian motion $\mathbf{B}_0$ may be taken to be any of the $\mathfrak{B}_t$ and we see that this flow is a Kolmogorov flow. We now return to the discussion of the flow of Brownian motion, the following being an immediate consequence of Proposition 4.4.

Corollary 1. *If we write $L_t^2 = L^2(\mathscr{S}^*, \mathfrak{B}_t, \mu)$, then*

i. $L_t^2 \supseteq L_s^2$ *if* $t > s$;
ii. $\bigvee_t L_t^2 = (L^2)$;
iii. $\bigcap_t L_t^2 = \{1\}$, *the space spanned by the constant function 1.*

In the above $\bigvee_t$ denotes the Hilbert subspace spanned by the union $\bigcup_t$. It follows from this corollary that we have

Corollary 2. *If $\varphi(x) \in L_a^2$ for some a, then the stationary process $\varphi(T_t x)$ is purely non-deterministic.*

PROOF. It follows from the assumptions stated that

$$\varphi(T_t x) \in L_{a+t}^2,$$

and hence a functional of the $\varphi(T_s x)$, $s \leq t$, that belongs to (L^2) is also a member of L_{a+t}^2. But by (iii) of Corollary 1 above,

$$\bigcap_t L_{a+t}^2 = \{1\},$$

and so $\{\varphi(T_t x): -\infty < t < \infty\}$ is purely non-deterministic. See §1.3,(ii). □

Now let us set $a = 0$ (for simplicity) and consider a $\varphi(x) \in L_0^2$ as in Corollary 2.

Proposition 4.5. *Suppose that $\varphi(x) \in L_0^2$ is also a member of $\mathscr{H}_n$. Then the kernel $F \in L^2(\mathbf{R}^n)$ of the integral representation of φ has the following property*

$$F(u_1, u_2, \ldots, u_n) = 0 \quad \text{if } u_j > 0 \text{ for some } j. \tag{4.38}$$

PROOF. Suppose that $\text{supp}(\xi) \subseteq [0, \infty)$. Then $\langle x, \xi \rangle$ is independent of $\mathfrak{B}_0$ and $\varphi(x) \in L_0^2 \cap \mathscr{H}_n$ is orthogonal to any Fourier-Hermite polynomial in $\langle x, \xi \rangle$ involving such a ξ. The inner product of such a polynomial and $\varphi(x)$ may be written

$$\int \cdots \int_{\mathbf{R}^n} F(u_1, u_2, \ldots, u_n) \xi_1(u_1) \xi_2(u_2) \cdots \xi_n(u_n) \, du_1 \, du_2 \cdots du_n,$$

where $\text{supp}(\xi_j) \subseteq [0, \infty)$ for some j, and this has to vanish. The proof is concluded by noting that the set of products of such ξ's is rich enough to give the conclusion (4.38). □

4.4 Stochastic Processes

We next apply U_t to a φ such as that in Proposition 4.5. Since we have

$$(\mathcal{T}\varphi)(\xi) = C(\xi)i^n \int \cdots \int_{\mathbf{R}^n} F(u_1, u_2, \ldots, u_n)\xi(u_1) \cdots \xi(u_n) \, du_1 \cdots du_n$$

then

$$(\mathcal{T}(U_t\varphi))(\xi) = \int_{\mathcal{S}^*} \exp[i\langle x, \xi\rangle]\varphi(T_t x) \, d\mu(x)$$

$$= \int_{\mathcal{S}^*} \exp[i\langle T_{-t}x, \xi\rangle]\varphi(x) \, d\mu(x)$$

$$= (\mathcal{T}\varphi)(S_{-t}\xi)$$

$$= C(\xi)i^n \int \cdots \int_{\mathbf{R}^n} F(u_1 - t, \ldots, u_n - t)\xi(u_1) \cdots \xi(u_n) \, du_1 \cdots du_n.$$

The kernel of the integral representation of $U_t \varphi$ is therefore of the form

$$F(u_1 - t, u_2 - t, \ldots, u_n - t) \tag{4.39}$$

and vanishes if $u_j > t$ for some j.

With these properties in mind we recall that U_t commutes with the projection operator P_n [relation (4.29)] and obtain the following result:

Proposition 4.6. *Let $\{X(t)\}$ be the stationary process defined by $\varphi \in L_0^2$ and U_t via*

$$X(t, x) = U_t\varphi(x), \quad -\infty < t < \infty, \tag{4.40}$$

and decompose it in the manner of (4.28). Then each of the stationary processes $\{X_n(t, x)\}$ given by $X_n(t, x) = P_n X(t, x)$ arising as a component, $n \geq 1$, has as kernel of its integral representation

$$F_n(u_1 - t, u_2 - t, \ldots, u_n - t),$$

where $F_n(u_1, u_2, \ldots, u_n)$ is the kernel associated with $\varphi_n = P_n\varphi$ and satisfies the condition (4.38).

EXAMPLE 1. Consider the simplest case $(n = 1)$ with $\varphi \in L_0^2 \cap \mathcal{H}_1$, and $X_1(t, x) = U_t(x)$, $-\infty < t < \infty$. The stationary process $\{X_1(t)\}$ is centred, Gaussian, mean-square continuous and purely non-deterministic. Further the kernel $F(u)$ of the integral representation of $X(0)$ is an $L^2(\mathbf{R})$-function vanishing on $[0, \infty)$. It will be shown in the next section that there is a stochastic integral expression for $X_1(t, x)$

$$X_1(t, x) = \int_{-\infty}^{t} F_1(u - t) \, dB(u, x), \tag{4.41}$$

where $\{B(u, x)\}$ is the Brownian motion given by (3.30):

$$B(u, x) = \langle x, \chi_{[u \wedge 0, u \vee 0]}\rangle.$$

Letting $\mathfrak{B}_t(X_1)$ be the σ-field generated by the $X_1(s, x)$, $s \leq t$, then the representation (4.41) proves that

$$\mathfrak{B}_t(X_1) \subseteq \mathfrak{B}_t, \qquad -\infty < t < \infty. \tag{4.42}$$

In particular, if equality holds for every t in (4.42), then (4.41) is called the *canonical representation* of $\{X_1(t)\}$ and F_1 is called the *canonical kernel*. Once the canonical representation is given, or at least the canonical kernel is known, then we can obtain sample path properties from this kernel, get the best predictor of future values of $X_1(t)$ using (4.41), and other interesting results.

Observations. The expression (4.41) tells us that $X_1(t)$ is obtained by applying the integral operator $F_1(t - u)$ of Volterra type to $dB(u)$. With this in mind we see that the condition $\mathfrak{B}_t(X_1) = \mathfrak{B}_t$ for (4.41) to be canonical involves, in some sense, the existence of an inverse to this integral operator. In addition, the representation is "causal" in the sense that the inverse operator acts linearly upon $\{X(s): s \leq t\}$ to recover $dB(t)$. Thus the theory of canonical representation is studied not only for its probabilistic interest, but also for its relevance to applications in communication theory and electrical engineering. See also P. Lévy (1956) and T. Hida (1960).

EXAMPLE 2. Another interesting example can be found in the case $n = 2$. For $\varphi(x) \in L_0^2 \cap \mathcal{H}_2$ we set

$$X_2(t, x) = U_t \varphi(x), \qquad -\infty < t < \infty.$$

The kernel of the integral representation of $X_2(t, x)$ can be written as

$$F_2(u_1 - t, u_2 - t),$$

and the process $X_2(t, x)$ can be dealt with analytically by using the fact that F_2 is an $\hat{L}^2(\mathbf{R}^2)$-kernel. Unlike Example 1 we should note that the inequality

$$\mathfrak{B}_t(X_2) \subsetneq \mathfrak{B}_t \tag{4.43}$$

holds for every $t \in (-\infty, \infty)$. We shall return to this functional again in §4.6.

For the present we will leave the flow $\{T_t\}$ of the Brownian motion, but the increasing family $\mathfrak{B}_t = \mathfrak{B}_t(B)$ of σ-fields will be used frequently.

Proposition 4.7. i. *If $\varphi(x)$ is a member of $\mathcal{H}_n$ ($n \geq 1$), then the conditional expectation $E(\varphi | \mathfrak{B}_t)$ also belongs to $\mathcal{H}_n$.*

ii. *Let the kernels of the integral representation of φ and $E(\varphi | \mathfrak{B}_t)$ be $F(u_1, u_2, \ldots, u_n)$ and $F(u_1, u_2, \ldots, u_n; t)$, respectively. Then*

$$F(u_1, u_2, \ldots, u_n; t) = F(u_1, u_2, \ldots, u_n) \chi_{(-\infty, t]^n}(u_1, u_2, \ldots, u_n). \tag{4.44}$$

Remark. In the assertion (i) the random variable $E(\varphi | \mathfrak{B}_t)$ might be 0 (the zero element of $\mathcal{H}_n$) even for a non-trivial φ.

4.4 Stochastic Processes

PROOF. i. is almost obvious. For the proof of (ii) we proceed as follows. The conditional expectation $E(\varphi | \mathfrak{B}_t)$ of φ is the orthogonal projection of φ onto L_t^2, so that by Proposition 4.5, $F(u_1, u_2, \ldots, u_n; t) = 0$ if $u_j > t$ for some j. Further, if ψ is an element of L_t^2, then

$$\int \varphi(x)\overline{\psi(x)}\, d\mu(x) = \int E(\varphi | \mathfrak{B}_t)(x)\overline{\psi}(x)\, d\mu(x).$$

We now let ψ vary over the space $L_t^2 \cap \mathscr{H}_n$, and in terms of the integral representation we get

$$\int_{-\infty}^{t} \cdots \int_{-\infty}^{t} F(u_1, u_2, \ldots, u_n)\overline{G(u_1, u_2, \ldots, u_n)}\, du_1\, du_2 \cdots du_n$$

$$= \int_{-\infty}^{t} \cdots \int_{-\infty}^{t} F(u_1, u_2, \ldots, u_n; t)\overline{G(u_1, u_2, \ldots, u_n)}\, du_1\, du_2 \cdots du_n,$$

where G runs through the entire space $\hat{L}^2((-\infty, t]^n)$. Therefore $F(u_1, u_2, \ldots, u_n; t)$ agrees with the restriction of $F(u_1, u_2, \ldots, u_n)$ to $(-\infty, t]^n$ and so (4.44) is proved. □

Now let us consider a stochastic process $\{X(t, x): -\infty < t < \infty\}$ on the probability space $(\mathscr{S}^*, \mathfrak{B}, \mu)$ such that

$$X(t, x) \in (L^2), \quad \text{for every} \quad t. \tag{4.45}$$

Theorem 4.6. *Let $\{X(t, x), \mathfrak{B}_t: -\infty < t < \infty\}$ be a zero-mean martingale such that $X(t, x)$ satisfies (4.45). If we denote by $X_n(t, x)$ the projection of $X(t, x)$ down onto $\mathscr{H}_n$, then*

i. *for each $n (> 0)$ $\{X_n(t, x), \mathfrak{B}_t: -\infty < t < \infty\}$ is a purely nondeterministic martingale, and*
ii. *there exists a symmetric function $F_n(u_1, u_2, \ldots, u_n)$ such that the kernel of the integral representation of $X_n(t, x)$ may be expressed in the form*

$$F_n(u_1, u_2, \ldots, u_n)\chi_{(-\infty, t]^n}(u_1, u_2, \ldots, u_n). \tag{4.46}$$

PROOF. i. One of the martingale assumptions for $\{X(t)\}$ is that asserting the $\mathfrak{B}_t$-measurability of $X(t, x)$, and this, together with (4.45), gives us the conclusion

$$X(t, x) \in L_t^2, \quad \text{for all} \quad t.$$

Since $E\{X(t)\} \equiv 0$ we see that $X(t, x)$ is always orthogonal to $\bigcap_t L_t^2 = \{1\}$, so $\{X(t)\}$, and hence $\{X_n(t)\}$, is purely nondeterministic. If we apply the projection operator P_n onto $\mathscr{H}_n$ to the relation

$$E(X(t) | \mathfrak{B}_s) = X(s), \quad t > s,$$

we obtain

$$P_n E(X(t) | \mathfrak{B}_s) = X_n(s).$$

Now the left-hand side of this equality may be expressed in the form

$$P_n \sum_{k=1}^{\infty} E(X_k(t)|\mathcal{B}_s)$$

which is just $E(X_n(t)|\mathcal{B}_s)$, since for every k, $E(X_k(t)|\mathcal{B}_s)$ belongs to $\mathcal{H}_k$ (Proposition 4.7, i). Thus we have

$$E(X_n(t)|\mathcal{B}_s) = X_n(s), \qquad s < t,$$

and hence $\{X_n(t, x), \mathcal{B}_t : -\infty < t < \infty\}$ is a martingale.

ii. Choose and fix an n and let $F_n(u_1, u_2, \ldots, u_n; t)$ be the kernel of the integral representation of $X_n(t)$. Proposition 4.7 (ii) and the relation $E(X_n(t)|\mathcal{B}_s) = X_n(s)$ for $t > s$ give the equation

$$F_n(u_1, u_2, \ldots, u_n; t)\chi_{(-\infty, s]^n}(u_1, u_2, \ldots, u_n) = F_n(u_1, u_2, \ldots, u_n; s), \qquad t > s.$$

Equivalently, for $t > s$, $F_n(u_1, u_2, \ldots, u_n; s)$ is the restriction of the function $F_n(u_1, u_2, \ldots, u_n; t)$ to $(-\infty, s]^n$. Thus we can let $t \to \infty$ and inductively define a symmetric function $F_n(u_1, u_2, \ldots, u_n)$ such that $F_n(u_1, u_2, \ldots, u_n; t)$ is of the form (4.46). □

Remarks. a. The function F_n determined by Theorem 4.6 (ii) is not always in $\hat{L}^2(\mathbf{R}^n)$; see example 3 below.

b. Even when t varies over $[0, \infty)$, or a finite interval, a similar result may be proved.

EXAMPLE 3. For t varying over $[0, \infty)$ and $\{B(t, x) : t \in [0, \infty)\}$ a Brownian motion we write

$$Y(t, x) = \exp\left[B(t, x) - \frac{1}{2}t\right]. \qquad (4.47)$$

Then we obtain a martingale $\{Y(t, x), \mathcal{B}_t : t \in [0, \infty)\}$ relative to $\mathcal{B}_t = \mathcal{B}\{B(u, x) : 0 \le u \le t\}$ as before. If we view $Y(t, x)$ as the generating function of the sequence of Hermite polynomials, then we have

$$Y(t, x) = \sum_{n=0}^{\infty} H_n(B(t, x); t),$$

so that $H_n(B(t, x); t)$ is a particular example of the $X_n(t, x)$ of Theorem 4.6. The kernel of its integral representation is proved in the next section to be $(n!)^{-1}\chi_{(0, t]^n}(u_1, u_2, \ldots, u_n)$ and the function F_n of Theorem 4.6 (ii) is thus given by

$$F_n(u_1, u_2, \ldots, u_n) = \frac{1}{n!}.$$

As F_n is constant it clearly cannot belong to $\hat{L}^2([0, \infty)^n)$.

4.5 Stochastic Integrals

As suggested by the name multiple Wiener integral, the elements of $\mathcal{H}_n$ may be regarded as members of (L^2) which can be expressed as n-fold integrals with respect to Brownian motion. In order to understand this fact we briefly summarise some background necessary for the material of this section.

For any fixed $\xi \in \mathcal{S}$, the function $\langle x, \xi \rangle$ of $x \in \mathcal{S}^*$ is a member of (L^2) with norm $\|\langle x, \xi \rangle\|_{(L^2)} = \|\xi\|$, the $L^2(\mathbf{R})$-norm of ξ. If we have the limit in $L^2(\mathbf{R})$:

$$\xi_n \to \chi_{[0, t]} \quad \text{as} \quad n \to \infty, \quad \xi_n \in \mathcal{S},$$

then we also have

$$\|\langle x, \xi_n \rangle - \langle x, \xi_m \rangle\|_{(L^2)} = \|\xi_n - \xi_m\|_{L^2(\mathbf{R})} \to 0$$

as $m, n \to \infty$. Thus there is a limit in (L^2) of the sequence $\{\langle x, \xi_n \rangle : n \geq 1\}$, and this limit does not depend upon the particular sequence $\{\xi_n\}$ used to approximate $\chi_{[0, t]}$. The (L^2)-norm of the mean-square limit of the sequence $\langle x, \xi_n \rangle$ is of course t, the $L^2(\mathbf{R})$-norm of $\chi_{[0, t]}$, and for simplicity we denote the limit by $\langle x, \chi_{[0, t]} \rangle$. It is a Gaussian random variable on $(\mathcal{S}^*, \mathfrak{B}, \mu)$ with zero mean and variance t, and the collection $\{\langle x, \chi_{[0, t]} \rangle : t \geq 0\}$ is a Gaussian process with zero mean and covariance function $t \wedge s$, i.e. a Brownian motion.

For a general element $f \in L^2(\mathbf{R})$ the random variable $\langle x, f \rangle$ is similarly defined, and is a zero-mean Gaussian random variable on $(\mathcal{S}^*, \mathfrak{B}, \mu)$ with variance $\|f\|^2$. Furthermore it belongs to $\mathcal{H}_1$. Now the random variable $\langle x, \chi_\Delta \rangle$ with $\Delta = (a, b]$ may be viewed as an increment of Brownian motion, i.e. $\langle x, \chi_\Delta \rangle = \Delta B$ where $\langle x, \chi_{[0, t]} \rangle = B(t)$. With this expression in mind the random variable $\langle x, f \rangle$ may be interpreted as

$$\int f(u) \, dB(u). \tag{4.48}$$

We know from Theorem 2.4, however, that for almost all x, $B(t, x)$ is of unbounded variation on any time interval. N. Wiener (1958, Lecture 1) was clever enough to get around this by introducing his integral with respect to $dB(t, \omega)$ in a particular way. His trick was to assume that the derivative f' belongs to $L^2([0, 1])$, and to use the hoped-for equality

$$\int_0^1 f(u) \, dB(u, \omega) = f(1) B(1, \omega) - \int_0^1 f'(u) B(u, \omega) \, du$$

to define the left-hand side in terms of the right-hand side, an expression which has the ordinary meaning. Then he was able to extend to the integral (4.48), where f is an arbitrary element of $L^2([0, 1])$.

Having made these observations we will define the *stochastic integral* below, where the integrand depends, in general, upon u as well as ω:

$$\int_0^t f(u, \omega) \, dB(u, \omega). \tag{4.49}$$

We will then go on to show how any non-constant member of (L^2) can be expressed as a stochastic integral (4.49), and further, we will establish a connection between the expression obtained and the integral representation of the previous section.

In what follows we define the stochastic integral (4.49) in the sense of K. Itô (1953c) and J. L. Doob (1953). For simplicity the time interval is taken to be $[0, 1]$.

Let $\{B(t, \omega): 0 \leq t \leq 1\}$ be a Brownian motion defined on the probability space $(\Omega, \mathbf{B}, P)$ and let $\mathbf{B}_s$ (equivalently, $\mathbf{B}_s(B)$) be the sub-σ-field of $\mathbf{B}$ generated by the $B(t, \omega)$, $t \leq s$. We further assume the existence of an increasing (in t) family $\{\mathbf{A}_t: t \in [0, 1]\}$ of sub-σ-fields of $\mathbf{B}$ such that for every $t \in [0, 1]$:

A.1. $\mathbf{A}_t \supseteq \mathbf{B}_t$;
A.2. $\mathbf{A}_t$ is independent of the system $\{B(t + s) - B(t): s \geq 0\}$ of random variables;

and that the integrand $f(t, \omega)$ satisfies:

f.1. $f(t, \omega)$ is $\mathcal{B}(\mathbf{R}) \times \mathbf{B}$-measurable;
f.2. $f(t, \cdot)$ is $\mathbf{A}_t$-measurable;
f.3. $\int_0^t f(u, \omega)^2 \, du < \infty$, a.e.

For such an f and system $\{\mathbf{A}_t\}$ the integral (4.49) can be defined in the following stages.

i. We begin with the case where f is a simple function of t. Let $\{\Delta_k\}$, $\Delta_k = [c_k, c_{k+1})$ be a partition of $[0, 1]$ into a finite number of intervals, where $c_0 = 0 < c_1 < c_2 < \cdots \leq 1$, and express f in the form

$$f(t, \omega) = \sum_k A_k(\omega) \chi_{\Delta_k}(t), \tag{4.50}$$

where $A_k(\omega)$ is $\mathbf{A}_{c_k}$-measurable. Then we define

$$\int_0^t f(u, \omega) \, dB(u, \omega) = \sum_{c_k < t} A_k(\omega) \, \Delta_k B(\omega) + A_l(\omega)[B(t, \omega) - B(c_l, \omega)], \tag{4.51}$$

$$c_l \leq t < c_{l+1}.$$

A few remarks are now in order. Firstly, there are many different expressions of the form (4.50) for f. However
a. the integral (4.51) does not depend on the way in which f is expressed;

4.5 Stochastic Integrals

b. the integral is additive in f: if f_1 and f_2 are simple functions of the form (4.50), then for constants λ_1, λ_2 we have

$$\int_0^t [\lambda_1 f_1(u, \omega) + \lambda_2 f_2(u, \omega)] \, dB(u, \omega)$$

$$= \lambda_1 \int_0^t f_1(u, \omega) \, dB(u, \omega) + \lambda_2 \int_0^t f_2(u, \omega) \, dB(u, \omega), \quad \text{a.e.;} \tag{4.52}$$

and c. for such a simple function f the integral

$$\int_0^t f(u, \omega) \, dB(u, \omega)$$

is, for almost all ω, a continuous function of t.

These properties follow directly from the definition.

ii. Given a function $f(t, \omega)$ satisfying (f.1), (f.2) and (f.3) we can approximate f by a suitable sequence $\{f_n\}$. Indeed we can define a function f_ε which is continuous in t by writing

$$\varepsilon^{-1} \int_{(t-\varepsilon) \vee 0}^t f(u, \omega) \, du = f_\varepsilon(t, \omega), \quad \varepsilon > 0.$$

If we then set

$$f_{\varepsilon, m}(t) = f_\varepsilon(k 2^{-m}), \quad t \in [k 2^{-m}, (k+1) 2^{-m}),$$

then we have

$$\lim_{\varepsilon \to 0+} \lim_{m \to \infty} \int_0^1 [f(u, \omega) - f_{\varepsilon, m}(u, \omega)]^2 \, du = 0, \quad \text{a.e.}$$

Choose a sequence $\varepsilon_n \to 0$ and sequence of integers $m_n \to \infty$ such that

$$P\left(\int_0^1 [f(u, \omega) - f_{\varepsilon_n, m_n}(u, \omega)]^2 \, du > 2^{-n} \right) \leq 2^{-n}.$$

Set $f_{\varepsilon_n, m_n} = f_n$. We can then apply the Borel-Cantelli Lemma to conclude that for almost all ω there exists an integer $n(\omega)$ such that for every $n \geq n(\omega)$

$$\int_0^1 [f(u, \omega) - f_n(u, \omega)]^2 \, du \leq 2^{-n}. \tag{4.53}$$

The sequence f_n is then exactly what we need.

iii. Returning now to the case of a simple function f given by (4.50), we set

$$X(t) = \exp\left[\int_0^t f(u, \omega) \, dB(u, \omega) - \frac{1}{2} \int_0^t f(u, \omega)^2 \, du \right]. \tag{4.54}$$

Then $\{X(t), \mathbf{A}_t: 0 \le t \le 1\}$ is a martingale with $E(X(t)) \equiv 1$. We prove this fact as follows. Suppose $s < t$. Then

$$\int_0^t f(u, \omega) \, dB(u, \omega) = \int_0^s f(u, \omega) \, dB(u, \omega) + A_j(\omega)[B(c_{j+1}, \omega) - B(s, \omega)]$$
$$+ \sum_{\Delta_k \subseteq [s, t]} A_k(\omega) \Delta_k B(\omega) + A_l(\omega)[B(t, \omega) - B(c_l, \omega)],$$

where $c_j \le s < c_{j+1}$, $c_l \le t < c_{l+1}$ and $\Delta_k B = B(c_{k+1}) - B(c_k)$, and also

$$\int_0^t f(u, \omega)^2 \, du = \int_0^s f(u, \omega)^2 \, du + A_j^2(\omega)(c_{j+1} - s)$$
$$+ \sum_{\Delta_k \subseteq [s, t]} A_k^2(\omega) |\Delta_k| + A_l^2(\omega)(t - c_l),$$

where $|\Delta_k| = c_{k+1} - c_k$ is the length of Δ_k. Using these formulae we obtain

$$E(X(t)|\mathbf{A}_s) = X(s) E \left\{ \exp \left[A_j\{B(c_{j+1}) - B(s)\} + \sum_{\Delta_k \subseteq [s, t]} A_k \Delta_k B \right. \right.$$
$$+ A_l\{B(t) - B(c_l)\} - \frac{1}{2} A_j^2(c_{j+1} - s)$$
$$\left. \left. - \frac{1}{2} \sum_{\Delta_k \subseteq [s, t]} A_k^2 |\Delta_k| - \frac{1}{2} A_l^2(t - c_l) \right] \middle| \mathbf{A}_s \right\}.$$

In order to compute the factor $E\{\exp[\cdot]|\mathbf{A}_s\}$ on the right-hand side, we first obtain the value of $E\{\exp[\cdot]|\mathbf{A}_{c_l}\}$, and then, successively, the same expression given $\mathbf{A}_{c_{l-1}}, \ldots$, so that we can apply the relation

$$E\{\exp[\cdot]|\mathbf{A}_s\} = E\{\cdots \{E\{\exp[\cdot]|\mathbf{A}_{c_l}\}|\mathbf{A}_{c_{l-1}}\} \cdots |\mathbf{A}_s\}.$$

Since $B(t) - B(c_l)$ is independent of $\mathbf{A}_{c_l}$, and since A_l is $\mathbf{A}_{c_l}$-measurable, $E\{\exp[\cdot]|\mathbf{A}_{c_l}\}$ turns out to be

$$\exp \left[A_j\{B(c_{j+1}) - B(s)\} + \sum_{\Delta_k \subseteq [s, t]} A_k \Delta_k B - \frac{1}{2} A_j^2(c_{j+1} - s) - \frac{1}{2} \sum_{\Delta_k \subseteq [s, t]} A_k^2 |\Delta_k| \right].$$

Taking the conditional expectation of this expression given $\mathbf{A}_{c_{l-1}}$, then given $\mathbf{A}_{c_{l-2}}, \ldots$, and so on, we finally obtain

$$E\left\{ \exp\left[A_j\{B(c_{j+1}) - B(s)\} - \frac{1}{2} A_j^2(c_{j+1} - s) \right] \middle| \mathbf{A}_s \right\} = 1.$$

Recalling that we have just proved that

$$E\{\exp[\cdot]|\mathbf{A}_s\} = 1,$$

we obtain the conclusion

$$E\{X(t)|\mathbf{A}_s\} = X(s), \qquad E\{X(t)\} \equiv 1.$$

4.5 Stochastic Integrals

We therefore deduce the inequality

$$P\left(\max_{0\le t\le 1}\left\{\int_0^t f(u,\omega)\,dB(u,\omega) - \frac{a}{2}\int_0^t f(u,\omega)^2\,du\right\} \ge b\right) \le e^{-ab}, \quad (4.55)$$

exactly as in §2.2 (iii).

iv. Let $\{f_n\}$ be the sequence of simple functions obtained in (ii). Then for any $n > n(\omega)$ we have the inequality

$$\int_0^1 \{f_n(u,\omega) - f_{n-1}(u,\omega)\}^2\,du \le 4\cdot 2^{-n}.$$

By taking $a = (2^{n-1}\log n)^{1/2}$ and $b = c(2^{-n+1}\log n)^{1/2}$, where $c > 1$, we apply the inequality (4.55) to $g_n = f_n - f_{n-1}$ to obtain

$$P\left(\max_{0\le t\le 1}\left\{\int_0^t g_n(u,\omega)\,dB(u,\omega) - \tfrac{1}{2}(2^{n-1}\log n)^{1/2}\int_0^t g_n(u,\omega)^2\,du\right\}\right.$$

$$\left. \ge c(2^{-n+1}\log n)^{1/2}\right) \le n^{-c}.$$

Again the Borel-Cantelli Lemma shows the existence of an integer $n_0(\omega) > n(\omega)$ such that for every $n > n_0(\omega)$ we have the inequality

$$\max_{0\le t\le 1}\int_0^t g_n(u,\omega)\,dB(u,\omega) \le \frac{1}{2}a\int_0^1 g_n(u,\omega)^2\,du + b$$

$$\le (1+c)(2^{-n+1}\log n)^{1/2}.$$

If g_n is replaced by $-g_n$ we see that the same inequality holds, and hence

$$\max_{0\le t\le 1}\left|\int_0^t \{f_n(u,\omega) - f_{n-1}(u,\omega)\}\,dB(u,\omega)\right| \le (1+c)(2^{-n+1}\log n)^{1/2} \quad (4.56)$$

holds for all sufficiently large n. This inequality guarantees the existence of the limit

$$\lim_{n\to\infty}\int_0^t f_n(u,\omega)\,dB(u,\omega) \equiv \int_0^t f(u,\omega)\,dB(u,\omega), \quad \text{a.e.} \quad (4.57)$$

In addition the convergence is seen to be uniform in t. It is also easily seen from the inequality (4.56) that the limit (4.57) does not depend on the choice of the sequence $\{f_n\}$ approximating f.

The discussion so far has enabled us to define the stochastic integral $\int_0^t f(u,\omega)\,dB(u,\omega)$ as the left-hand side of (4.57). One short remark on this is in order. If the time interval $[0,\infty)$ is of interest, the condition (f.3) on the integrand f should be replaced by

f.3'. $\quad \int_0^t f(u,\omega)^2\,du < \infty, \quad$ a.e. $\quad$ for every $\quad t < \infty.$

Then the integrals

$$\int_0^t f(u, \omega)\, dB(u, \omega), \qquad t \geq 0,$$

are all well-defined as above. Naturally if the stronger condition

$$\int_0^\infty f(u, \omega)^2\, du < \infty, \qquad \text{a.e.}$$

is supposed, then without any difficulty we can define the integral

$$\int_0^\infty f(u, \omega)\, dB(u, \omega).$$

Some important properties of the stochastic integral are now collected in the following proposition. All proofs are omitted, and we refer the reader to J. L. Doob (1953) or H. P. McKean Jr. (1969).

Proposition 4.8. *Let f, f_1 and f_2 satisfy the conditions (f.1), (f.2) and (f.3) or (f.3'). Then the following assertions hold true.*

i. *For constants λ_1 and λ_2*

$$\int_0^t \{\lambda_1 f_1(u, \omega) + \lambda_2 f_2(u, \omega)\}\, dB(u, \omega)$$

$$= \lambda_1 \int_0^t f_1(u, \omega)\, dB(u, \omega) + \lambda_2 \int_0^t f_2(u, \omega)\, dB(u, \omega).$$

ii. *The integral*

$$\int_0^t f(u, \omega)\, dB(u, \omega)$$

is continuous in t for almost all ω.

iii.

$$E\left[\left|\int_0^\infty f(u, \omega)\, dB(u, \omega)\right|^2\right] \leq E\left[\int_0^\infty f(u, \omega)^2\, du\right],$$

and if the right-hand side is finite, then

$$E\left[\int_0^\infty f(u, \omega)\, dB(u, \omega)\right] = 0$$

and

$$E\left[\left|\int_0^\infty f(u, \omega)\, dB(u, \omega)\right|^2\right] = E\left[\int_0^\infty f(u, \omega)^2\, du\right].$$

4.5 Stochastic Integrals

Remark. For a general (i.e. not necessarily simple) function f set

$$X(t, \omega) = \exp\left[\int_0^t f(u, \omega) \, dB(u, \omega) - \frac{1}{2} \int_0^t f(u, \omega)^2 \, du\right].$$

Then $\{X(t), \mathbf{A}_t : t \geq 0\}$ is a supermartingale, i.e. for $s < t$

$$E\{X(t) | \mathbf{A}_s\} \leq X(s), \quad \text{a.e.}$$

We also have

$$E\{X(t)\} \leq 1,$$

and there exists an example for which $E\{X(t)\} < 1$.

EXAMPLE 1. Let $F(t, u)$ be a non-random function of (u, t) which is square-integrable in u over $[0, t]$ for every $t \geq 0$. Then the stochastic integral

$$X(t, \omega) = \int_0^t F(t, u) \, dB(u, \omega), \quad t \geq 0, \tag{4.58}$$

can be defined as a particular case in which the integrand does not depend on ω. The process $\{X(t, \omega): t \geq 0\}$ is then Gaussian.

In particular if F is a function of u only, say $F(u)$, then

$$X(t, \omega) = \int_0^t F(u) \, dB(u, \omega), \quad t \geq 0, \tag{4.59}$$

defines an additive process whose sample functions are almost surely continuous. Further specialising to the case $F(u) \equiv 1$ gives $X(t, \omega) = B(t, \omega)$, $t \geq 0$, i.e.

$$B(t, \omega) = \int_0^t 1 \, dB(u, \omega). \tag{4.60}$$

There is an alternative definition of the stochastic integral available for the case (4.58) in which the integrand is non-random. For fixed t we think of the integrand F as a function of u only. If this function is a simple function, say f given by (4.50), with constants $\{A_k\}$, then the integral is defined by (4.51). We now choose a sequence of simple functions approximating the F under consideration, and find that the associated sequence of stochastic integrals has a limit in the strong topology on $L^2(\Omega, \mathbf{B}, P)$. The random variable so obtained depends on t and may be written in the form

$$X(t) = \int_0^t F(t, u) \, dB(u), \quad t \geq 0,$$

where we have omitted the parameter ω. The family $\{X(t): t \geq 0\}$ is then a version of the Gaussian process $\{X(t, \omega): t \geq 0\}$ defined by (4.58). The process $X_1(t)$ of the last section may be regarded as having been defined by either of the above methods.

EXAMPLE 2. The following equality is valid:

$$\int_0^t B(u, \omega) \, dB(u, \omega) = \frac{1}{2} B(t, \omega)^2 - \frac{1}{2} t \equiv H_2(B(t, \omega), t), \qquad 0 \le t \le 1. \quad (4.61)$$

To see this we take the sequence $\{B_n(u, \omega) : n \ge 1\}$ approximating $B(u, \omega)$ as

$$B_n(u, \omega) = B(k2^{-n}, \omega), \qquad u \in [k2^{-n}, (k+1)2^{-n}).$$

By definition, if $t \in [l2^{-n}, (l+1)2^{-n})$, we have

$$\int_0^t B_n(u, \omega) \, dB(u, \omega) = \sum_{k=0}^{l-1} B(k2^{-n}, \omega)\{B((k+1)2^{-n}, \omega) - B(k2^{-n}, \omega)\}$$
$$+ B(l2^{-n}, \omega)\{B(t, \omega) - B(l2^{-n}, \omega)\}$$
$$= \frac{1}{2}\left[\sum_0^{l-1} B((k+1)2^{-n}, \omega)^2 - \sum_0^{l-1} B(k2^{-n}, \omega)^2\right.$$
$$\left. - \sum_0^{l-1} \{B((k+1)2^{-n}, \omega) - B(k2^{-n}, \omega)\}^2\right]$$
$$+ B(l2^{-n}, \omega)\{B(t, \omega) - B(l2^{-n}, \omega)\}$$
$$= \frac{1}{2} B(t, \omega)^2 - \frac{1}{2}\sum_0^{l-1} \{\Delta_k B(\omega)\}^2$$
$$+ \frac{1}{2}\{B(l2^{-n}, \omega)^2 - B(t, \omega)^2\}$$
$$+ B(l2^{-n}, \omega)\{B(t, \omega) - B(l2^{-n}, \omega)\},$$

where $\Delta_k = [k2^{-n}, (k+1)2^{-n})$. Now let $n \to \infty$ and let $[l2^{-n}, (l+1)2^{-n})$ continue to denote the interval containing t. Then $B(l2^{-n}, \omega) \to B(t, \omega)$ follows, and Theorem 2.3 of §2.2 proves that the last expression approaches $\frac{1}{2}(B(t, \omega) - t)^2$ for almost all ω. The first equality in (4.61) has been proved, whilst the second is simply obtained by using the definition of $H_2(x; \sigma^2)$.

Incidentally if we recall formula (4.60), then the left-hand side of (4.61) may be rewritten in the form

$$\int_0^t \left\{\int_0^u dB(s, \omega)\right\} dB(u, \omega) = \int_0^t \int_0^u dB(s, \omega) \, dB(u, \omega),$$

and this is sometimes written

$$\frac{1}{2} \int_0^t \int_0^t dB(s, \omega) \, dB(u, \omega),$$

which may appropriately be termed a multiple, in this case double, Wiener integral.

4.5 Stochastic Integrals

Remark. We are *not* permitted to make use of Fubini's theorem in the formal way, for example to deduce that

$$\frac{1}{2}\int_0^t \left\{\int_0^t dB(s, \omega)\right\} dB(u, \omega) = \frac{1}{2} B(t)^2.$$

We should also keep in mind the crucial condition (f.2) necessary for the straightforward definition of a stochastic integral.

As the example of the foregoing remark illustrates, Fubini's theorem does not always hold for stochastic integrals. However there are certain conditions under which we may define multiple integrals with respect to dB.

For simplicity in what follows, let us suppose that the time parameter varies over $[0, 1]$. Take a function f satisfying (f.1), (f.2) and (f.3), and define the stochastic integral

$$f_1(u, \omega) = \int_0^u f(s, \omega)\, dB(s, \omega).$$

Clearly the definition of the integral implies that

a. $f_1(u, \omega)$ is (u, ω)-measurable; in fact
b. $f_1(u, \omega)$ is $\mathbf{A}_u$-measurable for every u; and
c. $f_1(u, \omega)$ is continuous in u for almost every ω.

This last property implies that, for almost all ω, $f_1(u, \omega)$ is bounded, and

$$\int_0^1 \{f_1(u, \omega)f(u, \omega)\}^2\, dB(u, \omega) \leq \max_{0 \leq u \leq 1} |f_1(u, \omega)|^2 \cdot \int_0^1 f(u, \omega)^2\, du < \infty.$$

Thus (a), (b) and (c) show that the product $f_1(u, \omega)f(u, \omega)$ satisfies the conditions (f.1), (f.2) and (f.3) above and hence we may define the stochastic integral

$$f_2(t, \omega) = \int_0^t f_1(u, \omega) f(u, \omega)\, dB(u, \omega). \tag{4.62}$$

Alternatively, it may be expressed in the form

$$f_2(t, \omega) = \int_0^t \int_0^u f(u_1, \omega) f(u, \omega)\, dB(u_1, \omega)\, dB(u, \omega). \tag{4.63}$$

We may define f_n inductively by

$$f_n(t, \omega) = \int_0^t f_{n-1}(u, \omega) f(u, \omega)\, dB(u, \omega) \tag{4.64}$$

and re-express this in the form

$$\int_0^t \int_0^{u_1} \cdots \int_0^{u_{n-1}} f(u_1, \omega) f(u_2, \omega) \cdots f(u_n, \omega)$$
$$\times dB(u_1, \omega)\, dB(u_2, \omega) \cdots dB(u_n, \omega). \tag{4.65}$$

The above integral can easily be extended to cover the case where the integrand is a product of distinct functions each satisfying the necessary conditions.

Later we will frequently meet simple but interesting cases in which the integrand $f(u, \omega)$ is independent of ω, say $f(u, \omega) = F(u)$. More particularly, let us consider the case $F(u) \equiv 1$. Then we have

$$f_n(t, \omega) = \int_0^t \int_0^{u_1} \cdots \int_0^{u_{n-1}} dB(u_1, \omega) \cdots dB(u_n, \omega). \tag{4.66}$$

If we take $\mathbf{A}_t$ to be $\mathbf{B}_t \, (= \mathbf{B}_t(B))$ and set

$$Y(t, \omega) = \exp\left[B(t) - \frac{1}{2}t\right], \tag{4.67}$$

then, as has already been proved, $\{Y(t, \omega), \mathbf{B}_t : t \in [0, 1]\}$ is a martingale. We can now state

Theorem 4.7. *Let* $\{Y(t, \omega): t \in [0, 1]\}$ *be the stochastic process defined by* (4.67). *Then the following relations hold*:

$$\int_0^t Y(u, \omega) \, dB(u, \omega) = Y(t, \omega) - 1, \quad \text{a.e.}, \tag{4.68}$$

and

$$\int_0^t H_n(B(u, \omega); u) \, dB(u, \omega) = H_{n+1}(B(t, \omega); t), \quad \text{a.e.}, \tag{4.69}$$

where this last expression also agrees with $f_{n+1}(t, \omega)$ *as defined by* (4.66).

Before giving the proof of this theorem, we give some relevant background. The basic Brownian motion will be taken to be $\{\langle x, \chi_{[0, t]} \rangle : t \in [0, 1]\}$ on the probability space $(\mathcal{S}^*, \mathfrak{B}, \mu)$, and for simplicity we suppose that all sample paths are continuous. We now introduce some notation. For a real number λ we denote by $E^{(k)}(\lambda)$ the operator acting on $\hat{L}^2(\mathbf{R}^k)$ by the definition

$$E^{(k)}(\lambda) f(x) = \chi_{(-\infty, \lambda]^k}(x) f(x), \quad f \in \hat{L}^2(\mathbf{R}^k)$$

This is a projection operator, and the system

$$\{E^{(k)}(\lambda): -\infty < \lambda < \infty\}$$

is seen to be a resolution of the identity I on $\hat{L}^2(\mathbf{R}^k)$.

For $f \in \hat{L}^2(\mathbf{R}^n)$ and $g \in L^2(\mathbf{R})$ we use $E^{(n)}(\lambda)$ and $E^{(1)}(\lambda)$ to define a product $f \rhd g$ in such a way that if $x_1 \leq x_2 \leq \cdots \leq x_n \leq y$, then

$$(f \rhd g)(x, y) = \int_{-\infty}^{\infty} \{E^{(n)}(\lambda) f \otimes dE^{(1)}(\lambda) g\}(x, y) \tag{4.70}$$

4.5 Stochastic Integrals

where $x = (x_1, x_2, \ldots, x_n)$, $\otimes$ denotes the tensor product, and for $d\lambda > 0$

$$dE^{(1)}(\lambda) = E^{(1)}(\lambda + d\lambda) - E^{(1)}(\lambda).$$

This product may be extended by symmetry to the entire space $\mathbf{R}^n \times \mathbf{R} = \mathbf{R}^{n+1}$, and we also denote this extension by $f \triangleright g$. We began in the domain $x_1 \leq x_2 \leq \cdots \leq x_n$ when defining the product, but as f is a symmetric function on $\mathbf{R}^n$, if we had begun the definition on a different domain given by permuting some of the x_i in the string of inequalities, the end result would have been the same. Thus the product $f \triangleright g$ is well-defined, and it is clearly bilinear in f and g.

Remark. The product $\triangleright$ can be defined similarly when $\hat{L}^2(\mathbf{R}^n)$ is replaced by $\hat{L}^2(T^n)$ for a finite or infinite subinterval T of $\mathbf{R}$.

EXAMPLE 3. In the case $n = 1$ we have

$$(f \triangleright g)(x, y) = f(x \wedge y)g(x \vee y). \tag{4.71}$$

EXAMPLE 4. If $f = \chi_{[0, 1]^n}$ $[\in \hat{L}^2(\mathbf{R}^n)]$ and $g = \chi_{[0, t]}$ $[\in L^2(\mathbf{R})]$ then

$$(f \triangleright g)(x, y) = \chi_{[0, t]^{n+1}}(x, y), \qquad x \in \mathbf{R}^n, \qquad y \in \mathbf{R}. \tag{4.72}$$

If we replace $f = \chi_{[0, t]^n}$ with $\chi_{[0, s]^n}$ where $s > t$, the result $f \triangleright g$ is still the same as before, i.e. it is $\chi_{[0, t]^{n+1}}$.

The result of the next lemma is established on the probability space $(\mathscr{S}^*, \mathfrak{B}, \mu)$ which is fixed for the moment, and instead of ω we will naturally use the notation x. A Brownian motion is then obtained as

$$\langle x, \chi_{[0, t]} \rangle = B(t, x), \qquad t \geq 0,$$

and other notation will be as before. The time parameter t is usually constrained to run over $[0, 1]$.

Lemma 4.1. Let $\{f(t, x), \mathfrak{B}_t: t \in [0, 1]\}$ be a martingale, with $f(u, x)$ an element of $\mathscr{H}_n$ satisfying the conditions (f.1), (f.2) and (f.3). Further let $F_u(s_1, s_2, \ldots, s_n)$ be the kernel of the integral representation of $f(u, x)$ and $g(u)$ a non-random function in $L^2([0, 1])$. Then

$$\int_0^1 f(u, x)g(u)\,dB(u, x) \tag{4.73}$$

belongs to $\mathscr{H}_{n+1}$, and the kernel of its integral representation is given by

$$(n + 1)^{-1} F_1 \triangleright g.$$

PROOF. The stochastic integral (4.73) is approximated by the sum

$$\sum_k f(u_k, x)g(u_k)\langle x, \chi_{\Delta_k}\rangle, \qquad \Delta_k = [u_k, u_{k+1}).$$

Now apply the transformation $\mathcal{T}$, formula (4.12) of §4.3, to this and obtain

$$\int_{\mathscr{S}^*} e^{i\langle x,\xi\rangle} \sum_k f(u_k, x) g(u_k) \langle x, \chi_{\Delta_k}\rangle \, d\mu(x) \qquad (\xi \in \mathscr{S})$$

$$= \sum_k \int_{\mathscr{S}^*} \{e^{i\langle x, \xi_k^+\rangle}\langle x, g(u_k)\chi_{\Delta_k}\rangle\}\{e^{i\langle x, \xi_k^-\rangle} f(u_k, x)\} \, d\mu(x),$$

where $\xi_k^+(u) = \xi(u)\chi_{[u_k, \infty)}$ and $\xi_k^-(u) = \xi(u) - \xi_k^+(u)$.

Each summand in this expression is the product of the two independent random variables in the braces, and so we obtain the sum of products of integrals:

$$= \sum_k i\langle g(u_k)\chi_{\Delta_k}, \xi_k^+\rangle \exp\left[-\frac{1}{2}\|\xi_k^+\|^2\right]$$

$$\times i^n \int_{[0,1]^n} \cdots \int F_{u_k}(s_1, \ldots, s_n) \xi_k^-(s_1) \cdots \xi_k^-(s_n) \, ds_1 \cdots ds_n \exp\left[-\frac{1}{2}\|\xi_k^-\|^2\right].$$

Since $\{f(u, x), \mathfrak{B}_t, t \in [0, 1]\}$ is a martingale, Theorem 4.6 assets that

$$F_u(s_1, \ldots, s_n) = F_1(s_1, \ldots, s_n)\chi_{[0, u]^n}(s_1, \ldots, s_n), \qquad 0 \le u \le 1,$$

and thus the above sum of integrals becomes

$$= C(\xi) i^{n+1} \sum_k \int_{[0,1]^{n+1}} \cdots \int g(u_k)\chi_{\Delta_k}(s_0) F_1(s_1, \ldots, s_n)\chi_{[0, u_k]^n}(s_1, \ldots, s_n)$$

$$\times \xi(s_0)\xi(s_1) \cdots \xi(s_n) \, ds_0 \, ds_1 \cdots ds_n.$$

Now we recall the definition of $\rhd$ and see that the sum converges to the integral

$$i^{n+1} C(\xi) \int \cdots \int (n + 1)^{-1} (F_1 \rhd g)(s_0, s_1, \ldots, s_n)$$

$$\times \xi(s_0)\xi(s_1) \cdots \xi(s_n) \, ds_0 \, ds_1 \cdots ds_n.$$

We have shown that the stochastic integral (4.73) does indeed belong to $\mathscr{H}_{n+1}$, and also that the kernel of its integral representation is given by $(n + 1)^{-1} F_1 \rhd g$. $\square$

PROOF OF THEOREM 4.7. Our probability space is still $(\mathscr{S}^*, \mathfrak{B}, \mu)$.

1. For a real constant λ set $Y_\lambda(t, x) = \exp[\lambda B(t, x) - \frac{1}{2}\lambda^2 t]$. We know that $\{Y_\lambda(t, x), \mathfrak{B}_t, t \in [0, 1]\}$ is a martingale and so, for $s < t$,

$$E(Y_\lambda(t, x)|\mathfrak{B}_s) = Y_\lambda(s, x), \qquad \text{a.e.}$$

The expression for $Y_\lambda(t)$ is recognised as the generating function of the sequence of Hermite polynomials with a parameter, and so may be expanded into a power series in λ. The martingale identity then becomes

$$\sum_n \lambda^n E(H_n(B(t, x); t)|\mathfrak{B}_s) = \sum_n \lambda^n H_n(B(s, x); s),$$

which, upon comparing coefficients of λ^n, gives us the result

$$E(H_n(B(t, x); t)|\mathfrak{B}_s) = H_n(B(s, x); s), \quad \text{a.e.},$$

i.e. for each n, $\{H_n(B(t, x); t), \mathfrak{B}_t, t \in [0, 1]\}$ is a martingale.

2. Our next step is to obtain the integral representation of $H_n(B(t, x); t)$. As in the case of (4.16) we have

$$(\mathcal{F} Y_\lambda(t))(\xi) = C(\xi) \exp\left[i\lambda \int \chi_{[0, t]}(u)\xi(u)\, du\right],$$

and again we view $Y_\lambda(t)$ as a generating function and expand the right-hand side as a power series. This proves that for $n \geq 0$

$$(\mathcal{F} H_n(B(t, x); t))(\xi) = C(\xi) \frac{i^n}{n!} \left\{\int \chi_{[0, t]}(u)\xi(u)\, du\right\}^n,$$

which shows that the kernel of the integral representation of $H_n(B(t, x); t)$ is $(1/n!)\chi_{[0, t]^n}$.

3. We are now in a position to use the lemma just proved. For every n the functional $H_n(B(t, x); t)$ is an element of $\mathcal{H}_n$ satisfying the conditions of Lemma 4.1, the exception being that this is now taken on $[0, t]$ rather than $[0, 1]$. Hence the stochastic integral

$$\int_0^t H_n(B(u, x); u)\, dB(u, x) \tag{4.74}$$

can be defined, is a member of $\mathcal{H}_{n+1}$, and, by Example 4, has as kernel of its integral representation the function

$$(n + 1)^{-1}(n!)^{-1}\chi_{[0, t]^n} \triangleright \chi_{[0, t]} = \{(n + 1)!\}^{-1}\chi_{[0, t]^{n+1}}.$$

Recalling that there is a one-to-one correspondence, indeed an isomorphism, between $\mathcal{H}_{n+1}$ and $\hat{L}^2(\mathbf{R}^{n+1})$ under $\mathcal{F}$, up to $(n!)^{-1/2}$, we see that the integral (4.74) must coincide with the (L^2)-functional $H_{n+1}(B(t, x); t)$ which is known to have the kernel $(n + 1)!^{-1}\chi_{[0, t]^{n+1}}$. We have actually proved that (4.69) holds for all values of t (more precisely: that the right-hand side of (4.69) is continuous in t for almost all ω, so that the equality holds for all t except on an ω-set of P-measure zero).

By summing this result over $n = 0, 1, 2, \ldots$, and using the generating function for Hermite polynomials, we obtain the equality (4.68). □

Remark. Example 2 above is a particular case of the formula (4.69) just proved.

Let us return once more to the expansion $Y_\lambda(t, x) = \exp[\lambda B(t, x) - \frac{1}{2}\lambda^2 t]$ $= \sum_0^\infty \lambda^n H_n(B(t, x); t)$ used in the first step of the proof of Theorem 4.7. Term-by-term integration with respect to dB is possible, and doing so we obtain

$$\int_0^t Y_\lambda(u, x) \, dB(u, x) = \lambda^{-1}[Y_\lambda(t, x) - 1]. \tag{4.75}$$

We now change the definition of $Y_\lambda(t, x)$ slightly, to the expression $Y_\lambda(t, x) = \exp[\lambda\{B(t + a, x) - B(a, x)\} - \frac{1}{2}\lambda^2 t]$. For this expression (4.75) still holds, and then Theorem 4.7 can be easily generalised. Take $f \in L^2([0, 1])$ and put

$$X(t, x) = \int_0^t f(u) \, dB(u, x). \tag{4.76}$$

The Wiener integral $X(t, x)$ may also be denoted by $\langle x, f \rangle$. If we introduce the notation

$$\|f\|_t^2 = \int_0^t f(u)^2 \, du$$

and set

$$Y(t, x) = \exp\left[X(t, x) - \frac{1}{2}\|f\|_t^2\right]. \tag{4.77}$$

Then $\{Y(t, x), \mathbf{B}_t, t \in [0, 1]\}$ is also a martingale, and we have the following generalisation of Theorem 4.7.

Proposition 4.9. *Let $X(t, x)$ and $Y(t, x)$ be given by (4.76) and (4.77) respectively. Then we have the relations*

$$\int_0^t Y(u, x) f(u) \, dB(u, x) = Y(t, x) - 1, \quad \text{a.e.,} \tag{4.78}$$

and

$$\int_0^t H_n(X(t, x); \|f\|_t^2) f(u) \, dB(u, x) = H_{n+1}(X(t, x); \|f\|_t^2), \quad \text{a.e.} \tag{4.79}$$

In addition $H_n(X(t, x); \|f\|_t^2)$ may be expressed as a multiple integral

$$H_n(X(t, x); \|f\|_t^2) = \int_0^t \int_0^{u_1} \cdots \int_0^{u_{n-1}} f^{n\otimes}(u_1, \ldots, u_n) \, dB(u_1, x) \cdots dB(u_n, x), \tag{4.80}$$

where $f^{n\otimes}(u_1, \ldots, u_n) = f(u_1) \cdots f(u_n)$.

Since the proposition can be proved in a manner similar to that of Theorem 4.7, the proof is omitted.

Before we close this section a few remarks are in order. Set $t = 1$ and consider an f in formula (4.80) with $\|f\| = 1$. Then the kernel of the integral representation of $H_n(\langle x, f \rangle; \|f\|^2)$ is seen to be $(n!)^{-1} f^{n\otimes}$, in agreement with §4.4.

Integration with respect to $dB(u, x)$ defines an operator which carries those elements of $\mathscr{H}_n$ that can be integrands to $\mathscr{H}_{n+1}$. For example in the case of (4.79) the element of $\mathscr{H}_n$ having $f^{n\otimes}$ as kernel is transformed into that element of $\mathscr{H}_{n+1}$ having kernel $f^{(n+1)\otimes}$. In the terminology of quantum mechanics such an integration defines one of the so-called creation-annihilation operators on the Fock space (4.22) of §4.3, in our case a creation operator. This operation corresponds to the formula given as (A.35) in §A.5 concerning Hermite polynomials, which in the present notation may be written as:

$$\langle x, f \rangle H_n(\langle x, f \rangle; \|f\|^2) = (n + 1) H_{n+1}(\langle x, f \rangle; \|f\|^2)$$
$$+ \|f\|^2 H_{n-1}(\langle x, f \rangle; \|f\|^2).$$

4.6 Examples of Applications

This section is devoted to several topics illustrating our theory, as well as to providing certain supplementary descriptions.

(i) Quadratic Functionals

Elements of the double Wiener integral $\mathscr{H}_2$ are quadratic functionals of white noise (or Brownian motion). These particular non-linear functionals have many interesting properties and the theory of integral representations provides a powerful tool for their analysis.

Let $F(u, v)$ be the integral representation of a real-valued $\varphi(x) \in \mathscr{H}_2$. Then F is a real function in $\hat{L}^2(\mathbf{R}^2)$ that defines an integral operator on $L^2(\mathbf{R}^2)$ of Hilbert-Schmidt type which, since F is real symmetric (i.e. $F(u, v) = F(v, u)$), is Hermitian. Denote the system of eigenvalues and normalised eigenfunctions of F by $\{\lambda_n : n \geq 1\} \subseteq \mathbf{R}$ and $\{\eta_n : n \geq 1\} \subseteq L^2(\mathbf{R})$ respectively. Then $\|\eta_n\| = 1$ for all n, and

$$\lambda_n \int F(u, v) \eta_n(v) \, dv = \eta_n(u).$$

The generalised Hilbert-Schmidt expansion theorem gives us the following development in $L^2(\mathbf{R}^2)$:

$$F(u, v) = \sum_{1}^{\infty} \lambda_n^{-1} \eta_n(u) \eta_n(v), \qquad (4.81)$$

where the $\{\lambda_n\}$ are arranged in order of increasing absolute value, multiplicity being taken into account. It is understood that $\sum_n \lambda_n^{-2} < \infty$.

Theorem 4.8. *Let $\varphi(x)$ be a real functional in $\mathcal{H}_2$ with associated kernel F. Then, following on from the expansion (4.81) of F, we can express $\varphi(x)$ as a sum of independent random variables*

$$\varphi(x) = \sum_1^\infty \lambda_n^{-1}(\langle x, \eta_n \rangle^2 - 1), \qquad \text{a.e.} \tag{4.82}$$

The characteristic function $\chi(z)$ of the random variable $\varphi(x)$ is given by

$$\chi(z) = \delta(2iz; F)^{-1/2}, \qquad z \in \mathbf{R}, \tag{4.83}$$

where $\delta(\lambda, F)$ is the modified Fredholm determinant of F.

PROOF. Through the transformation $\mathcal{T}$, there is a one-to-one correspondence between $\mathcal{H}_2$ and the set $\hat{L}^2(\mathbf{R}^2)$ of all kernels of integral representations. Thus the expansion (4.81) of F gives the corresponding expansion of $\varphi(x)$. It is known (see formula (4.17) of §4.3) that the $\mathcal{H}_2$-functional having kernel $\eta_n(u)\eta_n(v)$ for its integral representation is $\langle x, \eta_n \rangle^2 - 1$, and with this fact, (4.81) determines the expansion (4.82). Since $\{\eta_n\}$ is an orthonormal system in $L^2(\mathbf{R})$, we know that $\{\langle x, \eta_n \rangle : n \geq 1\}$ is an independent system of Gaussian random variables, and so $\{[\langle x, \eta_n \rangle^2 - 1] : n \geq 1\}$ is also an independent system.

The characteristic function of the random variable $\varphi(x)$ is also readily obtained from the expansion (4.82). Since $\langle x, \eta_n \rangle$ is a standard Gaussian random variable, $\langle x, \eta_n \rangle^2$ has a chi-squared distribution on one degree of freedom, so that for $n \geq 1$

$$\int_{\mathcal{S}^*} \exp[iz(\langle x, \eta_n \rangle^2 - 1)] \, d\mu(x) = (1 - 2iz)^{-1/2} \exp[-iz], \qquad z \in \mathbf{R}.$$

Evaluating this at $\lambda_n^{-1} z$ and taking the product over $n \geq 1$ we obtain the characteristic function of $\varphi(x)$ as

$$\int_{\mathcal{S}^*} \exp[iz\varphi(x)] \, d\mu(x) = \prod_n (1 - 2iz\lambda_n^{-1})^{-1/2} \exp[-iz\lambda_n^{-1}].$$

Finally we see that the right-hand side may be put into the form (4.83) using the modified Fredholm determinant of the kernel F which is $\delta(\lambda, F) = \prod_n (1 - \lambda_n^{-1}\lambda)\exp[\lambda_n^{-1}\lambda]$. $\square$

Corollary. *Let $\varphi(x)$ be the $\mathcal{H}_2$-functional given in Theorem 4.8. Then $\varphi(x)$ has moments of all orders, and its p-th cumulant γ_p is of the form*

$$\begin{cases} \gamma_1 = 0, \\ \gamma_p = 2^{p-1}(p-1)! \sum_1^\infty \lambda_n^{-p}, \qquad p \geq 2. \end{cases} \tag{4.84}$$

4.6 Examples of Applications

PROOF. This result involves the Taylor expansion of $\log \chi(z)$. Letting σ_p be the trace of the p-fold iterate $F^p(u, v)$ of the kernel F, then a formula for $\delta(\lambda; F)$ (see F. Smithies (1958) Chapter VI)

$$\frac{d}{d\lambda} \log \delta(\lambda; F) = -\sum_{p=1}^{\infty} \sigma_p \lambda^p, \qquad |\lambda| \text{ sufficiently small,}$$

together with (4.83), implies that

$$\log \chi(z) = -\frac{1}{2} \log \delta(2iz; F)$$

$$= \frac{1}{2} \sum_{2}^{\infty} p^{-1} \sigma_p (2iz)^p$$

$$= \frac{1}{2} \sum_{2}^{\infty} 2^p \sigma_p (p-1)! \frac{(iz)^p}{p!}.$$

On the other hand we know that

$$\sigma_p = \sum_{1}^{\infty} \lambda_n^{-p}$$

and hence the result (4.84) follows. $\square$

The existence of higher order moments follows from this result, or, alternatively, from the fact that $\{\langle x, \eta_n \rangle : n \geq 1\}$ is an independent system, and that $\sum_n \lambda_n^{-p}$ is convergent for $p \geq 2$.

Remark. The distribution of $\varphi(x) \in \mathcal{H}_2$ is completely determined by its cumulants, and the above theorem gives a method of calculating them.

We will also be interested in the particular case in which the kernel $F(u, v)$ of the integral representation of $\varphi(x)$ satisfies

$$F(-u, -v) = -F(u, v) \qquad (= -F(v, u)). \tag{4.85}$$

Such a quadratic functional φ appeared in the construction of a model of waves in the brain given by N. Wiener (1958). A characterisation of such functions is contained in the following proposition.

Proposition 4.10. *Suppose that the kernel $F(u, v)$ of the integral representation of a real-valued functional $\varphi(x) \in \mathcal{H}_2$ satisfies condition (4.85). Then:*
1. *If λ is an eigenvalue of F, so also is $-\lambda$.*
2. *The distribution of $\varphi(x)$ is symmetric about the origin with characteristic function*

$$\chi(z) = \delta(2iz, F)^{-1/2} = \prod_{\lambda_n > 0} (1 + 4z^2 \lambda_n^{-2})^{-1/2}, \qquad z \in \mathbb{R}, \tag{4.86}$$

where $\pm \lambda_n$, $n \geq 1$, are the eigenvalues of F.

3. *The cumulants of $\varphi(x)$ are given by*

$$\begin{cases} \gamma_{2p+1} = 0, \\ \gamma_{2p} = 2^{2p}(2p-1)! \sum_{\lambda_n > 0} \lambda_n^{-2p}, \quad p \geq 1. \end{cases} \quad (4.87)$$

The proof is straightforward and will be omitted.

We now return to consider a general real-valued functional $\varphi(x) \in \mathcal{H}_2$, for significant examples often arise in applications having the form $X(t, x) = \varphi(T_t x)$, $-\infty < t < \infty$, with $\varphi(x) \in \mathcal{H}_2$ and $\{T_t\}$ the flow of Brownian motion [see, for instance, N. Wiener (1958)]. As foreshadowed in Example 2 of §4.4, the use of the integral representation method will be shown to be a powerful tool in analysing such stochastic processes. In fact if the kernel $F(u, v)$ is associated with $\varphi(x)$, then $F(u - t, v - t)$, $-\infty < t < \infty$, can be used to describe the behaviour of $X(t, x)$ as t evolves. Incidentally, the distribution of a process $\{X(t): -\infty < t < \infty\}$ is known whenever that of $\{X(\xi, x): \xi \in \mathcal{S}\}$ is also given, where

$$X(\xi, x) = \int X(t, x)\xi(t)\, dt, \quad \xi \in \mathcal{S}.$$

In other words, it is enough for us to study the characteristic functional

$$C_X(\xi) = \int_{\mathcal{S}^*} \exp[iX(\xi, x)]\, d\mu(x). \quad (4.88)$$

The integrand is an exponential function of a quadratic functional, and such functionals will be discussed in (iii) below. However each random variable $X(\xi, x)$ can itself be considered, using the kernel

$$F_\xi(u, v) = \int_{-\infty}^{\infty} F(u - t, v - t)\xi(t)\, dt \quad (4.89)$$

to which Theorem 4.8 can be applied.

(ii) Stochastic Area

Let $\{B(t, \omega) = (B_1(t, \omega), B_2(t, \omega)): t \geq 0\}$ be a two-dimensional Brownian motion. P. Lévy wished to define the area $S(T, \omega)$ of the region enclosed by the Brownian curve in the time interval $[0, T]$ and the chord connecting the origin with the terminal point $B(T, \omega)$, and to this end he gave the integral

$$S(T, \omega) = \frac{1}{2} \int_0^T [B_1(t, \omega)\, dB_2(t, \omega) - B_2(t, \omega)\, dB_1(\omega)], \quad (4.90)$$

see P. Lévy (1940). If the curve $(B_1(t, \omega), B_2(t, \omega))$, $t \in [0, T]$, was sufficiently smooth, then $S(T, \omega)$ would express the area in the usual way. But it was

4.6 Examples of Applications

shown in §2.2 that the Brownian path is of unbounded variation and so this is unfortunately not possible. However the integral (4.90) does exist as a stochastic integral. Indeed the conditions (A.1), (A.2), (f.1), (f.2) and (f.3) given in §4.5 for the existence of the stochastic integral can be checked as follows. Firstly, let us consider the integral $\int_0^T B_1(t, \omega) \, dB_2(t, \omega)$. We set $A_t = \mathbf{B}((B_1(s), B_2(s)): s \leq t)$, $t \geq 0$, and readily check that for $\mathbf{B}_t = \mathbf{B}(B_2(s): s \leq t)$, the condition (A.1) $A_t \supseteq \mathbf{B}_t$ is obvious, whilst the requirement (A.2) that A_t be independent of the system $\{B_2(t + s) - B_2(t): s \geq 0\}$ comes directly from the definition of two-dimensional Brownian motion. The integrand $B_1(t, \omega)$ is measurable in (t, ω) jointly, and $B_1(t, \omega)$ is A_t-measurable for any t, and so the conditions (f.1) and (f.2) are satisfied. Thus the stochastic integral $\int_0^T B_1(t, \omega) \, dB_2(t, \omega)$ exists. If we recall the definition of A_t given above, and interchange the roles of $B_1(t)$ and $B_2(t)$ in the foregoing, then we see that $\int_0^T B_2(t, \omega) \, dB_1(t, \omega)$ can also be defined. We have therefore proved that the area $S(T, \omega)$ is well-defined as a stochastic integral, and the process $\{S(T, \omega): T \geq 0\}$ is called the *stochastic area*.

Notwithstanding the fact that $S(T, \omega)$ may not conform to any intuitive notion of area, we propose to study the stochastic process $\{S(T, \omega): T \geq 0\}$ given by (4.90). A concrete realisation of $\{B(t)\}$ may be obtained on the probability space $(\mathscr{S}^*, \mathfrak{B}, \mu)$ by setting

$$B_1(t, x) = \langle x, \chi_{[0, t]} \rangle, \quad \text{and}$$

$$B_2(t, x) = \langle x, \chi_{[-t, 0]} \rangle, \quad t \geq 0.$$

For then $B(t, x) = (B_1(t, x), B_2(t, x))$ is a two-dimensional Brownian motion. According to Lemma 4.1 of §4.5 the stochastic area $S(T, x)$ given by the stochastic integral (4.90) with respect to $B(t, x)$ belongs to $\mathscr{H}_2$. For the moment take $T = 1$. The kernel $F(u, v)$ of the integral representation of $S(1, x)$ can be illustrated by Fig. 8.

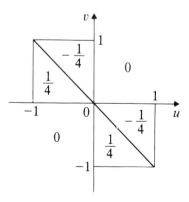

Figure 8

Namely,

$$F(u, v) = \begin{cases} -\dfrac{1}{4} & \text{if } uv \leq 0, \ u, v \leq 1, \ -v < u; \\ \dfrac{1}{4} & \text{if } uv \leq 0, \ u, v \geq -1, \ -v > u; \\ 0 & \text{otherwise.} \end{cases}$$

Theorem 4.9. *The characteristic function $\chi(z)$ and the cumulants of $S(1, x)$ may be written*

$$\chi(z) = \{\cosh(\tfrac{1}{2}z)\}^{-1}, \tag{4.92}$$

and

$$\begin{cases} \gamma_{2p+1} = 0, & p \geq 0, \\ \gamma_{2p} = (2^{2p} - 1)\dfrac{B_p}{4p}, & p \geq 1, \end{cases} \tag{4.93}$$

respectively, where B_p is the p-th Bernoulli number.

PROOF. The kernel $F(u, v)$ given by (4.91) satisfies condition (4.85) and its eigenvalues are $2(2n - 1)\pi$, $n = \pm 1, \pm 2, \ldots$, each of which has multiplicity two. We can thus use Proposition 4.10 to obtain

$$\chi(z) = \left[\prod_{n=1}^{\infty} \left(1 + \frac{4z^2}{4(2n-1)^2\pi^2}\right)^2\right]^{-1/2} = \left\{\cosh\left(\frac{1}{2}z\right)\right\}^{-1},$$

and the cumulants are easily expressible in the form (4.93). □

In addition the kernel of the integral representation of $S(T, x)$ may be written in terms of F as

$$F\left(\frac{u}{T}, \frac{v}{T}\right), \qquad T > 0.$$

The system $\{F(u/T, v/T): T > 0\}$ entirely describes the stochastic process $\{S(T, x): T > 0\}$, and in particular its behaviour in T is illustrated by Fig. 8.

(iii) Exponentials of Quadratic Functionals

We will begin with an exponential function $f(x)$ of a real-valued quadratic functional $\varphi(x)$ of the form $f(x) = \exp[i\varphi(x)]$, $\varphi(x) \in \mathscr{H}_2$. Clearly $f(x) \in (L^2)$, so that the integral representation theory may be applied. A good problem involving such an $f(x)$ was noted in (i) above, namely, the need to evaluate the characteristic function $C_X(\xi)$ of the process $\{X(t, x)\}$, see (4.88).

4.6 Examples of Applications

Let $F(u, v)$ be the kernel of the integral representation of $\varphi(x)$, and let the system of eigenvalues and eigenfunctions of the integral operator F be denoted by $\{\lambda_n : n \geq 1\}$ and $\{\eta_n : n \geq 1\}$, respectively.

Proposition 4.11. *The transformation $\mathcal{T}$ maps $f(x) = \exp[i\varphi(x)]$ to*

$$(\mathcal{T}f)(\xi) = C(\xi)\, \delta(2i; F)^{-1/2} \exp\left[i \sum_1^\infty \frac{(\eta_n, \xi)^2}{-\lambda_n + 2i} \right], \qquad (4.94)$$

where $(\cdot, \cdot)$ denotes the inner product in $L^2(\mathbf{R})$.

PROOF. By assumption F can be expanded in the form (4.81), and hence $\varphi(x)$ may be expressed as a sum (4.82) of independent random variables. Thus we may write

$$f(x) = \exp\left[i \sum_1^\infty \lambda_n^{-1}(\langle x, \eta_n \rangle^2 - 1) \right]. \qquad (4.95)$$

Now we use the fact that $\{\eta_n\}$ is an orthonormal system in $L^2(\mathbf{R})$, so that any ξ may be expressed in the form $\xi = \sum_n a_n \eta_n + \xi'$ where $a_n = (\xi, \eta_n)$ and ξ' is orthogonal to the $\{\eta_n\}$. We have the equality $\|\xi\|^2 = \sum_n a_n^2 + \|\xi'\|^2$ and this, together with (4.95), gives

$$(\mathcal{T}f)(\xi) = \int_{\mathcal{S}^*} \exp[i\langle x, \xi \rangle] f(x)\, d\mu(x)$$

$$= \int_{\mathcal{S}^*} \exp\left[i \sum_n a_n \langle x, \eta_n \rangle + i \sum_n \lambda_n^{-1}(\langle x, \eta_n \rangle^2 - 1) \right]$$

$$\times \exp[i\langle x, \xi' \rangle]\, d\mu(x).$$

Since $\{\langle x, \xi' \rangle, \langle x, \eta_n \rangle : n \geq 1\}$ is a system of independent random variables, the above integral turns out to be

$$= \int_{\mathcal{S}^*} \exp[i\langle x, \xi' \rangle]\, d\mu(x) \prod_n \int_{\mathcal{S}^*} \exp[ia_n \langle x, \eta_n \rangle + i\lambda_n^{-1}\langle x, \eta_n \rangle^2 - i\lambda_n^{-1}]\, d\mu(x)$$

$$= C(\xi') \prod_n \left\{ (2\pi)^{-1/2} \int_{-\infty}^\infty \exp[ia_n t + (i\lambda_n^{-1} - 2^{-1})t^2]\, dt \cdot \exp(-i\lambda_n^{-1}) \right\}$$

$$= C(\xi') \prod_n \left\{ (1 - 2i\lambda_n^{-1})^{-1/2} \exp(-i\lambda_n^{-1}) \cdot \exp\left[-\frac{1}{2} \frac{a_n^2}{1 - 2i\lambda_n^{-1}} \right] \right\}$$

$$= C(\xi')\, \delta(2i; F)^{-1/2} \exp\left[\sum_n \frac{-i\lambda_n^{-1} a_n^2}{1 - 2i\lambda_n^{-1}} \right] \exp\left[-\frac{1}{2} \sum_n a_n^2 \right]$$

$$= C(\xi)\, \delta(2i; F)^{-1/2} \exp\left[\sum_n \frac{-i\lambda_n^{-1} a_n^2}{1 - 2i\lambda_n^{-1}} \right].$$

This last formula is equal to (4.94). □

Given an $L^2(\mathbf{R}^2)$-kernel F by (4.81) we define a function $\tilde{F}$

$$\tilde{F}(u, v) = \sum_n \frac{i\lambda_n^{-1}}{-1 + 2i\lambda_n^{-1}} \eta_n(u)\eta_n(v). \tag{4.96}$$

The function $\tilde{F}$ is no longer real-valued, but it is an L^2-kernel and it shares the eigenfunctions $\{\eta_n\}$ with F. As an integral kernel $\tilde{F}$ may be characterised by the two conditions

a. F commutes with $\tilde{F}$, and they have the same range; and
b. $\tilde{F}(-I + 2iF) = iF$. $\hspace{2cm}$ (4.97)

Theorem 4.10. *Let $\varphi(x)$ be a real-valued functional in $\mathcal{H}_2$ and set $f(x) = \exp[i\varphi(x)]$. Also, denote the kernel of the integral representation of φ by F and define $\tilde{F}$ by (4.96). Then*

i. $(\mathcal{T}f)(\xi) = C(\xi) \, \delta(2i; F)^{-1/2} \exp[\iint \tilde{F}(u, v)\xi(u)\xi(v) \, du \, dv]$.
ii. *For the projection f_n of f onto $\mathcal{H}_n$ we have*

$$f_0(x) = \delta(2i; F)^{-1/2},$$

$$f_{2n+1}(x) = 0, \quad n \geq 0,$$

and the kernel of the integral representation of $f_{2n}(x)$ is given by

$$\delta(2i; F)^{-1/2}(-1)^n \frac{\hat{F}^{(2n)}}{n!}, \tag{4.98}$$

where $\hat{F}^{(2n)}$ is the symmetrization of $\tilde{F}^{n\otimes}$.

PROOF. i. is simply a rephrasing of Proposition 4.11 into the notation of (4.96).
  ii. The power series expansion of $(\mathcal{T}f)(\xi)$ is

$$(\mathcal{T}f)(\xi) = C(\xi) \, \delta(2i; F)^{-1/2} \sum_0^\infty \frac{1}{n!} \left[\iint \tilde{F}(u, v)\xi(u)\xi(v) \, du \, dv\right]^n.$$

Each term is a monomial in ξ of degree $2n$ and, by Theorem 4.5, must coincide with $(\mathcal{T}f_{2n})(\xi)$. In other words

$$(\mathcal{T}f_0)(\xi) = C(\xi) \, \delta(2i; F)^{-1/2}, \quad \text{and}$$

$$(\mathcal{T}f_{2n})(\xi) = \frac{1}{n!} C(\xi) \, \delta(2i; F)^{-1/2} (\tilde{F}^{n\otimes}, \xi^{2n\otimes})_{L^2(\mathbf{R}^{2n})}.$$

These relations imply (4.98) whilst the result $f_{2n+1}(x) = 0$ is obvious. $\square$

We will now similarly consider exponential functions of the form $g(x) = \exp[\varphi(x)]$ where, as before, $\varphi(x)$ is supposed to be a real-valued member of $\mathcal{H}_2$.

4.6 Examples of Applications

Proposition 4.12. *A necessary and sufficient condition for $g(x)$ of this form to belong to (L^2) is that the kernel of the integral representation of φ has no eigenvalue λ_n with $0 < \lambda_n \leq 4$. If this condition is satisfied then*

$$(\mathcal{T}g)(\xi) = C(\xi)\,\delta(2i;F)^{-1/2}\exp\left[\iint \tilde{G}(u,v)\xi(u)\xi(v)\,du\,dv\right], \quad (4.99)$$

where $\tilde{G}$ is the real-valued $\hat{L}^2(\mathbf{R}^2)$-function given by

$$\tilde{G}(u,v) = \sum_n (-\lambda_n + 2)^{-1}\eta_n(u)\eta_n(v). \quad (4.100)$$

Moreover, if $g_n(x)$ denotes the projection of $g(x)$ onto the space $\mathcal{H}_n$, then

$$g_0(x) = \delta(2;F)^{-1/2}$$

$$g_{2n+1}(x) = 0, \quad n \geq 0,$$

and

$$(\mathcal{T}g_{2n})(\xi) = \frac{1}{n!}C(\xi)\,\delta(2;F)^{-1/2}(\tilde{G}^{n\otimes},\xi^{2n\otimes})_{L^2(\mathbf{R}^{2n})}. \quad (4.101)$$

PROOF. The computation below is straightforward:

$$\int_{\mathcal{S}^*} g(x)^2\,d\mu(x) = \int_{\mathcal{S}^*}\exp\left[2\sum_n \lambda_n^{-1}(\langle x,\eta_n\rangle^2 - 1)\right]d\mu(x)$$

$$= \prod_n\left\{(2\pi)^{-1/2}\int_{-\infty}^{\infty}\exp[2\lambda_n^{-1}(x^2-1)]\exp[-\tfrac{1}{2}x^2]\,dx\right\}.$$

A necessary and sufficient condition for the integrals inside the braces $\{\cdot\}$ to exist is that $\lambda_n^{-1} < \tfrac{1}{4}$ for all n. If this inequality holds for every n, the above product of the integrals turns out to be $\prod_n \{(1 - 4\lambda_n^{-1})^{-1/2} \times \exp[-2\lambda_n^{-1}]\}$, which converges, and coincides with $\delta(4;F)$.

The proof of the second half of the proposition, including that of the relation (4.99) and the expression for $g_n(x)$, is similar to that of Theorem 4.10 and so is omitted. □

There is an interesting example in which a $g(x)$ such as the above arises quite naturally. In the problem of signal detection in electrical engineering, one uses the (Radon-Nikodym) density function of a Gaussian measure μ^X with respect to Wiener measure μ^W, where we assume that μ^X and μ^W are equivalent. Let $\{B(t,\omega): t \in [0,1]\}$ be a Brownian motion, and let $F_1(s,u)$ be an L^2-kernel of Volterra type (i.e. $F_1(s,u) = 0$ when $s < u$). If we define

$$X(t,\omega) = B(t,\omega) - \int_0^t \left\{\int_0^s F_1(s,u)\,dB(u,\omega)\right\}ds, \quad (4.102)$$

then $\{X(t): t \in [0,1]\}$ is a Gaussian process. Furthermore the distribution μ^X of $\{X(t)\}$ is known to be equivalent to Wiener measure μ^W; we refer to M.

Hitsuda (1968) and §5.7 for further discussion of this topic. The density function $\rho(\omega) = (d\mu^x/d\mu^W)(\omega)$ is an exponential function of a quadratic functional of Brownian motion, and in terms of stochastic integrals may be written in the form:

$$\rho(\omega) = \exp\left[\int_0^1 \int_0^s F_1(s, u) \, dB(u, \omega) \, dB(s, \omega)\right.$$
$$\left. - \frac{1}{2} \int_0^1 \left\{\int_0^s F_1(s, u) \, dB(u, \omega)\right\}^2 ds\right]. \quad (4.103)$$

We now proceed as before, realising $B(t)$, $X(t)$ and ρ as functionals on $(\mathscr{S}^*, \mathfrak{B}, \mu)$ so that our theory may be applied. Firstly, $B(t, \omega)$ is realised by $\langle x, \chi_{[0, t]}\rangle$, and $X(t, \omega)$ by $\langle x, \chi_{[0, t]}\rangle - \langle x, \int_0^t F_1(s, \cdot) \, ds\rangle$ in $\mathscr{H}_1$. Since the first term of the exponent of ρ in (4.103) is a stochastic integral, in fact an element of $\mathscr{H}_2$, its integral representation follows from results in §4.5. Also the term $\{\cdot\}^2$ is nothing but

$$2H_2\left(\int_0^s F_1(s, u) \, dB(u); \int_0^s F_1(s, u)^2 \, du\right) + \int_0^s F_1(s, u)^2 \, du.$$

Thus, again from §4.5, the sum of the exponent $[\cdot]$ in (4.103) and $\frac{1}{2}\int_0^1 \int_0^s F_1(s, u)^2 \, du \, ds$ is an element of $\mathscr{G}_2$, and the kernel $G(u, v)$ of its integral representation may be expressed in the form

$$G(u, v) = \frac{1}{2}\left\{F(u, v) - \int_{u \vee v}^1 F(u, s)F(s, v) \, ds\right\}, \quad (4.104)$$

where $F(u, v) = F_1(u \vee v, u \wedge v)$, $(u, v) \in [0, 1]^2$. Thus we have an expression $g(x)$ for $\rho(\omega)$ in the form

$$g(x) = \exp[\psi(x)], \quad \psi(x) = \psi_0 + \psi_2(x), \quad (4.105)$$

where $\psi_2(x)$ is the $\mathscr{H}_2$-functional with kernel $F(u, v)$ and ψ_0 is the constant $-\frac{1}{2}\int_0^1 \{\int_0^s F_1(s, u)^2 \, du\} \, ds$.

The following remark concerning the eigenvalues of the kernel G is worth making. If we denote by F_1^* the adjoint of the integral operator F_1, then F may be written as $F_1 + F_1^*$ and so $G = \frac{1}{2}\{F_1 + F_1^* - F_1^* F_1\}$. Since $I - 2G = (I - F_1^*)(I - F_1) > 0$, the eigenvalues of G can never be in $(0, 2]$, and thus by Proposition 4.12, a necessary and sufficient condition for $g(x)$ to belong to (L^2) is that there is no eigenvalues of G in the interval $(2, 4]$. If this condition is fulfilled, then the projection $g_n(x)$ of $\exp[\psi_2(x)]$ onto $\mathscr{H}_n$ can be defined, and is obtained via (4.101). Further properties may be obtained by studying the kernel $G^{n\otimes}$. In the particular case where $n = 0$ we have $\exp(\psi_0) = \delta(2; G)^{1/2}$ and we see that the expectation of $g(x)$ in (4.105), or of $\rho(\omega)$ in (4.103) is unity.

(iv) Stochastic Differential Equations

We will now consider the case in which a stochastic process $\{X(t, \omega): t \in [0, T]\}$ is defined by using a stochastic and an ordinary integral in the form

$$X(t, \omega) = X_0(\omega) + \int_0^t a(u, X(u, \omega)) \, du + \int_0^t b(u, X(u, \omega)) \, dB(u, \omega),$$

$$0 \le t \le T \quad (4.106)$$

The initial value $X_0(\omega)$ is taken to be a random variable independent of $\mathbf{B}_t = \mathbf{B}(B(u, \omega): u \in [0, t])$. If a solution $X(t, \omega)$ exists, then it is almost surely continuous in t. Our problem to obtain a solution $X(t, \omega)$, $t \in [0, T]$, to equation (4.106) such that $X(t, \omega)$ is $\mathbf{A}_t$-measurable for each t, where $\mathbf{A}_t = \mathbf{B}(X_0, B(u): 0 \le u \le t)$. To this end we make the following assumptions on a and b:

1. for every x, $a(u, x)$ and $b(u, x)$ are continuous in u; and
2. the Lipschitz condition: there exist constants K_1, K_2 such that for all $u \in [0, T]$ we have

$$|a(u, x) - a(u, y)| \le K_1 |x - y|, \quad \text{and} \quad |b(u, x) - b(u, y)| \le K_2 |x - y|.$$

Under these assumptions, a unique solution $\{X(t): t \in [0, T]\}$ exists to (4.106), and it is a Markov process. An alternative description of the equation (4.106), see K. Itô and H. P. McKean Jr. (1965), or H. P. McKean Jr. (1969), is in the form of a stochastic differential equation:

$$\begin{cases} dX(t, \omega) = a(t, X(t, \omega)) \, dt + b(t, X(t, \omega)) \, dB(t, \omega), & 0 \le t \le T, \\ X(0, \omega) = X_0(\omega). \end{cases} \quad (4.107)$$

In this subsection we use the integral representation theory to obtain a stochastic process on the probability space $(\mathscr{S}^*, \mathfrak{B}, \mu)$ which is a concrete solution to (4.106) or (4.107). As we saw in §4.4, this method works well with stochastic processes which may be realised in (L^2), and we continue the notations used there for Brownian motion $B(t, x) = \langle x, \chi_{[t \wedge 0, \, t \vee 0]} \rangle$, the σ-fields $\mathfrak{B}_t$, the flow $\{T_t\}$ of Brownian motion, the unitary group $\{U_t\}$, the subspace $\{L_t^2\}$, and so on. With these notational preliminaries we may describe our position as follows:

$$\begin{cases} X(t) \text{ may be expressed as } X(t) = U_t \varphi \text{ for some } \varphi \in L_0^2; \\ a \text{ and } b \text{ in (4.107) are independent of } t \text{ and } a(X(0)), b(X(0)) \in (L^2), \end{cases} \quad (4.108)$$

(see §A.4 concerning N. Wiener's non-linear circuit theory). From these assumptions we see that both $a(X(0))$ and $b(X(0))$ belong to L_0^2 and further, $a(X(t)) = U_t a(X(0))$ and $b(X(t)) = U_t b(X(0))$, so that these both belong to

L_t^2. Thus if $b_n(X(t))$ is the projection of $b(X(t))$ onto $\mathcal{H}_n$, then $b_n(X(t)) = U_t b_n(X(0))$. Letting $F_n(u_1, u_2, \ldots, u_n)$ be the kernel of the integral representation of $b_n(X(0))$, Proposition 4.5 of §4.4 tells us that $F_n(u_1, u_2, \ldots, u_n) = 0$ if $u_j > 0$ for some j, and Proposition 4.6 that the kernel for $b_n(X(t))$ is $F_n(u_1 - t, u_2 - t, \ldots, u_n - t)$. Also we can associate $dB(t)$ with $\chi_{dt}(u - t)$, where χ_{dt} denotes the indicator function of $[0, dt]$. Furthermore, the proof of Lemma 4.1 of §4.5 shows that the kernel of the integral representation of $b_n(X(t)) dB(t)$ ($\in \mathcal{H}_{n+1}$) may be obtained by symmetrising the product $\chi_{dt}(u_1 - t) F_n(u_2 - t, u_3 - t, \ldots, u_{n+1} - t), u_1 \geq u_2 \geq \cdots \geq u_{n+1}$. We proceed similarly for $a(X(t))$. As for $a_n(X(t))$, the projection $a_n(X(t))$ of $a(X(t))$ onto $\mathcal{H}_n$ satisfies $a_n(X(t)) = U_t a_n(X(0))$, and $a_n(X(t)) dt$ is an element of $\mathcal{H}_n$ the kernel of whose integral representation is obtained simply by multiplying that of $a_n(X(t))$ by dt. Thus the stochastic differential equation (4.107) has been expressed in terms of a system of functional equations in $\hat{L}^2(\mathbf{R}^n)$, $n = 0$, $1, 2, \ldots$.

We are now going to present the explicit solutions to three stochastic differential equations using the method just explained.

EXAMPLE 1. This is the so-called *Langevin equation*

$$dX(t) = -\lambda X(t) dt + dB(t), \qquad \lambda > 0, \tag{4.109}$$

where the time interval is taken to be $(-\infty, \infty)$. Our result can be extended to the cases of other time intervals with some obvious modifications. It is quite easy to check that the coefficient functions in this equation satisfy all the conditions necessary for the existence and uniqueness of a (purely non-deterministic) solution. We now give an explicit formula for $X(t)$ satisfying (4.109). Denote the projection of $X(t)$ onto $\mathcal{H}_n$ by $X_n(t)$, and let the kernel of the integral representation of $X_n(0)$ be $F_n(u_1, u_2, \ldots, u_n)$. Then the system $\{d_t F_n(u_1 - t, u_2 - t, \ldots, u_n - t): n \geq 0\}$ is associated with $dX(t)$, whilst the system $\{-\lambda F_n(u_1 - t, u_2 - t, \ldots, u_n - t) dt : n \neq 1\}$ and $-\lambda F_1(u_1 - t) dt + \chi_{dt}(u_1 - t)$ ($n = 1$) is associated with $-\lambda X(t) dt + dB(t)$. In terms of these kernels, (4.109) can be expressed

$$d_t F_0 = -\lambda F_0 dt;$$

$$d_t F_1(u_1 - t) = -\lambda F_1(u_1 - t) dt + \chi_{dt}(u_1 - t);$$

$$d_t F_n(u_1 - t, u_2 - t, \ldots, u_n - t) = -\lambda F_n(u_1 - t, u_2 - t, \ldots, u_n - t) dt,$$

$$n = 2, 3, \ldots . \tag{4.110}$$

Since $F_n \in \hat{L}^2(\mathbf{R}^n)$ the only possible solution to (4.110) is the following:

$$\begin{cases} F_n(u_1, u_2, \ldots, u_n) = 0, & n \neq 1, \\ F_1(u_1) = e^{\lambda u_1} \chi_{(-\infty, 0]}(u_1), \end{cases} \tag{4.111}$$

4.6 Examples of Applications

and so the solution $X(t)$ to (4.109) is given by

$$X(t, x) = e^{-\lambda t} \int_{-\infty}^{t} e^{\lambda u} \, dB(u, x). \tag{4.112}$$

This is the particular case of Example 1 in §4.4 in which $\{X(t): -\infty < t < \infty\}$ is a stationary Gaussian process with mean zero and covariance function $(2\lambda)^{-1} \exp[-\lambda|t|]$. The process is called the *Ornstein-Uhlenbeck process*, or the Ornstein-Uhlenbeck Brownian motion. We note in passing that the representation (4.112) is canonical in the sense of Remark 2, §2.6 (iii).

EXAMPLE 2. As above the time interval is taken to be $(-\infty, \infty)$. Under the assumption (4.108) we will obtain the solution of the stochastic differential equation

$$dX(t) = (aX(t) + a') \, dt + (bX(t) + b') \, dB(t). \tag{4.113}$$

Letting $X_n(0)$ be the projection of $X(0)$ onto $\mathscr{H}_n$, and denoting its associated kernel by $F_n \in \hat{L}^2(\mathbf{R}^n)$, the sort of argument used in Example 1 allows us to replace equation (4.113) by the system

$$d_t F_0 = aF_0 \, dt + a'$$

$$d_t F_n(u_1 - t, u_2 - t, \ldots, u_n - t)$$
$$= \{aF_n(u_1 - t, u_2 - t, \ldots, u_n - t) + a'\} \, dt$$
$$+ \chi_{dt}(u_1 - t)\{bF_{n-1}(u_2 - t, \ldots, u_n - t) + b' \, \delta_{n,1}\},$$
$$n \geq 1, \quad t \geq u_1 \geq u_2 \geq \cdots \geq u_n. \tag{4.114}$$

Assuming now that $E(X(t)) = 0$, we obtain $F_0 = 0$ and hence $a' = 0$. For the case $n = 1$ (4.114) can be solved explicitly as in Example 1 to yield

$$F_1(u_1) = b'\{\exp[-au_1]\}\chi_{(-\infty, 0]}(u_1).$$

Moreover the condition $F_1 \in L^2(\mathbf{R})$ forces $a < 0$. With this F_1 we can move on to the case $n = 2$ in (4.114). For $t > 0$ we get the equation

$$F_2(u_1 - t, u_2 - t) - F_2(u_1, u_2)$$
$$= a \int_{u_1 \vee 0}^{t} F_2(u_1 - s, u_2 - s) \, ds + bb' \int_{u_2 \vee 0}^{t} \exp[-a(u_2 - s)]\chi_{ds}(u_1 - s).$$

This equation proves that F_2 is independent of u_1, and the solution to it is

$$F_2(u_1, u_2) = bb' \exp(-au_2)\chi_{(-\infty, 0]}(u_1), \quad u_1 \geq u_2.$$

Using equation (4.114) we can obtain F_n when F_{n-1} is known and so, by induction, we may form all the $F_n(u_1, u_2, \ldots, u_n)$ with $u_1 \geq u_2 \geq \cdots \geq u_n$, $n \geq 1$. After symmetrisation we obtain

$$F_n(u_1, u_2, \ldots, u_n) = (n!)^{-1} b^{n-1} b' \exp\left[-a \min_{1 \leq j \leq n} u_j\right] \chi_{(-\infty, 0]^n}(u_1, u_2, \ldots, u_n),$$

$$n \geq 1, \quad (4.115)$$

and the process itself may be expressed in the form

$$X(t) = \sum_1^\infty X_n(t), \qquad (4.116)$$

where $X_n(t) = U_t X_n(0)$, and the kernel associated with $X_n(0)$ is F_n given by (4.115) with $b^2 < -2a$.

Remark. We can obtain an explicit expression for the non-linear predictor with this example. To this end we first prove that $\mathfrak{B}_t(X) = \mathfrak{B}(X(s): s \leq t)$ coincides with $\mathfrak{B}_t$, and therefore when $\{X(s): s \leq 0\}$ is known, the optimal (non-linear) predictor $E(X(t)|\mathfrak{B}_0(X)) = E(X(t)|\mathfrak{B}_0) = \hat{X}(t)$ say, of $X(t)$, $t > 0$, can be obtained from the integral representation. Indeed if $\hat{X}_n(t)$ denotes the projection of $\hat{X}(t)$ to $\mathscr{H}_n$, then by (4.115) kernel of its integral representation is of the form

$$F_n(u_1 - t, u_2 - t, \ldots, u_n - t)\chi_{(-\infty, 0]^n}(u_1, u_2, \ldots, u_n)$$

$$= (n!)^{-1} b^{n-1} b' \exp(at)\exp\left[-a \min_{1 \leq j \leq n} u_j\right] \chi_{(-\infty, 0]^n}(u_1 - t, u_2 - t, \ldots, u_n - t)$$

$$\times \chi_{(-\infty, 0]^n}(u_1, u_2, \ldots, u_n)$$

$$= (n!)^{-1} b^{n-1} b' \exp(at)\exp\left[-a \min_{1 \leq j \leq n} u_j\right] \chi_{(-\infty, 0]^n}(u_1, u_2, \ldots, u_n).$$

There is an example of a stochastic differential equation in which the solution $X(t)$ is a $\mathfrak{B}_t$-measurable (L^2)-function not of the form $U_t \varphi$, but the integral representation theory can still be used effectively to obtain the solution.

EXAMPLE 3. The solution to the stochastic differential equation

$$\begin{cases} dX(t) = f(t)X(t) \, dB(t), \\ X(0) = 1, \end{cases} \qquad (4.117)$$

where $f \in \mathscr{C}([0, \infty))$ is given by $X(t) = \exp[\int_0^t f(u) \, dB(u) - \frac{1}{2}\|f\|_t^2]$. We will show that this solution can be obtained by solving the system of equations in terms of kernels equivalent to (4.117). As before $X_n(t)$ denotes the projection of $X(t)$ onto $\mathscr{H}_n$, and $F_n(u_1, u_2, \ldots, u_n; t)$ the associated kernel. Since $X(t)$ is

$\mathfrak{B}_t$-measurable by assumption, so also is $X_n(t)$ and hence $F_n(u_1, u_2, \ldots, u_n; t)$ vanishes off $[0, t]^n$.

We are now ready to form the solution. The first equation of (4.117) gives $dX_0(t) = 0$ for $X_0(t)$, and this together with the initial conditions, gives $X_0(t) \equiv 1$. When $n \geq 1$ and we project (4.117) onto $\mathscr{H}_n$, we obtain

$$dX_n(t) = f(t)X_{n-1}(t)\, dB(t), \tag{4.118}$$

and in terms of kernels this turns out to be

$$d_t F_n(u_1, u_2, \ldots, u_n; t) = f(t)\chi_{dt}(u_1 - t)F_{n-1}(u_2, \ldots, u_n; t),$$
$$u_1 \geq u_2 \geq \cdots \geq u_n. \tag{4.119}$$

Since $F_n(u_1, u_2, \ldots, u_n; 0)$ is clearly 0, if we integrate (4.119) over $[0, t]$ we obtain

$$F_n = n^{-1} F_{n-1} \triangleright f, \quad n \geq 1,$$

where $\triangleright$ is the product introduced in §4.5. Hence $F_n(u_1, u_2, \ldots, u_n; t)$ coincides with $(n!)^{-1}(f^{n\otimes} \cdot \chi_{[0,\,t]^n})(u_1, u_2, \ldots, u_n)$, and so by Proposition 4.9 in §4.5, $X_n(t) = H_n(\int_0^t f(u)\, dB(u); \|f\|_t^2)$. Thus we have proved that $X(t) = \sum_n X_n(t) = \sum_n H_n(\int_0^t f(u)\, dB(u); \|f\|_t^2)$ is given by

$$X(t) = \exp\left[\int_0^t f(u)\, dB(u) - \frac{1}{2}\|f\|_t^2\right].$$

We note that equation (4.117) is the particular case of equation (4.107) in which $a(t, x) = f(t)x$ and $b(t, x) = 0$, so that the existence and uniqueness of the solution to (4.117) was guaranteed. The process $X(t)$ given above is that solution, and no other can be found.

The reader might have noticed that the above example is simply a paraphrasing of Proposition 4.9 of the previous section. The process $Y(t)$ given by (4.77) there satisfies equation (4.78), which is equivalent to (4.117), and as the continuity of $f(t)$ was supposed there, the uniqueness of the solution was assured. Thus the construction of such a process $Y(t)$ was really solving the equation under discussion.

4.7 The Fourier-Wiener Transform

We turn now to the search for a linear transformation on $(L^2) = L^2(\mathscr{S}^*, \mathfrak{B}, \mu)$ which would be an infinite-dimensional analogue of the Fourier transform on $L^2(\mathbf{R}^n)$. Recall the transformation $\mathscr{T}$ introduced in §4.3 and defined at $\varphi(x) \in (L^2)$ to be

$$(\mathscr{T}\varphi)(\xi) = \int_{\mathscr{S}^*} e^{i\langle x,\, \xi\rangle} \varphi(x)\, d\mu(x), \quad \xi \in \mathscr{S}. \tag{4.120}$$

Although similar in form to the Fourier transform, the transformation $\mathscr{T}$ maps (L^2) to a space of functionals on $\mathscr{S}$, in fact the reproducing kernel Hilbert space with reproducing kernel $C(\xi - \eta)$, $(\xi, \eta) \in \mathscr{S} \times \mathscr{S}$, where $C(\xi)$ is the characteristic functional of μ. Therefore $\mathscr{T}$ cannot be a unitary operator.

There is another transformation on (L^2) introduced by Cameron and Martin (1945), (1947a) which is different from $\mathscr{T}$ and more like the Fourier transform. The present section is devoted to this transformation, now known to be significant because of its rôle in the harmonic analysis on $(\mathscr{S}^*, \mathfrak{B}, \mu)$.

For $\varphi(x) \in (L^2)$ we take the following two steps:

i. $\sqrt{2}$-complexification of the variable: $\varphi(x) \to \varphi(\sqrt{2}\,x + iy)$, $y \in \mathscr{S}^*$;
ii. integration with respect to $d\mu$

$$(\mathfrak{J}\varphi)(y) = \int_{\mathscr{S}^*} \varphi(\sqrt{2}\,x + iy)\, d\mu(x). \tag{4.121}$$

We now have to show that the transformation $\mathfrak{J}$ of $\varphi(x)$ to the functional of y given by (4.121) is well-defined and, as usual, we begin by applying $\mathfrak{J}$ to the basic functionals of §4.1.

Proposition 4.13. *The transformation $\mathfrak{J}$ is well-defined for the classes of polynomials and exponential functionals on $\mathscr{S}^*$, respectively, and each class is mapped into itself by $\mathfrak{J}$.*

PROOF. Let $\varphi(x)$ be a Fourier-Hermite polynomial based on the sequence $\{\xi_k\}$ and expressed in the form

$$\varphi(x) = \prod_k H_{n_k}\!\left(\frac{\langle x, \xi_k \rangle}{\sqrt{2}}\right). \tag{4.122}$$

The first step, the $\sqrt{2}$-complexification, gives us

$$\varphi(\sqrt{2}\,x + iy) = \prod_k H_{n_k}\!\left(\langle x, \xi_k \rangle + \frac{i\langle y, \xi_k \rangle}{\sqrt{2}}\right),$$

and since the $\langle x, \xi_k \rangle$ are mutually independent random variables on $(\mathscr{S}^*, \mathfrak{B}, \mu)$, we have

$$\int_{\mathscr{S}^*} \varphi(\sqrt{2}\,x + iy)\, d\mu(x) = \prod_k \int_{\mathscr{S}^*} H_{n_k}\!\left(\langle x, \xi_k \rangle + \frac{i\langle y, \xi_k \rangle}{\sqrt{2}}\right) d\mu(x).$$

Making use of the addition formula for H_n, §A.5 formula (A.21), with $a = \sqrt{2}$, each factor on the right-hand side can be expanded to

$$\prod_k \sum_{j=0}^{n_k} \binom{n_k}{j} 2^{(n_k - j)/2} i^j H_j\!\left(\frac{\langle y, \xi_k \rangle}{\sqrt{2}}\right) \int_{\mathscr{S}^*} H_{n_k - j}\!\left(\frac{\langle x, \xi_k \rangle}{\sqrt{2}}\right) d\mu(x).$$

4.7 The Fourier-Wiener Transform

Since each integral vanishes except when $j = n_k$, this sum is simply $i^{n_k} H_{n_k}(\langle y, \xi_k \rangle / \sqrt{2})$ and we arrive at the conclusion: for φ given by (4.122)

$$\int_{\mathscr{S}^*} \varphi(\sqrt{2}\,x + iy)\, d\mu(x) = \prod_k i^{n_k} H_{n_k}\left(\frac{\langle y, \xi_k \rangle}{\sqrt{2}}\right).$$

Thus the transformation $\mathfrak{J}$ is indeed well-defined for a Fourier-Hermite polynomial φ, and if φ is given by (4.122) then we have

$$(\mathfrak{J}\varphi)(y) = i^n \varphi(y), \qquad n = \sum_k n_k. \tag{4.123}$$

Furthermore $\mathfrak{J}$ may be defined for general polynomials in $\langle x, \xi_k \rangle$, for such polynomials are expressible as linear combinations of Fourier-Hermite polynomials based on $\{\xi_k\}$, and the results will also be polynomials.

We now consider exponential functionals, say

$$\varphi(x) = \exp[a\langle x, \xi \rangle], \qquad a \in \mathbf{C}, \qquad \|\xi\| = 1. \tag{4.124}$$

Carrying out steps (i) and (ii) in this case we obtain

$$\varphi(\sqrt{2}\,x + iy) = \exp[ia\langle y, \xi \rangle]\exp[\sqrt{2}\,a\langle x, \xi \rangle]$$

and

$$\int_{\mathscr{S}^*} \varphi(\sqrt{2}\,x + iy)\, d\mu(x) = \exp[ia\langle y, \xi \rangle + a^2],$$

which shows that

$$(\mathfrak{J}\varphi)(y) = \exp[ia\langle y, \xi \rangle + a^2]. \tag{4.125}$$

Thus the proposition is completely proved. □

Theorem 4.11. *The transformation $\mathfrak{J}$ defined above for polynomials and exponential functions can be uniquely extended to a unitary operator on (L^2). For $\varphi(x) \in \mathscr{H}_n$ we have*

$$(\mathfrak{J}\varphi)(x) = i^n \varphi(y). \tag{4.126}$$

PROOF. Recall that for any complete orthonormal system $\{\xi_k\} \subseteq \mathscr{S}$ of $L^2(\mathbf{R})$, the collection of all Fourier-Hermite polynomials of degree n based upon $\{\xi_k\}$ forms a basis for $\mathscr{H}_n$. By using the fact that for any $\varphi(x)$ of the form (4.122) the relation (4.123) holds, we can easily see that $\mathfrak{J}$ is defined for all Fourier-Hermite polynomials of degree n, and that the relation (4.126) is valid. Consequently we have the isometry $\|\mathfrak{J}\varphi\| = \|\varphi\|$ for any orthonormal base, and we see that $\mathfrak{J}$ can be extended to a unitary operator on $\mathscr{H}_n$ satisfying (4.126).

If we wished to start with $\mathfrak{J}$ defined for exponential functions, we could obtain the above result for polynomials. For if we multiply any $\varphi(x)$ of the form (4.124) by $\exp(-\frac{1}{2}a^2)$, we obtain the generating function of the Hermite polynomials $H_n(\langle x, \xi\rangle/\sqrt{2})$, $n \geq 0$ [§A.5 (i)]. Then we may apply the transformation $\mathfrak{J}$, and the known result for exponentials gives the action of $\mathfrak{J}$ on polynomials. Of course it turns out to be the same as that obtained above starting from Hermite polynomials, and we can then extend the definition of $\mathfrak{J}$ to the whole space (L^2). Thus the theorem is proved. □

Definition 4.5. The unitary operator $\mathfrak{J}$ defined on (L^2) by Theorem 4.11 is called the *Fourier-Wiener transform*.

Just like the Fourier transform on $L^2(\mathbf{R}^n)$, the Fourier-Wiener transform satisfies:

$$\mathfrak{J}^4 = I, \quad (I = \text{the identity operator}). \tag{4.127}$$

This comes directly from (4.126), and another property which follows easily from the theorem is:

$$P_n \mathfrak{J} = \mathfrak{J} P_n, \quad n \geq 0, \tag{4.128}$$

where P_n is the projection onto the subspace $\mathcal{H}_n$.

Remark. With the usual notion of Fourier transform in mind, the expression (4.121) defining $\mathfrak{J}$ seems to be rather strange. To help with this and to give an intuitive interpretation to $\mathfrak{J}$ we note a fact and make an observation below.

For any complex number a we may define an operator $\mathfrak{J}_a$ on (L^2) by writing

$$(\mathfrak{J}_a \varphi)(y) = \int_{\mathscr{S}^*} \varphi(ax + iy) \, d\mu(x), \quad \varphi \in (L^2). \tag{4.129}$$

This operator is well-defined and it is interesting to note that $\mathfrak{J}_a$ is unitary if and only if $a = \sqrt{2}$. When $a = 1$ we refer to $\mathfrak{J}_1$ as the *Gauss transform*, and a monomial in $x \in \mathscr{S}^*$ is transformed by $\mathfrak{J}_1$ into a Fourier-Hermite polynomial. The details of this are omitted here, but we refer to formula (A.30) of §A.5.

Observation. Let $\{\xi_k\}$ be a complete orthonormal system in $L^2(\mathbf{R})$ and let φ be a function of finitely many of the $\langle x, \xi_k\rangle$, say $\langle x, \xi_1\rangle, \ldots, \langle x, \xi_n\rangle$. Such a function will be called a "tame" function, and there exists a function f on $\mathbf{R}^n$ for which

$$\varphi(x) = f(\langle x, \xi_1\rangle, \ldots, \langle x, \xi_n\rangle). \tag{4.130}$$

The condition $\varphi \in (L^2)$ is equivalent to the one on f requiring that f be an element of $L^2(\mathbf{R}^n, \exp(-\frac{1}{2}\sum_j t_j^2) \, dt_1 \cdots dt_n)$. If we rewrite this as

$$f(\sqrt{2}\,t_1, \ldots, \sqrt{2}\,t_n) \cdot \exp\left[-\frac{1}{2}\sum_1^n t_j^2\right] \in L^2(\mathbf{R}^n),$$

4.7 The Fourier-Wiener Transform

we can proceed to some formal computations:

$$(\mathfrak{J}\varphi)(y) = \int_{\mathscr{S}^*} f(\sqrt{2}\langle x, \xi_1\rangle + i\langle y, \xi_1\rangle, \ldots, \sqrt{2}\langle x, \xi_n\rangle + i\langle y, \xi_n\rangle)\, d\mu(x)$$

$$= (2\pi)^{-n/2} \int_{\mathbf{R}^n} \cdots \int f(\sqrt{2}\,t_1 + is_1, \ldots, \sqrt{2}\,t_n + is_n)$$

$$\times \exp\left[-\frac{1}{2}\sum_1^n t_j^2\right] dt_1 \cdots dt_n,$$

$$= (2\pi)^{-n/2} \int_{\mathbf{R}^n} \cdots \int \exp\left[i\sum_1^n u_j \frac{s_j}{\sqrt{2}}\right] f(\sqrt{2}\,u_1, \ldots, \sqrt{2}\,u_n)$$

$$\times \exp\left[-\frac{1}{2}\sum_1^n u_j^2\right] \cdot \exp\left[\frac{1}{2}\sum_1^n \left(\frac{s_j}{\sqrt{2}}\right)^2\right] du_1 \cdots du_n,$$

where $s_j = \langle y, \xi_j\rangle$ and $u_j = t_j + is_j/\sqrt{2}$, $j = 1, 2, \ldots, n$. The idea behind these computations is to show that if for a functional φ given by (4.130), we form the $L^2(\mathbf{R}^n)$-function $f(\sqrt{2}\,t_1, \ldots, \sqrt{2}\,t_n) \cdot \exp[-\frac{1}{2}\sum_1^n t_j^2]$, take the usual Fourier transform to obtain $g(v_1, \ldots, v_n)$, multiply this g by $\exp[\frac{1}{2}\sum_1^n v_j^2]$ and then replace v_j by $\langle y, \xi_j\rangle/\sqrt{2}$, we obtain the Fourier-Wiener transform of $\varphi(x)$.

The above calculations were, of course, done in a purely formal way. For example, f should have been specified in such a way that $f(\sqrt{2}\,t_1 + is_1, \ldots, \sqrt{2}\,t_n + is_n)$ is well-defined, we should give a more precise meaning to the integration in du_j where u_j is a complex variable, and so on. These points aside, it was hoped that the above formal computations give some idea why the transformation (4.121) was called a Fourier transform.

Let $\{T_t: -\infty < t < \infty\}$ be the "flow" of Brownian motion (see §4.4), and let $\{U_t: -\infty < t < \infty\}$ be the one-parameter unitary group on (L^2) defined via $\{T_t\}$ by $(U_t\varphi)(x) = \varphi(T_t x)$, $-\infty < t < \infty$, $\varphi \in (L^2)$.

Proposition 4.14. *The Fourier-Wiener transform $\mathfrak{J}$ commutes with $\{U_t\}$:*

$$\mathfrak{J}U_t = U_t\mathfrak{J}, \quad -\infty < t < \infty. \tag{4.131}$$

PROOF. For $\varphi(x) \in (L^2)$ the relation (4.131) follows from the computations:

$$(\mathfrak{J}U_t\varphi)(y) = \int_{\mathscr{S}^*} \varphi(T_t(\sqrt{2}\,x + iy))\, d\mu(x)$$

$$= \int_{\mathscr{S}^*} \varphi(\sqrt{2}\,x' + iT_t y)\, d\mu(x'), \quad \text{since } \mu \text{ is } T_t\text{-invariant,}$$

$$= (\mathfrak{J}\varphi)(T_t y) = U_t(\mathfrak{J}\varphi)(y). \qquad \square$$

We will find further significant properties similar to these enjoyed by the usual Fourier transform. For example the operators consisting of differentiation and of multiplication by the independent variable respectively, are interchanged by $\mathfrak{J}$. Properties such as these will be discussed in §5.7 below when we consider harmonic analysis (on an infinite-dimensional space!) in the context of the rotation group.

The Rotation Group 5

5.1 Transformations of White Noise (I): Rotations

Once again we begin with the triple $E \subseteq L^2(\mathbf{R}) \subseteq E^*$, where as in §3.1, E is a real nuclear space with dual space E^*, and the two spaces are linked by the canonical bilinear form $\langle x, \xi \rangle$, $x \in E^*$, $\xi \in E$.

The measure μ of white noise can be introduced into the space E^* via the characteristic functional $C(\xi) = \exp[-\frac{1}{2}\|\xi\|^2]$, and with this background we introduce the infinite-dimensional rotation group [first done by H. Yoshizawa (1970)].

Definition 5.1. A *rotation* of E is transformation g on $L^2(\mathbf{R})$ satisfying

i. g is an orthogonal transformation, i.e. a linear transformation such that $\|gf\| = \|f\|$, $f \in L^2(\mathbf{R})$; and
ii. the restriction of g to E is a homeomorphism of E.

The collection of all rotations of E is denoted $O(E)$ and a product may be defined so that $O(E)$ forms a group. For any g_1, g_2 in $O(E)$ the product $g_1 g_2$ is defined by the relation

$$(g_1 g_2)\xi = g_1(g_2 \xi), \quad \xi \in E. \tag{5.1}$$

Clearly the product $g_1 g_2$ belongs to $O(E)$, and since every rotation is a linear homeomorphism, an inverse g^{-1} with respect to this product can be found in $O(E)$ for every $g \in O(E)$. With the identity operator as a unit, we easily prove that $O(E)$ forms a group when equipped with the product (5.1).

Either of the following two topologies may be used to topologise $O(E)$:
a. The compact-open topology. This is reasonable since $O(E)$ is a collection of transformations on E, and a sub-base for this topology is given by the

family of all sets of the form $W(K, U) = \{g \in O(E): gK \subseteq U\}$ where K is a compact and U an open subset of E, respectively.

b. The so-called μ-topology. Associated with every $g \in O(E)$ is its adjoint g^* acting on E^* and (by Theorem 5.1 below) g^* preserves the measure μ of white noise. Thus we can form the unitary operator U_g defined by $(U_g \varphi)(x) = \varphi(g^*x)$, $\varphi \in (L^2)$, $x \in E^*$. The μ-topology is then defined to be the weakest topology relative to which the regular unitary representation $(U_g, (L^2))$ of $O(E)$ is continuous.

It is easily proved that $O(E)$ is a topological group with either of these topologies.

Remark. The μ-topology will be better understood after the results of this section. However, it makes no significant difference in this chapter which of these topologies is chosen.

Definition 5.2. The group $O(E)$ of all rotations of E equipped with the product (5.1) and either of the topologies (a) or (b) above is referred to as the *rotation group* of E. If E is not specified, it is simply called an *infinite-dimensional rotation group*.

For any $g \in O(E)$ and fixed $x \in E^*$, the map $\xi \to \langle x, g\xi \rangle$ is a continuous linear functional on E by (i) of Definition 5.1, and so there exists an element of E^*, denoted by g^*x, such that

$$\langle x, g\xi \rangle = \langle g^*x, \xi \rangle. \tag{5.2}$$

The map $x \to g^*x$ defines the linear automorphism g^* on E^* adjoint to g, and the collection of all such maps g^* is written $O^*(E^*)$:

$$O^*(E^*) = \{g^*: g \in O(E)\}.$$

A product on $O^*(E^*)$ is derived naturally from (5.1), namely,

$$(g_1^* g_2^*)x = g_1^*(g_2^*x), \quad g_1^*, g_2^* \in O^*(E^*), \tag{5.3}$$

and with this product $O^*(E^*)$ forms a group.

Noting that $(g^*)^{-1} = (g^{-1})^*$ we immediately see the relationship between these two groups, given in

Proposition 5.1. *The mapping that associates* $(g^*)^{-1}$ *of* $O^*(E^*)$ *with* g *in* $O(E)$

$$g \to (g^*)^{-1} \in O^*(E^*),$$

defines an algebraic isomorphism between $O(E)$ *and* $O^*(E^*)$.

It follows that as far as algebraic structure is concerned, we may take either of these groups. However we prefer to use $O(E)$ because E is more easily visualised than E^*, although the importance of $O^*(E^*)$ in our analysis is readily illustrated by the following theorem. Indeed $O^*(E^*)$ describes the

5.1 Transformations of White Noise (I): Rotations

properties of μ so well that we could hardly study white noise without it. There is no need to describe a topology on $O^*(E^*)$ for we simply suppose it is such that the resulting topological group is isomorphic to $O(E)$.

At this point we recall the measurable space $(E^*, \mathfrak{B})$ constructed in §3.2.

Theorem 5.1. i. *Any element g^* of $O^*(E^*)$ is a $\mathfrak{B}$-measurable mapping of E^* onto itself.*

ii. *If we define the product $g^* \circ \mu$ by*

$$g^* \circ \mu(B) = \mu(g^*B), \quad B \in \mathfrak{B}, \quad (5.4)$$

then the equality

$$g^* \circ \mu = \mu \quad (5.5)$$

is valid for all $g^* \in O^*(E^*)$, i.e. μ is $O^*(E^*)$-invariant.

PROOF. i. As before let $\mathfrak{A}$ be the field of all cylinder subsets of E^*. By (5.2) we have

$$g^*\mathfrak{A} = \mathfrak{A}$$

for every g^*, where $g^*\mathfrak{A} = \{g^*A : A \in \mathfrak{A}\}$. Since g^* is an isomorphism of E^*, we see that $g^*\mathfrak{B}$ is again a σ-field, and because $g^*\mathfrak{B}$ includes $\mathfrak{A} = g^*\mathfrak{A}$, we deduce that

$$g^*\mathfrak{B} \supset \mathfrak{B}.$$

Now the same conclusion holds for $(g^*)^{-1}$ and hence $g^*\mathfrak{B} = \mathfrak{B}$, i.e. g^* is $\mathfrak{B}$-measurable.

ii. Let us calculate the characteristic functional of the measure $g^* \circ \mu$. It is

$$\int_{E^*} e^{i\langle x, \xi\rangle} dg^* \circ \mu(x) = \int_{E^*} e^{i\langle x, \xi\rangle} d\mu(g^*x) = \int_{E^*} e^{i\langle g^{*-1}x, \xi\rangle} d\mu(x)$$

$$= \int_{E^*} e^{i\langle x, g^{-1}\xi\rangle} d\mu(x) = \exp\left[-\frac{1}{2}\|g^{-1}\xi\|^2\right].$$

By condition (i) of Definition 5.1 this last expression coincides with the characteristic functional of μ, and since a probability measure on the measurable space $(E^*, \mathfrak{B})$ is uniquely determined by its characteristic functional, we have proved (5.5). □

As remarked earlier, we now see that any investigation of $O^*(E^*)$ [or $O(E)$] will have implications for the measure μ. In the remainder of this chapter we will study the group $O(E)$ as much as possible, and, at the same time, throw indirect light on μ.

Speaking intuitively, the group $O(E)$ is a very large group, and as we will see in the following sections, it has an extremely rich structure. Of course it is neither compact nor locally compact. Our immediate aim is to elucidate the structure of several interesting subgroups of $O(E)$, as well as to examine the significance of the results obtained for the theory of white noise.

5.2 Subgroups of the Rotation Group

The subgroups of $O(E)$ which we discuss below fall naturally into two classes: [I] Finite-dimensional rotation groups, and their limits, and [II] Subgroups of rotations of E derived from time changes. The present section is devoted to a quick description of these two classes.

I. Let g be a linear transformation on E. If there exists a finite-dimensional subspace F of E such that
a. The restriction g_F of g to F is an orthogonal transformation of F, and
b. the restriction $g_{F^\perp}$ of g to $F^\perp = \{\xi \in E : \xi \perp F\}$ is the identity, then g belongs to $O(E)$ and is called a *finite-dimensional rotation*.

For a fixed finite-dimensional subspace F of E, the collection $G(F)$ of all such finite-dimensional rotations forms a subgroup of $O(E)$ which, if F is n-dimensional, satisfies

$$G(F) \cong O(n), \quad \text{the } n\text{-dimensional orthogonal group.}$$

Consequently $O(E)$ has a subgroup isomorphic to the n-dimensional rotation group $SO(n)$ for every n. We are naturally led to the formation of a limit (in a suitable sense) of a sequence of subgroups the n-th of which is isomorphic to $O(n)$. With every increasing sequence $\{F_n\}$ of subspaces

$$F_1 \subseteq \cdots \subseteq F_n \subseteq F_{n+1} \subseteq \cdots,$$

where F_n is n-dimensional, we associate the sequence

$$G(F_1) \subseteq \cdots \subseteq G(F_n) \subseteq G(F_{n+1}) \subseteq \cdots$$

where $G(F_n) \cong O(n)$ for every $n \geq 1$. The limit G_∞ of the sequence

$$G_\infty \equiv \lim_{n \to \infty} G(F_n) = \bigcup_1^\infty G(F_n) \tag{5.6}$$

becomes an algebraic subgroup of $O(E)$. Further discussion of this subgroup is omitted, because if we wanted to discuss the closure of G_∞ in $O(E)$, or the topological subgroup of $O(E)$ generated by G_∞, we would first have to specify the topology on E. Clearly the subgroup G_∞ depends upon the choice of the sequence $\{F_n\}$ and so when G_∞ is discussed, it is convenient to fix in advance a complete orthonormal system $\{\xi_n\} \subseteq E$ of $L^2(\mathbf{R})$, and then let F_n be the span of $\{\xi_i : 1 \leq i \leq n\}$.

In the course of his analysis of functionals on the Hilbert space $L^2([0, 1])$ P. Lévy (1951, Part III, §§36–37) introduced a group of permutations of the coordinates of $L^2([0, 1])$. At the back of his mind would no doubt have been the support of a measure such as μ, so that permutations of the coordinates should necessarily be heavily constrained. An example from Lévy's discussion may be viewed in a light similar to that of the subgroup G_∞ of $O(E)$, and it is to this which we now turn.

5.2 Subgroups of the Rotation Group

EXAMPLE 1. The basic nuclear space E is taken to be $\hat{\mathscr{D}}(\pi)$ given in Example 1 of §A.3. We also take the complete orthonormal system $\{\xi_k : k \geq 1\}$ introduced there by (A.12). Elementary calculations show that the $\{\xi_k\}$ can be rearranged in such a way that for every n, $\|\xi_k\|_n$ is non-decreasing in k.

Suppose that we have a permutation $\pi = \{\pi(n)\}$ of the positive integers satisfying the following condition:

$$(*) \begin{cases} \text{there exists a sequence } \{n_p : p \geq 1\} \text{ of positive integers} \\ \text{with } n_1 = 1 \text{ such that } \lim_{p \to \infty} \left(\dfrac{n_{p+1}}{n_p} \right) = 1, \text{ and} \\ n_p \leq k < n_{p+1} \text{ implies } n_p \leq \pi(k) < n_{p+1}. \end{cases}$$

Any such permutation, π, defines a transformation g_π by

$$\xi = \sum_1^\infty a_k \xi_k \to g_\pi \xi = \sum_1^\infty a_k \xi_{\pi(k)}, \qquad \xi \in \hat{\mathscr{D}}(\pi). \tag{5.7}$$

Clearly g_π is linear and satisfies the condition

$$\|g_\pi \xi\| = \|\xi\|.$$

We prove the continuity of g_π in the following way. By assumption the ratios n_{p+1}/n_p are bounded above, i.e. there exists a constant C such that for every p, $n_{p+1}/n_p \leq C$, and therefore for every n

$$\|g_\pi \xi\|_n^2 = \sum_{p=1}^\infty \sum_{k=n_p}^{n_{p+1}-1} a_k^2 \|\xi_{\pi(k)}\|_n^2$$

$$\leq \sum_{p=1}^\infty \sum_{k=n_p}^{n_{p+1}-1} a_k^2 \|\xi_{n_{p+1}}\|_n^2$$

$$\leq C_n \sum_{p=1}^\infty \left(\frac{1}{2} n_{p+1}\right)^{2n} \sum_{k=n_p}^{n_{p+1}-1} a_k^2 \qquad C_n \text{ a constant depending only on } n, \text{ see (A.13);}$$

$$\leq C^{2n} C_n \sum_{p=1}^\infty \left(\frac{1}{2} n_p\right)^{2n} \sum_{k=n_p}^{n_{p+1}-1} a_k^2$$

$$\leq D_n \|\xi\|_n^2$$

where D_n is another constant depending only on n. This proves that g_π is continuous on $\hat{\mathscr{D}}(\pi)$.

Let π^{-1} be the permutation inverse to π. Then we see that π^{-1} also satisfies condition (*), so that $g_{\pi^{-1}}$ is also continuous, and moreover we can prove that

$$g_\pi^{-1} = g_{\pi^{-1}}.$$

Thus g_π is a member of $O(\hat{\mathscr{D}}(\pi))$, and for a fixed sequence $\{n_p\}$ of positive integers, the collection

$$G_{\{n_p\}} = \{g_\pi : \pi \text{ is such that } n_p \leq k < n_{p+1} \text{ implies } n_p \leq \pi(k) < n_{p+1}\}$$

forms an algebraic subgroup of $O(\hat{\mathscr{D}}(\pi))$ isomorphic to the direct product $\prod_{p=1}^{\infty} \mathfrak{S}_{[n_p, n_{p+1})}$, where $\mathfrak{S}_{[n_p, n_{p+1})}$ is the group of all permutations of the integers in the interval $[n_p, n_{p+1})$.

One could devise many other types of permutations of coordinates, but we will not go into details since our aim was solely to show a subgroup like G_∞ with a direct connection to the analysis of functionals on (L^2); see also §5.7.

Remark. There is an intrinsic difference between the two subgroups G_∞ and $G_{\{n_p\}}$ which it is worth pointing out, describable in terms of "average power". If we set

$$\gamma(x, g) = \limsup_{n \to \infty} n^{-1} \sum_{1}^{n} \langle x, \xi_k - g\xi_k \rangle^2$$

then we find that $\gamma(x, g) = 0$, a.e. if $g \in G_\infty$, whilst there exists elements $g \in G_{\{n_p\}}$ such that $\gamma(x, g) = 2$, a.e. For example we could take $g = g_\pi$ where π is a product of cyclic permutations of $(n_p, n_p + 1, \ldots, n_{p+1} - 1)$, $p \geq 1$.

II. When the measure μ of white noise is defined on the space E^*, each member $x \in E^*$ may be regarded as the time derivative $B'(t)$ of a Brownian sample path $B(t)$. Our interest now focusses on linear transformations of x which correspond to scale changes of the time parameter t of $B'(t)$. Such transformations are naturally required to be automorphisms of the class E^* of all sample functions of $B'(t)$ and, using the canonical bilinear form $\langle x, \xi \rangle$, we can discuss them as linear transformations of $\xi \in E$.

Let the new time scale be expressed in terms of the monotone real-valued function $\psi(u)$, $u \in \mathbf{R}$. The scale change may be written $\xi(u) \to \xi(\psi(u))$, but in order that μ be invariant under the induced transformation on E^*, we need a further factor. Indeed we must consider

$$\xi(u) \to \xi(\psi(u)) \cdot |\psi'(u)|^{1/2}. \tag{5.8}$$

Related topics, as well as background to this material, can be found in H. Sato (1971). With the factor $|\psi'(u)|^{1/2}$ we see that the $L^2(\mathbf{R})$-norm of ξ, and hence the value $C(\xi) = \exp[-\frac{1}{2}\|\xi\|^2]$ of the characteristic functional, is unchanged. If the transformed function again belongs to E, and if the transformation is a continuous automorphism, then the transformation induced on E^* leaves the measure μ invariant.

These observations raise the problem of studying those members of $O(E)$ that are defined by (5.8). The choice of $\psi(u)$ has to be compatible with the structure of E and, more importantly, it should be of probabilistic interest if possible. There are indeed several important one-parameter subgroups of $O(E)$ involving such interesting transformations of the time scale, and the following three sections are devoted to the study of just these one-parameter subgroups. Our approach will be more in a heuristic vein than a systematic discussion.

5.2 Subgroups of the Rotation Group

The function ψ will now be permitted to depend on a parameter t, $-\infty < t < \infty$, and will be written $\psi_t(u)$, and the transformation of ξ it defines will be written g_t:

$$(g_t \xi)(x) = \xi(\psi_t(u)) \cdot |\psi'_t(u)|^{1/2}, \qquad \psi'_t(u) = \frac{\partial}{\partial u} \psi_t(u). \tag{5.9}$$

The requirement that $\{g_t: -\infty < t < \infty\}$ forms a one-parameter group

$$g_t g_s = g_{t+s}, \qquad -\infty < t, s < \infty,$$

$$g_0 = I \quad \text{(the identity)},$$

is expressed in terms of ψ_t as:

$$\begin{cases} \psi_s(\psi_t(u)) = \psi_{t+s}(u), & -\infty < t, s < \infty, \\ \psi_0(u) = u. \end{cases} \tag{5.10}$$

The following example, known as "the shift", is our model for such one-parameter subgroups of $O(E)$.

EXAMPLE 2. The shift $\{S_t\}$. If we take

$$\psi_t(u) = u - t, \qquad -\infty < t < \infty, \tag{5.11}$$

then ψ_t satisfies (5.10). The g_t defined by (5.9) with this ψ_t is denoted by S_t, and $\{S_t : -\infty < t < \infty\}$ is just the *shift* introduced in §3.3. As we proved there, if the space E is taken to be $\mathscr{S}$, then $\{S_t\}$ is a continuous one-parameter subgroup of $O(\mathscr{S})$. The same is also true for the case $E = D_0$ (notation as in §A.3). Although the shift is the simplest one-parameter subgroup, we will see later that it is also one of the most important.

Returning to the general case, we now look for a formula for such $\{\psi_t\}$. Since $\psi_t(u)$ denotes a family of changes of time-scale, it appears reasonable to suppose (a) that $\psi_t(u)$ is strictly monotone in both t and u, and (b) that $\psi_t(u)$ is continuous in u. Under these two assumptions, $\psi_t(u)$ may be written (see J. Aczél (1966)) as

$$\psi_t(u) = f(f^{-1}(u) + t), \tag{5.12}$$

where f is a strictly monotone function on $(-\infty, \infty)$, uniquely determined up to an additive constant. Intuitively, (5.12) says that ψ_t is conjugate to the shift, and the shift itself, with $f(u) = -u$, is seen to occupy a central place in the class of all such ψ_t. We make use of (5.12) by expressing the conditions for $\{g_t\}$ given by (5.9) to be a one-parameter subgroup of $O(E)$ in terms of the function f, although these will, of course, also depend on the structure of the basic nuclear space E. The one-parameter subgroups of the next section will illustrate these ideas as well as describing certain profound properties of Brownian motion.

5.3 The Projective Transformation Group

Our aim in this section is to find subgroups of $O(E)$ which are in class [II] of the classification of the previous section. Such subgroups will not only be treated individually, but also jointly, and the central role of the shift in defining the relationships between them is specifically taken into account. A probabilistic interpretation of each subgroup will be given later, as will also the manner in which these were discovered, see §8, T. Hida (1970b), and T. Hida, I. Kubo, H. Nomoto and H. Yoshizawa (1968).

Take a one-parameter $\{g_t\}$ given as in (5.12) by ψ_t, and denote its infinitesimal generator by α:

$$\frac{d}{dt} g_t \bigg|_{t=0} = \alpha,$$

where in forming this derivative we use the topology of strong convergence. The generator α is a first-order differential operator involving the f of (5.12). Indeed if we apply d/dt to (5.12) at $t=0$ and set

$$a(u) = \left\{\frac{d[f^{-1}(u)]}{du}\right\}^{-1} (= f'[f^{-1}(u)]),$$

then α may be expressed in the form

$$\alpha = a(u)\frac{d}{du} + \frac{1}{2}a'(u).$$

For the shift we have $a(u) \equiv -1$ and its generator is denoted by s:

$$s = -\frac{d}{du}.$$

One method of describing the relationship between the shift and other one-parameter subgroups is via the commutation relations between their generators. For a general α and s the commutator $[\alpha, s] = \alpha s - s\alpha$ is

$$[\alpha, s] = a'(u)\frac{d}{du} + \frac{1}{2}a''(u). \tag{5.13}$$

Initially we are interested in possible special cases of the result (5.13).

i. $[\alpha, s] = 0$.

This is equivalent to $a'(u) \equiv 0$, i.e. $a(u) = c$ (constant), so that

$$\alpha = -cs.$$

Thus α is essentially the same as the generator of the shift.

ii. $[\alpha, s] = \lambda s$, λ a constant.

5.3 The Projective Transformation Group

Again using (5.13) we find that $a(u) = -\lambda u + d$, d constant, and since we are free to choose α modulo s, the constant d may be taken to be 0. Thus we consider

$$\alpha = -\lambda u \frac{d}{du} - \frac{1}{2}\lambda.$$

By solving this equation we have $a(u)^{-1} = \{f^{-1}(u)\}' = -1/\lambda u$, and so

$$f^{-1}(u) = -\lambda^{-1} \log\left(\frac{u}{c}\right), \qquad \text{if } u > 0 \text{ and } c \text{ is a positive constant.}$$

But this implies that

$$f(u) = c \cdot \exp[-\lambda u], \qquad u > 0,$$

and so

$$\psi_t(u) = c \cdot \exp\left[-\lambda\left\{-\lambda^{-1}\log\left(\frac{u}{c}\right) + t\right\}\right]$$
$$= u \exp[-\lambda t], \qquad u > 0.$$

This last expression makes sense for $u \leq 0$; indeed in this case the discussion above only needs changes in certain intermediate steps to lead to the same expression. The transformation of ξ defined by this $\psi_t(u)$ will be denoted by τ_t:

$$\tau_t \xi(u) = \xi(ue^{-\lambda t})e^{-\lambda t/2}. \tag{5.14}$$

Thus we have a one-parameter group $\{\tau_t: -\infty < t < \infty\}$ of transformations on E and this group is called the *dilation* or tension (group). For simplicity in what follows we take $\lambda = -1$, and the associated infinitesimal generator is denoted by τ, having the expression

$$\tau = u\frac{d}{du} + \frac{1}{2}.$$

The commutation relation $[\tau, s] = -s$ is easily obtained, and implies the following relation between the corresponding subgroups:

$$S_t \tau_s = \tau_s S_{t \exp[s]}. \tag{5.15}$$

In words, $\{S_t\}$ is transversal to $\{\tau_t\}$.

iii. $[\alpha, s] = \mu\alpha$, μ a constant.

This relation implies that $a'(u) = \mu a(u)$, which implies $f^{-1}(u) = -c\mu^{-1}\exp[-\mu u] + c'$, but the function $f(u)$ in this case does not define a $\psi_t(u)$ carrying $(-\infty, \infty)$ onto itself.

Changing our viewpoint slightly, let us now focus on τ. The commutation relation between τ and a general α is

$$[\tau, \alpha] = \{ua'(u) - a(u)\}\frac{d}{du} + \frac{1}{2}ua''(u),$$

and possible special cases include:

i. $[\tau, \alpha] = 0$,
ii. $[\tau, \alpha] = \lambda s$,
iii. $[\tau, \alpha] = \mu\tau$,
iv. $[\tau, \alpha] = \nu\alpha$,

where λ, μ and ν are all non-zero constants. We state without proofs the end results in each of the cases (i) to (iv) above.

Apart from the trivial solution, the only α satisfying (i) is $\alpha = c\tau$ where c is a constant.

In case (ii) we find that α is a constant multiple of s modulo τ, and no new generator appears.

There are no suitable solutions in case (iii).

Finally, relation (iv) gives the equation

$$\alpha = u^{\nu+1}\frac{d}{du} + \frac{1}{2}(\nu + 1)u^{\nu},$$

and this is the only generator which has been discovered using τ. It satisfies the commutation relation

$$[\alpha, s] = (\nu + 1)\left\{u^{\nu}\frac{d}{du} + \frac{1}{2}\nu u^{\nu-1}\right\}.$$

A consideration of the expression inside the braces $\{\cdot\}$ suggests the idea of a family of generators involving s, τ and α which is closed under the commutator product. In fact if we take $\nu = 1$ and introduce

$$\kappa = u^2\frac{d}{du} + u,$$

we see that $[\kappa, s] = 2\tau$. We also see that κ is the generator of the one-parameter group $\{\kappa_t\}$ of transformations given by

$$(\kappa_t\xi)(u) = \xi\left(\frac{u}{-tu+1}\right)|-tu+1|^{-1}, \quad -\infty < t < \infty. \quad (5.16)$$

In this case, f in (5.12) is discontinuous.

So far we have only discussed algebraic relations between generators and we now go on to study the one-parameter groups with these generators. For this purpose we take the basic nuclear space E to be the space D_0 introduced and discussed in Example 3 of §A.3.

5.3 The Projective Transformation Group

Proposition 5.2. *Let E be the space D_0. Then $\{S_t\}$, $\{\tau_t\}$ and $\{\kappa_t\}$ are all continuous one-parameter subgroups of $O(D_0)$. Their infinitesimal generators satisfy the commutation relations*

$$[\tau, s] = -s,$$
$$[\tau, \kappa] = \kappa, \qquad (5.17)$$

and
$$[s, \kappa] = -2\tau.$$

The assertions have all been proved apart from those concerning $\{\kappa_t\}$, but these are straightforward consequences of (5.16). □

Now let the generators $\{s, \tau, \kappa\}$ span a vector space $\mathfrak{R}$ on $\mathbf{R}$, and introduce the product $[\cdot, \cdot]$ into $\mathfrak{R}$, forming a Lie algebra. The commutation relations (5.17) show that $\mathfrak{R}$ is isomorphic to the Lie algebra of the linear Lie group $PGL(2, \mathbf{R})$. That this is no accident can be seen by considering the following mapping from $GL(2, \mathbf{R})$ into $O(D_0)$. For an element $X = \begin{pmatrix} a & b \\ c & d \end{pmatrix}$ of $GL(2, \mathbf{R})$, we define a transformation g_X of the ξ by

$$(g_X \xi)(u) = \xi\left(\frac{au+b}{cu+d}\right) \cdot \frac{|X|^{1/2}}{|cu+d|}, \qquad (5.18)$$

where $|X|$ denotes the absolute value of the determinant of X. As we noted in the case of $\{\kappa_t\}$, it can be proved that g_X belongs to $O(D_0)$ for every $X \in GL(2, \mathbf{R})$. Moreover the mapping

$$X \to g_X \in O(D_0) \qquad (5.19)$$

is seen to be a homomorphism whose kernel is the centre of $GL(2, \mathbf{R})$. Thus we obtain

Proposition 5.3. *The subgroup G_p of $O(D_0)$ generated by $\{S_t\}$, $\{\tau_t\}$ and $\{\kappa_t\}$ is isomorphic to $PGL(2, \mathbf{R})$ under the mapping (5.19).*

Observation. A general element of the factor group of $GL(2, \mathbf{R})$ modulo the centre may be expressed as a product

$$\zeta(t)\delta(u)z(v) \quad \text{or} \quad s\zeta(t)\delta(u)z(v),$$

where

$$\zeta(t) = \begin{pmatrix} 1 & t \\ 0 & 1 \end{pmatrix}, \; \delta(u) = \begin{pmatrix} e^{u/2} & 0 \\ 0 & e^{-u/2} \end{pmatrix}, \; z(v) = \begin{pmatrix} 1 & 0 \\ v & 1 \end{pmatrix} \quad \text{and} \quad s = \begin{pmatrix} 0 & 1 \\ 1 & 0 \end{pmatrix}.$$

In our notation S_t, τ_t and κ_t correspond to $\zeta(-t)$, $\delta(t)$ and $z(-t)$, respectively.

5.4 Projective Invariance of Brownian Motion

The purpose of this section is to examine the probabilistic significance of the subgroup G_p of Proposition 5.3, as well as that of the three one-parameter subgroups generating G_p. In particular we see that the so-called projective invariance of Brownian motion may be described in terms of G_p.

Given a one-parameter subgroup $\{g_t\}$ of $O(E)$, we may (by Theorem 5.1) form a one-parameter group $\{g_t^*\}$ of measure preserving transformations (i.e. a "flow") on (E^*, μ).

Firstly, let us consider the shift $\{S_t\}$. As has already been discussed in §4.4, $\{S_t^*\}$ is the "flow" of Brownian motion, although there it was considered on a general E^*. To enable a contrast with other flows to be made below, let us quickly review the facts concerning $\{S_t^*\}$. The canonical bilinear form $\langle x, \xi \rangle$ defines a Gaussian random variable on (E^*, μ) for each $\xi \in E$, and this extends to the case when $\xi \in L^2(\mathbf{R})$. For $\xi \in L^2(\mathbf{R})$, the random variable $\langle x, \xi \rangle$ has mean 0 and variance $\|\xi\|^2$. If we now take for ξ the indicator function $\chi_{[t \wedge 0, t \vee 0]}$, then the Gaussian system

$$\{\langle x, \chi_{[t \wedge 0, t \vee 0]} \rangle \equiv B(t, x): -\infty < t < \infty\} \quad (5.20)$$

can be shown to be a Brownian motion with time parameter space $(-\infty, \infty)$. For $b > a$ we may take its increment $B(b, x) - B(a, x)$ and let S_t^* act on x, to find that with probability 1

$$B(b, S_t^* x) - B(a, S_t^* x) = B(b + t, x) - B(a + t, x).$$

In this sense $\{S_t^*\}$ may be regarded as inducing the passage of time on the increments of Brownian motion, and therefore (see Definition 4.3 of §4.4) it is referred to as the "flow" of Brownian motion.

We turn now to $\{\tau_t^*\}$. If we write

$$U(t, x) = \langle \tau_t^* x, \chi_{[0, 1]} \rangle, \quad -\infty < t < \infty, \quad (5.21)$$

then $\{U(t, x): -\infty < t < \infty\}$ is a stationary Gaussian process on the probability space (D_0^*, μ). Its mean is clearly 0 and its covariance function is given by

$$\gamma(h) = \int_{D_0^*} U(t + h, x) U(t, x) \, d\mu(x)$$

$$= \int_{-\infty}^{\infty} \exp\left[t + \frac{1}{2} h\right] \chi_{[0, 1]}(u \exp[t + h]) \chi_{[0, 1]}(u \exp[t]) \, du$$

$$= \exp\left[-\frac{1}{2} |h|\right].$$

Thus we see that $\{U(t)\}$ is the Ornstein-Uhlenbeck process introduced in §4.6. (iv) Example 1, and so $\{\tau_t^*\}$ is referred to as the flow of this process.

5.4 Projective Invariance of Brownian Motion

Before the role of the flow $\{\kappa_t^*\}$ can be discussed, some further background is necessary. For a fixed closed interval $[a, b] \subset [0, \infty)$ we introduce a family of functions of $u \in (a, b)$ parameterised by t varying over the same interval (a, b):

$$f(u; t) = \left|\frac{b-t}{t-a}\right|^{1/2} \cdot \chi_{[a, t]}(u) \cdot \frac{(b-a)^{1/2}}{b-u}, \qquad a < t < b.$$

Also let $X \in GL(2, \mathbf{R})$ define a projective transformation of (a, b) onto itself:

$$X: t \to X(t) \in (a, b), \; X(a) = a, \; X(b) = b, \qquad t \in (a, b). \tag{5.22}$$

The transformation g_X defined by (5.18) extends to one acting on $L^2(\mathbf{R})$, and with this in mind, we can see that

$$g_X f(u; t) = f(u; X^{-1}(t)). \tag{5.23}$$

Moreover, since $g_X \in O(D_0)$, the following theorem can be proved.

Theorem 5.2 (Projective invariance of Brownian motion). *For a projective transformation X satisfying (5.22), the following two Gaussian processes coincide:*

a. $\{\langle x, f(\,\cdot\,; t)\rangle : a < t < b\}$,
b. $\{\langle x, f(\,\cdot\,; X^{-1}(t))\rangle : a < t < b\}$.

It is of some interest to give an alternative proof of this result which is more straightforward, but which does not use the group $O(D_0)$. The two Gaussian processes both have means which are identically zero, and the covariance of the process (b) is

$$\int_{D_0^*} \langle x, f(\,\cdot\,, X^{-1}(t))\rangle \langle x, f(\,\cdot\,, X^{-1}(s))\rangle \, d\mu(x)$$

$$= \int_a^b f(u, X^{-1}(t)) f(u, X^{-1}(s)) \, du$$

$$= (a, b; X^{-1}(s \wedge t), X^{-1}(s \vee t))^{1/2},$$

where $(\,\cdot\,,\,\cdot\,;\,\cdot\,,\,\cdot\,)$ denotes the anharmonic ratio. Since X^{-1} is also a projective transformation, the anharmonic ratio is unchanged by X^{-1}, and so the last expression coincides with $(a, b; s \wedge t, s \vee t)^{1/2}$. Now the inner product of $f(\,\cdot\,, t)$ and $f(\,\cdot\,, s)$, which is the covariance function of the process (a), also coincides with $(a, b; s \wedge t, s \vee t)^{1/2}$, thus proving our assertion. □

Observation. Take a Brownian bridge $X(t) = X_{(a, 0)}^{(b, 0)}(t)$ [§2.6 (iii)] vanishing at $t = a$ and $t = b$ and normalise it, dividing, at each time point t, by the standard deviation of $X(t)$. The resulting Gaussian process $\{Y_{a, b}(t) : a < t < b\}$ has a covariance function which is the same as that of the process of (a) or (b) above. This means that if we apply a projective transformation of the time parameter t to $\{Y_{a,b}(t)\}$, we obtain the same process again. This result is one of the more important properties of Brownian motion, and was referred to as the property of *projective invariance* in §2.6. For further details we refer the reader to T. Hida, I. Kubo, H. Nomoto and H. Yoshizawa (1968).

Four remarks are now in order.

Remark 1. The statement of Theorem 5.2 tells us more than just the fact that two Gaussian processes have the same distribution. Indeed, it shows that each can be transformed into the other by the measure-preserving transformation g_X^* or g_X^{*-1} acting on D_0^*.

Remark 2. It should be noted that in order to describe the projective invariance in the sense of Theorem 5.2, g_X has to be outside the subgroup of $O(D_0)$ generated by $\{S_t\}$ and $\{\tau_t\}$. In other words, the entire subgroup G_p involving $\{\kappa_t\}$ plays an essential role in the description of projective invariance in terms of the rotations of D_0.

Remark 3. A significant open problem is the finding of further subgroups of $O(D_0)$ with associated probabilistic interpretations, not using the heuristic method of §5.3 based on the shift $\{S_t\}$. In particular, attention should be restricted to those that are finite-dimensional Lie groups, since then the powerful techniques of group theory are available.

Remark 4. The reader should observe that the construction of Brownian motion explained in §2.3 (i) rests heavily upon the property established in Theorem 5.2.

5.5 Spectral Type of One-Parameter Subgroups

It will be shown in this section that each of the flows derived from the three one-parameter subgroups of $O(D_0)$ discussed so far (§§5.3, 5.4) has countable Lebesgue spectrum.

Once more we recall that the flow $\{T_t: -\infty < t < \infty\}$ of Brownian motion is obtained by taking the adjoint $T_t = S_t^*$ of the shift $\{S_t: -\infty < t < \infty\}$ on the measure space $(D_0^*, \mathfrak{B}, \mu)$ of white noise. It is known (§4.4) that $\{T_t\}$ is a Kolmogorov flow and thus, according to Sinai's theorem (see Ya. G. Sinai (1961)) in the general theory of flows, it has countable Lebesgue spectrum. However, in our case $\{T_t\}$ has been formed quite explicitly and so a concrete proof can be given.

5.5 Spectral Type of One-Parameter Subgroups

The Hahn-Hellinger theorem on unitary groups is the key theorem in any discussion of spectral type, and we therefore begin with this theorem, see K. Yosida (1951) §6.8. Given a flow $\{T_t: -\infty < t < \infty\}$ on a probability space $(\Omega, \mathbf{B}, P)$, we define a one-parameter group $\{U_t: -\infty < t < \infty\}$ of unitary operators on $L^2(\Omega, \mathbf{B}, P)$ by

$$(U_t f)(\omega) = f(T_t \omega), \quad f \in L^2(\Omega, \mathbf{B}, P). \tag{5.24}$$

If $\{U_t\}$ is strongly continuous in t, then the well-known theorem of M.H. Stone gives us the spectral decomposition:

$$U_t = \int e^{it\lambda}\, dE(\lambda), \tag{5.25}$$

where $\{E(\lambda): -\infty < \lambda < \infty\}$ is a resolution of the identity, that is the $\{E(\lambda)\}$ are projection operators on $L^2(\Omega, \mathbf{B}, P)$ satisfying

i. $E(\lambda)E(\mu) = E(\lambda \wedge \mu)(= E(\mu)E(\lambda))$;
ii. $\text{s-lim}_{\lambda \to \mu + 0} E(\lambda) = E(\mu)$;
iii. $E(-\infty) = 0$, the zero operator, and $E(\infty) = I$.

We are now in a position to state the Hahn-Hellinger theorem giving the unitary invariants of a general group of unitary operators [not necessarily one of the form (5.24)] on a separable Hilbert space $\mathbf{H}$. Suppose that U_t has a spectral decomposition of the form (5.25).

Theorem 5.3 (Hahn-Hellinger Theorem).
 i. *There exists a sequence $\{f_n\}$ of elements of $\mathbf{H}$ such that the Hilbert space $\mathbf{H}$ has a direct sum decomposition*:

$$\mathbf{H} = \sum_{n \geq 1} \oplus \mathbf{H}_n, \tag{5.26}$$

where

$$\mathbf{H}_n = \left\{\int g(\lambda)\, dE(\lambda) f_n : g \in L^2(\mathbf{R}, d\rho_n)\right\}, \tag{5.27}$$

in the notation

$$\rho_n(\lambda) = \|E(\lambda) f_n\|^2,$$

and the sequence $\{d\rho_n\}$ of measures on $\mathbf{R}$ is non-increasing:

$$d\rho_1 \gg d\rho_2 \gg \cdots. \tag{5.28}$$

 ii. *The type of the sequence $\{d\rho_n\}$ is independent of the choice of sequence $\{f_n\}$.*

Remark 1. The subspace $\mathbf{H}_n$ given by (5.27) coincides with the closed subspace spanned by $\{U_t f_n : -\infty < t < \infty\}$, and for this reason it is referred to as a cyclic subspace, and f_n a cyclic vector.

Remark 2. The statement (ii) of Theorem 5.3 is to be understood in the following sense. If there is another decomposition $\mathbf{H} = \sum_{n \geq 1} \oplus \mathbf{H}'_n$, where $\mathbf{H}'_n$ is a cyclic subspace with cyclic vector f'_n, then

$$d\rho_n \sim d\rho'_n \quad \text{(equivalence) for every } n,$$

where $\rho'_n(\lambda) = \|E(\lambda)f'_n\|^2$.

Remark 3. The number of cyclic subspaces in the decomposition (5.26) may be finite or countably infinite.

Remark 4. If we denote by P_n the projection operator onto the subspace $\mathbf{H}_n$, then

$$U_t P_n = P_n U_t.$$

An immediate consequence of Theorem 5.3 is

Corollary. *There exists an isometry V from $\mathbf{H}$ onto $\sum_{n \geq 1} \oplus L^2(\mathbf{R}, d\rho_n)$ such that*

$$V\mathbf{H}_n = L^2(\mathbf{R}, d\rho_n),$$
$$(VU_t V^{-1} g)(\lambda) = e^{i\lambda t} g(\lambda), \quad g \in L^2(\mathbf{R}, d\rho_n).$$

As a further consequence of the theorem we assert the following. If $\{U_t\}$ and $\{U'_t\}$ are one-parameter groups of unitary operators acting on $\mathbf{H}$ and $\mathbf{H}'$ respectively, and if they are unitarily equivalent, i.e. if there exists an isometry V of $\mathbf{H}$ onto $\mathbf{H}'$ such that $U'_t = VU_t V^{-1}$, then the corresponding sequences of measures $\{d\rho_n\}$ and $\{d\rho'_n\}$ are of the same type. Conversely, if these two sequences are of the same type, then we can construct an isometry which makes $\{U'_t\}$ and $\{U_t\}$ unitarily equivalent. In other words, the sequence $\{d\rho_n\}$ is a *unitary invariant*.

Let the support of the measure $d\rho_n$ be denoted by Λ_n. [The set Δ_n does not depend on the decomposition (5.26) of $\mathbf{H}$.] The integer $m(\lambda) = \max\{n : \lambda \in \Lambda_n\}$ is referred to as the *multiplicity* of λ, and if we denote the largest measure $d\rho_1$ by simply $d\rho$, the pair $\{\rho, m\}$ is called the *spectral type* of $\{U_t\}$. We now see that there is a one-to-one correspondence between classes of unitarily equivalent one-parameter unitary groups $\{U_t\}$ and spectral types, i.e. we have a classification of such groups.

Returning now to flows, we can define the spectral type of $\{T_t\}$ to be that of $\{U_t\}$ defined by (5.24), and equivalence of flows can be defined similarly. More fully, two flows $\{T_t\}$ and $\{T'_t\}$ are said to be *spectrally equivalent* if the one-parameter unitary groups, $\{U_t\}$ and $\{U'_t\}$ respectively, which they define, have the same spectral type. We now mention some examples of typical spectral types. The spectral type of a flow $\{T_t\}$ is said to be *countable Lebesgue* or *σ-Lebesgue* if $d\rho$ is equivalent to Lebesgue measure, and if $m(\lambda) \equiv \infty$. Similarly if $d\rho$ is (equivalent to) Lebesgue measure but $m(\lambda) \equiv 1$, then the spectral type is said to be *simple Lebesgue*. Another important particular case is that in which $d\rho$ is discrete, and in this case we speak of a *discrete spectrum* or *pure point spectrum*.

With this background we return once more to the flow $\{T_t\}$ of Brownian motion, that is, the flow obtained from the shift $\{S_t\}$ on the measure space of white noise by putting $T_t = S_t^*$.

Theorem 5.4. *The spectral type of the flow of Brownian motion on the measure space* $(D_0^*, \mathfrak{B}, \mu)$ *of white noise is countable Lebesgue on the Hilbert space* $(L^2) \ominus \{1\}$, *the orthogonal complement in* (L^2) *of the 1-dimensional subspace consisting of the constant functions.*

PROOF. Define U_t by $(U_t\varphi)(x) = \varphi(T_t x)$, $\varphi \in (L^2)$, and let us find the spectral type of $\{U_t\}$ on the space $(L^2) \ominus \{1\}$. Since spectral type is a unitary invariant, this may be reduced to a search for the spectral type of the unitary group $\{V_t\}$ given by $V_t = \mathcal{T} U_t \mathcal{T}^{-1}$ on the Hilbert space (see (4.22))

$$\sum_{n=1}^{\infty} \oplus (n!)^{1/2} \hat{L}^2(\mathbf{R}^n) \left[\cong (L^2) \ominus \{1\} = \sum_{n=1}^{\infty} \oplus \mathcal{H}_n \right],$$

where V_t acts in the following way:

$$V_t F(u_1, u_2, \ldots, u_n) = F(u_1 - t, u_2 - t, \ldots, u_n - t), \quad F \in \hat{L}^2(\mathbf{R}^n),$$

and $-\infty < t < \infty$, (see §4.4).

Since $V_t \hat{L}^2(\mathbf{R}^n) = \hat{L}^2(\mathbf{R}^n)$, it is enough to know the spectral type of V_t on each $\hat{L}^2(\mathbf{R}^n)$, $n \geq 1$, where we denote the restriction of V_t to this space by the same symbol.

If $n = 1$, and if we take an $F \in L^2(\mathbf{R})$ such that its Fourier transform $\hat{F}$ is almost everywhere non-zero, then $\{e^{i\lambda t}\hat{F}(\lambda): -\infty < t < \infty\}$ spans the whole of $L^2(\mathbf{R})$. By taking Fourier transforms we thus see that the closed subspace spanned by $\{F(u - t): -\infty < t < \infty\}$ also coincides with $L^2(\mathbf{R})$. Noting that

$$(V_t F)(u) = F(u - t) = (2\pi)^{-1/2} \int_{-\infty}^{\infty} e^{-iu\lambda} \{e^{i\lambda t}\hat{F}(\lambda)\} d\lambda,$$

we see that F is a cyclic vector and that $L^2(\mathbf{R})$ itself is a cyclic (sub) space. Moreover, comparing this with (5.27), we see that $d\rho$ is a constant multiple of Lebesgue measure, and so the spectral type is simple Lebesgue on $L^2(\mathbf{R})$.

The cases when $n \geq 2$ are all similar, so that for simplicity we shall only discuss the case $n = 2$. The equation

$$V_t F(u_1, u_2) = F(u_1 - t, u_2 - t), \quad F \in \hat{L}^2(\mathbf{R}^2)$$

is already known, and since F is symmetric, we may restrict our attention to a half-plane, say $u_1 \geq u_2$. If we introduce the change of variables

$$v_1 = \frac{1}{2}(u_1 + u_2), \quad v_2 = u_1 - u_2 \geq 0 \qquad (5.29)$$

and set $F(u_1, u_2) = \tilde{F}(v_1, v_2)$, then we have

$$(V_t \tilde{F})(v_1, v_2) = \tilde{F}(v_1 - t, v_2). \tag{5.30}$$

In terms of a complete orthonormal system $\{\eta_n : n \geq 1\} \subseteq L^2([0, \infty))$ the Hilbert space $L^2(\mathbf{R}_+^2)$ formed by the square-integrable functions defined on the upper half-space, $v_2 \geq 0$, admits the following direct sum decomposition

$$L^2(\mathbf{R}_+^2) = \sum_{n=1}^{\infty} \oplus L_n, \tag{5.31}$$

where $L_n = \{f(v_1)\eta_n(v_2) : f \in L^2(\mathbf{R})\}$. Using the property (5.30) the operation of V_t on L_n may be expressed as $V_t(f \otimes \eta_n)(v_1, v_2) = f(v_1 - t)\eta_n(v_2)$. Clearly $V_t L_n = L_n$, so that when it is restricted to L_n, V_t acts exactly as it did in the case $n = 1$ above. More fully, take f such that $\hat{f}(\lambda) \neq 0$ a.e. Then $\{V_t(f \otimes \eta_n) : -\infty < t < \infty\}$ spans a subspace of $L^2(\mathbf{R}_+^2)$ which coincides with L_n. The function $f \otimes \eta_n$ is a cyclic vector of the cyclic subspace L_n, and $d\rho$ is equivalent to Lebesgue measure. Now as given in (5.31), $L^2(\mathbf{R}_+^2)$ is the direct sum of a countably infinite number of such cyclic subspaces, so that the spectral type of $\{V_t\}$ is countable Lebesgue on $\hat{L}^2(\mathbf{R}^2)$.

It is quite easy to extend the above proof to the case $n > 2$, and we note simply the change of variables that generalises (5.29). For example we might take $v_1 = n^{-1} \sum_1^n u_j$ and the other variables are $v_2, v_3, \ldots, v_n$ chosen in such a way that they are invariant under the shifts $u_j \to u_j - t$, $j = 1, 2, \ldots, n$, and that the Jacobian $\partial(v_1, \ldots, v_n)/\partial(u_1, \ldots, u_n)$ is identically 1 or -1. Thus the theorem is proved. □

One consequence of the above result, which follows from the fact that the spectral measure is continuous, is the weakly mixing property of $\{T_t\}$. For any $\varphi(x), \psi(x) \in (L^2)$, we can prove:

$$\lim_{T-S \to \infty} (T-S)^{-1} \int_S^T \{U_t \varphi(x)\}\overline{\psi(x)}\, dt = \int \varphi(x)\, d\mu(x) \int \overline{\psi(x)}\, d\mu(x). \tag{5.32}$$

The *ergodic property*, a much weaker condition, follows automatically: if $\mu(T_t B \triangle B) = 0$ for every t, then $\mu(B) = 0$ or 1. [The notation $A \triangle B$ denotes the symmetric difference $(A \backslash B) \cup (B \backslash A)$.] With this ergodic property we can show that for any $\varphi(x) \in (L^2)$, which we may suppose to be μ-integrable, the time-average of $(U_t \varphi)(x)$ coincides a.e. with its phase average, i.e.

$$\lim_{T-S \to \infty} (T-S)^{-1} \int_S^T (U_t \varphi)(x)\, dt = \int_{D_0^*} \varphi(x)\, d\mu(x), \text{ a.e.} \tag{5.33}$$

Incidentally it is worth noting that the general theory of Kolmogorov flows shows that our $\{T_t\}$ enjoys a higher order mixing property which is much stronger than the weak mixing property. Of course the proof would be quite different from our proof in both spirit and method.

Our next question is to ask about the spectral type of the flow of the

5.5 Spectral Type of One-Parameter Subgroups

Ornstein-Uhlenbeck process $\{\tau_t^*: -\infty < t < \infty\}$ derived from the dilation $\{\tau_t: -\infty < t < \infty\}$ given by

$$(\tau_t \xi)(u) = \xi(u e^t) e^{t/2}. \tag{5.34}$$

Let us set

$$U_t^\tau(x) = \varphi(\tau_t^* x), \qquad \varphi \in (L^2). \tag{5.35}$$

The equality $U_t^\tau \mathcal{H}_n = \mathcal{H}_n$ holds by the general theory, so that U_t^τ can be restricted to $\mathcal{H}_n$, and we retain the same notation. Having done this, we can use the integral representation to study the spectral type of U_t^τ on $\mathcal{H}_n$.

Let $F \in \hat{L}^2(\mathbf{R}^n)$ be the kernel of the integral representation of $\varphi \in \mathcal{H}_n$. Since μ is τ_t^*-invariant, we have

$$\mathcal{T}(U_t^\tau \varphi)(\xi) = \int_{\mathscr{S}^*} \exp[i\langle \tau_{-t}^* x, \xi\rangle]\varphi(x)\,d\mu(x)$$

$$= C(\xi)i^n \int \cdots \int F(u_1, \ldots, u_n)\xi(u_1 e^{-t}) \cdots \xi(u_n e^{-t}) e^{-nt/2}\,du_1 \cdots du_n$$

$$= C(\xi)i^n \int \cdots \int F(u_1 e^t, \ldots, u_n e^t) e^{nt/2}\xi(u_1) \cdots \xi(u_n)\,du_1 \cdots du_n.$$

Thus the kernel associated with $U_t^\tau \varphi$ must be $F(u_1 e^t, \ldots, u_n e^t) e^{-nt/2}$. If we introduce an operator V_t^τ on $\hat{L}^2(\mathbf{R}^n)$ by writing

$$(V_t^\tau F)(u_1, \ldots, u_n) = F(u_1 e^t, \ldots, u_n e^t) e^{nt/2}, \tag{5.36}$$

then $\{V_t^\tau: -\infty < t < \infty\}$ becomes a one-parameter unitary group on $\hat{L}^2(\mathbf{R}^n)$ linked with $\{U_t\}$ as follows:

$$\mathcal{T} U_t \mathcal{T}^{-1} = V_t^\tau. \tag{5.37}$$

Now $\mathcal{T}$ restricted to $\mathcal{H}_n$ is, apart from the multiplicative constant $(n!)^{1/2}$, an isometry onto $\hat{L}^2(\mathbf{R}^n)$, and so it is enough for our purposes to study the spectral type of $\{V_t^\tau\}$.

We begin with the case $n = 1$. Let us define the subspaces

$$L^2(\mathbf{R}_+) = \{f \cdot \chi_{[0, \infty)}: f \in L^2(\mathbf{R})\}, \qquad L^2(\mathbf{R}_-) = \{f \cdot \chi_{(-\infty, 0]}: f \in L^2(\mathbf{R})\}.$$

Both of these subspaces of $L^2(\mathbf{R})$ are V_t^τ-invariant for each t, and their direct sum is $L^2(\mathbf{R})$:

$$L^2(\mathbf{R}) = L^2(\mathbf{R}_+) \oplus L^2(\mathbf{R}_-). \tag{5.38}$$

We next take one of these two components, say $L^2(\mathbf{R}_+)$, and introduce the isometry W defined by

$$(Wf)(u) = f(e^u) e^{u/2}, \qquad f \in L^2(\mathbf{R}_+). \tag{5.39}$$

The group $\{V_t^\tau\}$ and the group $\{V_t\}$ on $L^2(\mathbf{R})$ derived from the shift $\{S_t\}$ are conjugate with each other under the isometry W:

$$V_t^\tau = W^{-1} V_{-t} W. \tag{5.40}$$

Thus $\{V_t^\tau\}$ on $L^2(\mathbf{R}_+)$ and $\{V_{-t}\}$ on $L^2(\mathbf{R})$ are unitarily equivalent, and so are $\{V_t^\tau\}$ and $\{S_t\}$. We already know the spectral type of $\{S_t\}$ and so can assert that $\{V_t^\tau\}$ has simple Lebesgue spectrum on $L^2(\mathbf{R}_+)$. Exactly the same is true for $L^2(\mathbf{R}_-)$, and we have therefore proved that $\{V_t^\tau\}$ has Lebesgue spectrum with multiplicity 2 ($m(\lambda) \equiv 2$) on the space $\mathcal{H}_1$.

We turn now to the case $n = 2$. As a particular case of (5.36) we see that V_t^τ acts as follows:

$$(V_t^\tau F)(u_1, u_2) = F(u_1 e^t, u_2 e^t) e^t, \qquad F \in \hat{L}^2(\mathbf{R}^2).$$

Since F is symmetric, we need only consider its values when $u_1 \geq u_2$, and we are thus led to the change of variables

$$v_1 = u_1 u_2, \qquad v_2 = \frac{u_1}{u_2}, \tag{5.41}$$

cf. (5.29). From (5.41) we see that the domains

$$D_+ = \{(u_1, u_2): u_1 \geq u_2, u_1 u_2 > 0\} \text{ and } D_- = \{(u_1, u_2): u_1 \geq u_2, u_1 u_2 < 0\}$$

transform into the 1st and 3rd orthant, denoted by $\tilde{D}_+$ and $\tilde{D}_-$ respectively, of the (v_1, v_2)-plane. We now concentrate our attention on functions defined on D_+. Obviously $V_t^\tau L^2(D_+) = L^2(D_+)$ is valid. Associated with the change (5.41) of variables, we have an isometry which carries a function $F(u_1, u_2) \in L^2(D_+)$ to the function $\tilde{F}(v_1, v_2)$ given by

$$\tilde{F}(v_1, v_2) = F\left((v_1 v_2)^{1/2}, \left(\frac{v_1}{v_2}\right)^{1/2}\right)(2v_2)^{-1/2}.$$

Under this isometry $(V_t^\tau F)(u_1, u_2)$ is transformed into $\tilde{F}(v_1 e^t, v_2) e^{t/2}$, and thus V_t^τ has been reduced to a dilation of only one variable. The next step is to obtain a direct sum decomposition of $L^2(D_+)$ into countably many cyclic subspaces, similar to that involving $\{T_t\}$, where the decomposition (5.31) was obtained. When we have this decomposition we can use the relation (5.40) to show that the spectral type of $\{V_t^\tau\}$ on $L^2(D_+)$ is countable Lebesgue. Exactly the same result can be obtained for $L^2(D_-)$ and we can conclude that the flow $\{\tau_t^*\}$ has countable Lebesgue spectrum on $\mathcal{H}_2$.

The generalisation of the above proof to the case $n > 2$ is almost obvious. Summing up the whole discussion, we have proved that the spectral type of the flow $\{\tau_t^*\}$ is countable Lebesgue on the Hilbert space $(L^2) \ominus \{1\}$.

We turn now to the flow $\{\kappa_t^*: -\infty < t < \infty\}$, first recalling the definition of κ_t (which is sometimes called the special conformal transformation):

$$(\kappa_t \xi)(u) = \xi\left(\frac{u}{-tu + 1}\right)|-tu + 1|^{-1}, \qquad \xi \in D_0.$$

Introduce a linear transformation σ on D_0 by

$$(\sigma \xi)(u) = \xi\left(\frac{1}{u}\right)|u|^{-1}. \tag{5.42}$$

Then we recall the definition of D_0 and see that $\sigma \in O(D_0)$. Furthermore,

$$\kappa_t = \sigma^{-1} S_t \sigma, \qquad -\infty < t < \infty, \tag{5.43}$$

(note that $\sigma^{-1} = \sigma$), and hence $\kappa_t^* = \sigma^* T_t \sigma^{*-1}$. Now σ^* is a measure-preserving transformation on $(D_0^*, \mathfrak{B}, \mu)$, so that the two flows $\{\kappa_t^*\}$ and $\{T_t\}$ are isomorphic, and consequently have the same spectral type. Summing up, we have the following

Theorem 5.5. *The flows $\{\tau_t^*: -\infty < t < \infty\}$ and $\{\kappa_t^*: -\infty < t < \infty\}$ both have countable Lebesgue spectrum on the Hilbert space $(L^2) \ominus \{1\}$.*

Remark 1. The proof of Theorem 5.5 consisted of reducing the discussion to that concerning the flow of Brownian motion, this being possible by virtue of the formula (5.12) above.

Remark 2. The properties of ergodicity and mixing are enjoyed by $\{\tau_t^*\}$ and by $\{\kappa_t^*\}$ as they are in the case of $\{T_t\}$.

As a continuation of the discussion of §4.4 we now mention an interesting application of the ergodic property of $\{T_t\}$. It is an actual example proposed by N. Wiener (1958, Lectures 10, 11) as a method of analysing a non-linear circuit. The model involves a Brownian motion (or white noise) input, and the internal structure of the system is assumed to be entirely unknown. In addition we make some reasonable practical assumptions such as (loosely) the fact that the present values of the circuit never depend on future values of the input, that is, we make an assumption of causality. Thus the output $X(t)$ at the instant t is viewed as a functional of the Brownian motion (or white noise) input up to time t. An important property which should be noted is the fact that both the input and the output share the time *shift*. We now formulate the problem under consideration as the elucidation of the dynamical structure of the circuit from observations on the stochastic process $\{X(t)\}$. The following diagram illustrates the situation.

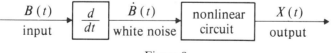

Figure 9

In terms of our analysis on $(D_0^*, \mathfrak{B}, \mu)$ the problem may be re-expressed as follows: a sample function of the white noise input is given by $x \in D_0^*$, and the output $X(t)$ is a non-linear functional of x, $\mathfrak{B}_t$-measurable since it is determined by the input up to the instant t. Thus we may assume that $X(t, x) = (U_t \varphi)(x)$ for $\varphi \in L_0^2$, where square-integrability of φ is assumed simply for convenience. (If this was not so, we could choose a one-to-one bounded measurable function f and take $f(\varphi(x))$ instead of $\varphi(x)$ in what follows.)

Now we know that an application of the Hahn-Hellinger theorem (Theorem 5.3) to $\{U_t\}$ on (L^2) gives the direct sum decomposition

$$(L^2) = \mathcal{H}_0 \oplus \sum_{n,k} \oplus L_{n,k}, \quad \mathcal{H}_n = \sum_k \oplus L_{n,k}, \qquad (5.44)$$

where each $L_{n,k}$ is a cyclic subspace relative to $\{U_t\}$ and $\mathcal{H}_n$ is the multiple Wiener integral of degree n. Further, we know that in the case $n = 1$ a single cyclic subspace is involved, whilst when $n \geq 2$, there are a countably infinite number of $L_{n,k}$. As we saw on the proof of Theorem 5.4, each $L_{n,k}$ contains a cyclic vector $\psi_{n,k}$ belonging to L_0^2 such that

$$L_{n,k} = \mathfrak{S}\{U_t \psi_{n,k} : -\infty < t < \infty\}, \qquad (5.45)$$

where $\mathfrak{S}\{\cdot\}$ denotes the closed subspace spanned by the elements between the braces. Let $Q_{n,k}$ denote the projection operator onto the subspace $L_{n,k}$ and let $Q_{n,k} X(t) = X_{n,k}(t)$. Since $Q_{n,k} U_t = U_t Q_{n,k}$, we have $U_t X_{n,k}(s) = X_{n,k}(s+t) = U_{s+t} X_{n,k}(0) = U_{s+t} Q_{n,k} \varphi$. By the corollary to the Hahn-Hellinger theorem with $d\rho_n$ taken as Lebesgue measure for every n, there are $L^2(\mathbf{R})$-functions $f(\lambda)$ ($\neq 0$, a.e.) and $g(\lambda)$, corresponding to $\psi_{n,k}$ and $X_{n,k}(0)$, respectively, such that the results of the application of U_t to $\psi_{n,k}$ and $X_{n,k}(0)$ correspond to $e^{i\lambda t} f(\lambda)$ and $e^{i\lambda t} g(\lambda)$, respectively. Because it is determined by $\psi_{n,k}$, we may now assume that $f(\lambda)$ is known, and its actual form can be obtained using the technique in the proof of Theorem 5.4. [See particularly (5.31); N. Wiener (1958, Lecture 10) also describes how $f(\lambda)$ can be obtained.] In order to determine the unknown function $g(\lambda)$ by using the known function $f(\lambda)$, it suffices for us to know the covariance function

$$\gamma_{n,k}(h) = \int_{-\infty}^{\infty} e^{i\lambda h} f(\lambda) \overline{g(\lambda)} \, d\lambda. \qquad (5.46)$$

(Note that $f(\lambda) \neq 0$, a.e.). Since $\{T_t\}$ is ergodic, $\gamma_{n,k}(h)$ can be obtained from the sample function of $X(t, x)$ by the following procedure:

$$\lim_{T-S \to \infty} (T-S)^{-1} \int_S^T (U_{t+h} \psi_{n,k})(x) \overline{\{U_t \varphi(x)\}} \, dt$$

$$= \lim_{T-S \to \infty} (T-S)^{-1} \int_S^T U_t \{U_h \psi_{n,k}\}(x) \cdot \overline{\varphi(x)}\} \, dt$$

$$= \int_{D_0^*} (U_h \psi_{n,k})(x) \cdot \overline{\varphi(x)} \, d\mu(x) \quad \text{a.e. } (\mu)$$

$$= \int_{D_0^*} (U_h \psi_{n,k})(x) \cdot \overline{(Q_{n,k} \varphi)(x)} \, d\mu(x) \quad (\text{since } Q_{n,k} U_h = U_h Q_{n,k})$$

$$= \int_{-\infty}^{\infty} e^{ih\lambda} f(\lambda) \overline{g(\lambda)} \, d\lambda.$$

5.6 Derivation of Properties of White Noise Using the Rotation Group

The first limit is the sample covariance of $(U_t \psi_{n,k})(x)$, constructed artificially from the observed process $X(t, x)$, x fixed, and is therefore the quantity that can actually be computed. The following diagram shows an algorithm to get $\gamma_{n,k}(h)$:

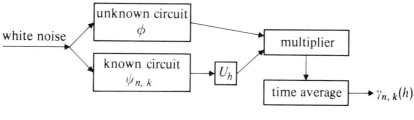

Figure 10

Thus if we could construct the infinitely many circuits described by the $\psi_{n,k}$, then each $\gamma_{n,k}$ would be obtained and therefore the structure of $X_{n,k}(t, x)$ could be found. Summing up the results concerning the $X_{n,k}(t, x)$ we obtain those concerning $X(t, x)$, assumed in advance to be unknown. Before we close this discussion of circuit theory, we should note that non-linear operations did appear in the construction of non-linear circuits being described by $\psi_{n,k}$, and in the multiplier in Fig. 10. Also we note that, as shown in Appendix §A.4, these non-linear circuits can be constructed mechanically.

5.6 Derivation of Properties of White Noise Using the Rotation Group

The purpose of this section is to characterise the white noise $(E^*, \mathfrak{B}, \mu)$ using the rotation group $O(E)$, where the basic nuclear space is taken to be $\mathscr{S}$ or D_0. The main idea behind the discussion which follows is due to Y. Umemura (1965).

Before we come to the main topic, the general notion of G-ergodicity has to be explained. Let $(\Omega, \mathbf{B}, P)$ be a probability space and G a group of automorphisms of the measurable space $(\Omega, \mathbf{B})$. The measure P is said to be *G-ergodic* if (i) it is G-invariant, i.e. if P is g-invariant for every $g \in G$, and (ii) if every $\mathbf{B}$-measurable G-invariant set is either $\emptyset$ or Ω mod 0. In particular, if G is the flow $\{T_t\}$ of a Brownian motion, then the property of G-ergodicity is the same as the property of ergodicity introduced in §5.5. Hence the results of the last section for $\{T_t\}$, coupled with the fact that $\{T_t\} \subset O^*(E^*)$ imply the following assertion:

Proposition 5.4. *Let μ be the measure of white noise. Then μ is $O^*(E^*)$-ergodic.*

When a coordinate system is fixed, the group G_∞ consisting of all finite-dimensional rotations of E can be introduced as in §5.2. We can proceed further than this to define the bigger subgroup G_f of $O(E)$ generated by the G_∞ obtained as we let the coordinate system, i.e. the complete orthornormal system $\{\xi_n\}$, vary in all possible ways. Set $G_f^* = \{g^* : g \in G_f\}$.

Proposition 5.5. *The measure μ of white noise is G_f^*-ergodic.*

PROOF. Suppose that $\varphi(x)$ is bounded and U_g-invariant:
$$(U_g \varphi)(x) (\equiv \varphi(g^*x)) = \varphi(x)$$
for every $g \in G_f$. We will show that $\varphi(x)$ is a.e. constant. Let $\{\xi_n\}$ be a complete orthornormal system in $L^2(\mathbf{R})$ and set $\mathfrak{B}_n = \mathfrak{B}\{\langle x, \xi_1\rangle, \ldots, \langle x, \xi_n\rangle\}$. The conditional expectation $E\{\varphi | \mathfrak{B}_n\}$ of φ given $\mathfrak{B}_n$ is denoted by φ_n, and we have the decomposition
$$\varphi = \varphi_n + \varphi_n^\perp, \qquad \varphi_n^\perp \text{ orthogonal to } L^2(E^*, \mathfrak{B}_n, \mu).$$
Since $\varphi = U_g \varphi = U_g \varphi_n + U_g \varphi_n^\perp$, we have
$$\varphi_n - U_g \varphi_n = -\varphi_n^\perp + U_g \varphi_n^\perp \quad \text{for every} \quad g \in G_f. \tag{5.47}$$

Thus (L^2)-norm of the right-hand side of (5.47) is at most $\|\varphi_n^\perp\| + \|U_g \varphi_n^\perp\| = 2\|\varphi_n^\perp\|$ which tends to 0 as $n \to \infty$. Further, when n is fixed, we can find an element $g \in G_f$ such that the vector space spanned by $\{\xi_1, \ldots, \xi_n\}$ is orthogonal to that spanned by $\{g\xi_1, \ldots, g\xi_n\}$. Then $g\mathfrak{B}_n$ is independent of $\mathfrak{B}_n$ with this g, and hence φ_n and $U_g \varphi_n$ are independent, for they are $\mathfrak{B}_n$- and $g^*\mathfrak{B}_n$-measurable, respectively. The squared (L^2)-norm of the left-hand side of (5.47) thus comes out as

$$\|\varphi_n\|^2 + \|U_g\varphi_n\|^2 - 2\,\mathrm{Re}\int_{E^*} \varphi_n(x)\overline{U_g\varphi_n(x)}\,d\mu(x)$$
$$= 2\|\varphi_n\|^2 - 2\,\mathrm{Re}\left|\int_{E^*} \varphi_n(x)\,d\mu(x)\right|^2 \leq 2\|\varphi_n^\perp\|^2 \qquad \text{Re real part.} \tag{5.48}$$

The last inequality implies that
$$\lim_{n\to\infty}\left|\int_{E^*} \varphi_n(x)\,d\mu(x)\right|^2 = \|\varphi\|^2.$$
On the other hand, we know that
$$\lim_{n\to\infty}\int_{E^*} \varphi_n(x)\,d\mu(x) = \int_{E^*} \varphi(x)\,d\mu(x) \equiv m$$
and hence we have
$$m^2 = \left(\int \varphi(x)\,d\mu(x)\right)^2 = \int \varphi(x)^2\,d\mu(x),$$
i.e. $\varphi(x) = m$, a.e. (μ). $\square$

5.6 Derivation of Properties of White Noise Using the Rotation Group   209

In §3.3 we introduced the notation μ_σ to denote the measure of a white noise with variance σ^2.

Theorem 5.6. *A necessary and sufficient condition for a probability measure v on the measurable space $(E^*, \mathfrak{B})$ to be $O^*(E^*)$-invariant is that v be expressible in the form*

$$v = a\delta_0 + \int_{(0,\infty)} \mu_\sigma \, dm(\sigma), \qquad a \geq 0, \tag{5.49}$$

where δ_0 is the Dirac measure, and m is a bounded Borel measure on $(0, \infty)$.

For the proof we need two lemmas.

Lemma 5.1. *A probability measure v on the measurable space $(E^*, \mathfrak{B})$ is $O^*(E^*)$-invariant if and only if the characteristic functional $C(\xi)$ of v is a function only of the $L^2(\mathbf{R})$-norm $\|\xi\|$ of ξ.*

PROOF. Sufficiency. If the characteristic functional is expressible as $C(\xi) = h(\|\xi\|)$ with a function h on $[0, \infty)$, then that of $g^* \circ v$ has the form

$$C(g^{-1}\xi) = h(\|g^{-1}\xi\|) = h(\|\xi\|) = C(\xi).$$

Thus (Theorem 3.2 of §3.2) we must have $g^* \circ v = v$.

Necessity. If $g^* \circ v = v$ for every $g^* \in O^*(E^*)$, then the characteristic functional of v satisfies $C(g\xi) = C(\xi)$, $g \in O(E)$. Hence $C(\xi)$ is constant on the sets $E_\xi = \{g\xi : g \in O(E)\}$. But $E_\xi = \{\eta : \|\eta\| = \|\xi\|\}$ since for any η with $\|\eta\| = \|\xi\|$, there is a $g \in O(E)$, in fact in G_f, such that $\eta = g\xi$. Thus we must have $C(\xi) = f(\|\xi\|)$ for a suitable function f on $[0, \infty)$. □

We now give a definition. A function $f(t)$ on $[0, \infty)$ is said to be *completely monotonic* if for any $t \geq 0$, $\alpha \geq 0$ and non-negative integer n, we have

$$\sum_{k=0}^{n} (-1)^k \binom{n}{k} f(t + k\alpha) \geq 0. \tag{5.50}$$

There is a well known theorem, due to Bernstein, which gives a representation for all completely monotonic functions, see e.g. D. V. Widder (1946). We state it as

Lemma 5.2 (S. Bernstein). *A function $f(t)$ on $[0, \infty)$ is completely monotonic if and only if there exists a bounded Borel measure m on $[0, \infty)$ such that*

$$f(t) = \int_0^\infty e^{-t\lambda} \, dm(\lambda). \tag{5.51}$$

PROOF OF THEOREM 5.6. i. Since the measure v is $O^*(E^*)$-invariant, Lemma 5.1 asserts that the characteristic functional C of v is a function of $\|\xi\|$ only, $C(\xi) = f(\|\xi\|^2)$, say.

ii. We will prove that the function f just introduced is completely monotonic, first proving that

$$f(t) \geq 0. \tag{5.52}$$

Let $\{\xi_j: 1 \leq j \leq n\}$ be an orthonormal system in $L^2(\mathbf{R})$ and set $\tilde{\xi}_j = (\tfrac{1}{2}t)^{1/2}\xi_j$, $1 \leq j \leq n$. Since $C(\xi) = f(\|\xi\|^2)$ is positive definite, we have

$$\sum_{j,k=1}^{n} f(\|\tilde{\xi}_j - \tilde{\xi}_k\|^2) = nf(0) + \sum_{j \neq k} f(\|\tilde{\xi}_j - \tilde{\xi}_k\|^2) = n + n(n-1)f(t) \geq 0,$$

since $f(0) = 1$ and $\|\tilde{\xi}_j - \tilde{\xi}_k\|^2 = \|\tilde{\xi}_j\|^2 + \|\tilde{\xi}_k\|^2 = t$. Since n is arbitrary, inequality (5.52) follows.

We proceed to the next step. Fixing $\alpha > 0$ and setting

$$-\{f(\|\xi\|^2 + \alpha) - f(\|\xi\|^2)\} = f_1(\|\xi\|^2),$$

we will now prove the positive-definiteness of

$$f_1(\|\xi\|^2), \qquad \xi \in E. \tag{5.53}$$

For an arbitrary set $\xi_1, \xi_2, \ldots, \xi_n$ of linearly independent vectors, we can find an $\xi_0 \in E$ with $\|\xi_0\|^2 = \alpha$ which is orthogonal to this set. Next set

$$\eta_j = \begin{cases} \xi_j & 1 \leq j \leq n, \\ \xi_{j-n} + \xi_0 & n+1 \leq j \leq 2n, \end{cases}$$

and for a given sequence $a_1, a_2, \ldots, a_n$ of complex numbers, we define

$$b_j = \begin{cases} a_j & 1 \leq j \leq n, \\ -a_{j-n} & n+1 \leq j \leq 2n. \end{cases}$$

Once again recalling that $f(\|\xi\|^2)$ is positive definite, we have

$$\sum_{j,k=1}^{2n} b_j \bar{b}_k f(\|\eta_j - \eta_k\|^2) \geq 0,$$

which can be written as

$$\sum_{j,k=1}^{n} a_j \bar{a}_k f(\|\xi_j - \xi_k\|^2) - \sum_{j,k=1}^{n} a_j \bar{a}_k f(\|\xi_j - \xi_k - \xi_0\|^2)$$

$$- \sum_{j,k=1}^{n} a_j \bar{a}_k f(\|\xi_j + \xi_0 - \xi_k\|^2) + \sum_{j,k=1}^{n} a_j \bar{a}_k f(\|\xi_j - \xi_k\|^2)$$

$$= 2 \sum_{j,k=1}^{n} a_j \bar{a}_k f(\|\xi_j - \xi_k\|^2) - 2 \sum_{j,k=1}^{n} a_j \bar{a}_k f(\|\xi_j - \xi_k\|^2 + \alpha)$$

$$= 2 \sum_{j,k=1}^{n} a_j \bar{a}_k f_1(\|\xi_j - \xi_k\|^2) \geq 0.$$

Thus we have proved equation (5.53).

5.6 Derivation of Properties of White Noise Using the Rotation Group 211

iii. In the proof of (5.53) just given, the property of positive-definiteness of $f(\|\xi\|^2)$ was all that was used, and so the same argument can be applied to the functional $f_1(\|\xi\|^2)$, to prove that $f_2(\|\xi\|^2) = -\{f_1(\|\xi\|^2 + \alpha) - f_1(\|\xi\|^2)\}$ is positive definite. Repeating this procedure successively, and noting (5.52), we can prove that $f(t)$ is completely monotonic. Thus, by Lemma 5.2, there exists a bounded Borel measure m on $[0, \infty)$ such that $C(\xi)$ can be expressed in the form

$$C(\xi) = f(\|\xi\|^2) = \int_0^\infty \exp[-\lambda\|\xi\|^2] \, dm(\lambda). \tag{5.54}$$

The mass $m(\{0\}) = a$ at 0 may be regarded as an exceptional part, and the expression (5.54) may be rewritten

$$C(\xi) = a + \int_{(0, \infty)} \exp[-\lambda\|\xi\|^2] \, dm(\lambda). \tag{5.55}$$

The constant a and the functional $\exp[-\lambda\|\xi\|^2]$ are the characteristic functionals of $a\delta_0$ and a white noise $\mu_{\sqrt{2\lambda}}$ with variance 2λ, respectively, and therefore the conclusion (5.49) of the theorem follows [with the notational change $m(\tfrac{1}{2}\sigma^2)$ to $m(\sigma)$]. □

Proposition 5.6. *If an $O^*(E^*)$-invariant measure v on the measurable space $(E^*, \mathfrak{B})$ is $O^*(E^*)$-ergodic, then either $v = \delta_0$ or $v = \mu_\sigma$ for some $\sigma > 0$.*

PROOF. An $O^*(E^*)$-invariant measure v is $O^*(E^*)$-ergodic if and only if any $O^*(E^*)$-invariant measure v'

$$v' \ll v \quad \text{implies} \quad v' \sim v \quad \text{or} \quad v' = 0. \tag{5.56}$$

This follows because if v is $O^*(E^*)$-ergodic, the support A of the measure v' has to be either the entire set ($=$ the support of v mod 0) or $\emptyset$ mod 0. The former implies that $v' \sim v$ whilst the latter implies that $v' = 0$. The converse is almost obvious.

Under the assumption that v is $O^*(E^*)$-invariant it can be expressed in the form (5.49). The further assumption that v is $O^*(E^*)$-ergodic shows that the support of m (which contains 0 if $a \neq 0$) in the expression (5.49) reduces to a single point. Indeed if the support of m contains more than a single point, it can be divided into two non-void subsets A and B. Take A, say, and form an $O^*(E^*)$-invariant measure

$$v' = \int_A \mu_\sigma \, dm(\sigma).$$

Obviously $v' \ll v$ and $v' \neq v$ so that $v' = 0$ must follow from the $O^*(E^*)$-ergodicity of v, a contradiction. Thus the support of m consists of precisely one point and the assertion is proved. □

Remark. We can easily perceive the finite-dimensional analogue of Proposition 5.6. For it is an obvious assertion that a probability measure (distribution) on $\mathbf{R}^n$ which is invariant under all n-dimensional rotations, and which is ergodic, is (up to a scalar multiple) either a Dirac measure at the origin or the uniform measure on a sphere $S^{n-1}(r)$ with radius r having the origin as centre. Uniform measures on spheres seem quite different from Gaussian distributions, however it is known that the uniform measure on $S^n(\sqrt{n})$ tends in a certain sense to the measure μ of white noise as $n \to \infty$, see T. Hida and H. Nomoto (1964). It is also of interest to compare Proposition 5.6 above with Proposition 1.19, both being characterisations of measures using rotations.

5.7 Transformations of White Noise (II): Translations

Before we discuss the main topic of this section we make an observation which should provide some motivation, at least intuitively.

Observation. The result of Example 2 of §1.3(iii) obtained as an application of the Kakutani theorem says the following: a necessary and sufficient condition for white noise on $\mathbf{R}^\infty$ to transform to an equivalent measure under the translation of $\mathbf{R}^\infty$ by a vector m is that m be an element of l^2. This result can be rephrased to refer to E^* instead of $\mathbf{R}^\infty$. For if we take a complete orthonormal system $\{\xi_n\}$ of $L^2(\mathbf{R})$, we obtain a natural mapping from E^* into $\mathbf{R}^\infty$ by forming

$$x \to x_n = \langle x, \xi_n \rangle, \quad x \in E^*, \quad n = 1, 2, \ldots.$$

This mapping defines an isomorphism between E^* and a subset of $\mathbf{R}^\infty$, so that $\{x_n\}$ may be viewed as a coordinate set for $x \in E^*$. In addition $\{\langle x, \xi_n \rangle\}$ is an independent system of standard Gaussian random variables on the measure space $(E^*, \mathfrak{B}, \mu)$, so that white noise on $\mathbf{R}^\infty$ arises as the probability distribution of $\{\langle x, \xi_n \rangle\}$. (Note that with probability 1, the sequence $\{x_n\}$ given by $x_n = \langle x, \xi_n \rangle$ lies outside l^2.)

Now introduce a translation T_m, $m \in E^*$, acting on E^* by

$$T_m: x \to x + m, \quad x \in E^*.$$

The measure μ is transformed by T_m to another measure which we denote by μ_m. This discussion may be phrased in terms of $\mathbf{R}^\infty$, for m and x may be written as $\{m_n\}$ and $\{x_n\}$ respectively, where $m_n = \langle m, \xi_n \rangle$, and $m \in L^2(\mathbf{R})$ is equivalent to $\{m_n\} \in l^2$. Then we can say that μ_m is equivalent to μ if and only if $m \in L^2(\mathbf{R})$ (i.e. $\{m_n\} \in l^2$). In the case of equivalence, the density function is

$$\frac{d\mu_m}{d\mu}(x) = \exp\left[-\langle x, m \rangle - \frac{1}{2}\|m\|^2\right] \quad \text{(in } E^*\text{)},$$

$$= \exp\left[-\sum_n x_n m_n - \frac{1}{2}\sum_n m_n^2\right] \quad \text{(in } \mathbf{R}^\infty\text{)}.$$

5.7 Transformations of White Noise (II): Translations

With these simple observations as our inspiration, we turn now to our main discussion.

As before the basic nuclear space E is taken to be either $\mathscr{S}$ or D_0, and the translation by $\eta \in E$ acting on E^* is denoted by T_η:

$$T_\eta: x \to x + \eta, \qquad x \in E^*. \tag{5.57}$$

Obviously T_η is a $\mathfrak{B}$-measurable automorphism of E^* and the composition $T_\eta \circ \mu$ satisfies

$$(T_\eta \circ \mu)(B) = \mu(B + \eta), \qquad B \in \mathfrak{B}, \tag{5.58}$$

where $B + \eta = \{x + \eta : x \in B\}$.

The characteristic functional $C_\eta(\xi)$ of the probability distribution $T_\eta \circ \mu$ on $(E^*, \mathfrak{B})$ is given by

$$C_\eta(\xi) = \int_{E^*} \exp[i\langle x, \xi \rangle] \, dT_\eta \circ \mu(x) = \int_{E^*} \exp[i\langle x, \xi \rangle] \, d\mu(x + \eta). \tag{5.59}$$

Let E_2 be the 2-dimensional subspace spanned by ξ and η and recall the discussion of §3.2. From the results there we know that $E_2^* \cong E^*/E_2^a = \rho_{E_2} E^*$ and $E_2^{\perp *} \cong E^*/E_2^{\perp a}$. Any element x of E^* is expressible as

$$x = x' + x'', \qquad x' \in E_2^*, \, x'' \in E_2^{\perp *}.$$

Relative to this decomposition the measures μ and $T_\eta \circ \mu$ are in turn expressible as product measures

$$\begin{aligned} d\mu(x) &= d\mu'(x') \times d\mu''(x'') \\ d\mu(x + \eta) &= d\mu'(x' + \eta) \times d\mu''(x''), \end{aligned} \tag{5.60}$$

where μ' and μ'' are probability measures on E_2^* and $E_2^{\perp *}$ respectively. If the restriction of T_η to E_2^* is denoted by the same symbol T_η, then $d\mu'(x' + \eta)$ may be written as $dT_\eta \circ \mu'(x')$.

Let $\{\xi_1, \xi_2\}$ be a pair of orthogonal vectors in the two-dimensional space E_2, and take a vector $\eta = b_1 \xi_1 + b_2 \xi_2$. With this coordinate system and the observation that E_2^* is isomorphic to E_2, we see that a member x' of E_2^* may be expressed as $x' = (x_1, x_2)$, $x_1, x_2 \in \mathbf{R}$. With respect to the measure μ', $x_1 = \langle x', \xi_1 \rangle$ and $x_2 = \langle x', \xi_2 \rangle$ are independent standard Gaussian random variables, and whilst they remain independent Gaussian random variables with unit variances with respect to $T_\eta \circ \mu'$, their expectations in this case are $-b_1$ and $-b_2$ respectively. The two measures μ' and $T_\eta \circ \mu'$ are of course equivalent, and the density $dT_\eta \circ \mu'/d\mu'$ is

$$\begin{aligned} \frac{dT_\eta \circ \mu'}{d\mu'}(x') &= \exp\left[-\frac{1}{2}\{(x_1 + b_1)^2 + (x_2 + b_2)^2 - x_1^2 - x_2^2\}\right] \\ &= \exp\left[-\langle x', \eta \rangle - \frac{1}{2}\|\eta\|^2\right], \qquad (x' \equiv (x_1, x_2)). \end{aligned} \tag{5.61}$$

Returning to (5.59) we obtain

$$C_\eta(\xi) = \int_{E_2^*} \int_{E_2^{\perp *}} \exp[i\langle x, \xi\rangle]\, d\mu'_\eta(x' + \eta)\, d\mu''(x'') \qquad (x = x' + x'')$$

$$= \int_{E_2^{\perp *}} \exp[i\langle x'', \xi\rangle]\, d\mu''(x'')$$

$$\times \int_{E_2^*} \exp[i\langle x', \eta\rangle] \exp\left[-\langle x', \eta\rangle - \frac{1}{2}\|\eta\|^2\right] d\mu'(x')$$

$$= \int_{E^*} \exp[i\langle x, \xi\rangle] \exp\left[-\langle x, \eta\rangle - \frac{1}{2}\|\eta\|^2\right] d\mu(x),$$

since $\langle x, \eta\rangle = \langle x', \eta\rangle$.

Comparing the last integral with the definition (5.59) of $C_\eta(\xi)$ we see that

$$\frac{dT_\eta \circ \mu}{d\mu}(x) = \exp\left[-\langle x, \eta\rangle - \frac{1}{2}\|\eta\|^2\right]. \tag{5.62}$$

In general if a measure μ is transformed into an equivalent measure under a measurable transformation T, then μ is said to be *quasi-invariant* relative to T or simply *T-quasi-invariant*. The definition of quasi-invariance can be generalised to the case of a group G of measurable transformations. If μ is quasi-invariant relative to each member of G, then μ is said to be *quasi-invariant* relative to G or to be *G-quasi-invariant*.

As we saw above, the measure μ of white noise is quasi-invariant relative to T_η, $\eta \in E$, defined by (5.57). The set $\mathbf{T} = \{T_\eta : \eta \in E\}$ of such operators admits a natural product

$$(aT_{\eta_1})(bT_{\eta_2}) = T_{a\eta_1 + b\eta_2}, \qquad a, b \in \mathbf{R}.$$

Thus $\mathbf{T}$ becomes an abelian group with operator domain $\mathbf{R}$ (i.e. a real vector space) and it is isomorphic to E. We summarise our results in terms of $\mathbf{T}$.

Theorem 5.7. i. *The measure μ of white noise is $\mathbf{T}$-quasi-invariant.*

ii. *For each $T_\eta \in \mathbf{T}$, the density $\dfrac{dT_\eta \circ \mu}{d\mu}$ is given by (5.62).*

Corollary. *For each $\eta \in E$ the mapping U_η on (L^2) defined by*

$$(U_\eta \varphi)(x) = \varphi(x + \eta) \exp\left[-\frac{1}{2}\langle x, \eta\rangle - \frac{1}{4}\|\eta\|^2\right], \qquad \varphi \in (L^2), \quad (5.63)$$

is a unitary operator on (L^2).

Clearly $\{U_\eta : \eta \in E\}$ forms a unitary group on (L^2) and the continuity of U_η in η follows from the next proposition.

5.7 Transformations of White Noise (II): Translations

Proposition 5.7. *Let η tend to 0 in E. Then for any $\varphi \in (L^2)$ we have*

$$(U_\eta \varphi)(x) \to \varphi(x), \text{ strongly in } (L^2). \tag{5.64}$$

PROOF. We consider the algebra **A** introduced in §4.2, a dense linear subspace of (L^2). The proposition will be proved if we can prove (5.64) for any element φ of **A**, because for a general φ and $\varepsilon > 0$ we can find $\psi \in \mathbf{A}$ such that $\|\varphi - \psi\| < \varepsilon$. Then it follows from the unitarity of U_η that

$$\|U_\eta \varphi - \varphi\| \leq \|U_\eta \varphi - U_\eta \psi\| + \|U_\eta \psi - \psi\| + \|\psi - \varphi\|$$
$$\leq 2\|\varphi - \psi\| + \|U_\eta \psi - \psi\| \leq 2\varepsilon + \|U_\eta \psi - \psi\|.$$

Moreover elements ψ of **A** are functionals of the form $\psi(x) = \sum_j a_j \exp[i\langle x, \xi_j \rangle]$ (finite sum), so that it is enough to show (5.64) holds for $\psi(x) = \exp[i\langle x, \xi \rangle]$. For such a ψ we have:

$$\int_{E^*} \left| \exp[i\langle x, \xi \rangle + i\langle \eta, \xi \rangle] \exp\left[-\frac{1}{2}\langle x, \eta \rangle - \frac{1}{4}\|\eta\|^2\right] - \exp[i\langle x, \xi \rangle] \right|^2 d\mu(x)$$

$$= \int_{E^*} \left| \exp\left[i\langle \xi, \eta \rangle - \frac{1}{4}\|\eta\|^2\right] \exp\left[-\frac{1}{2}\langle x, \eta \rangle\right] - 1 \right|^2 d\mu(x)$$

$$= 2 - 2\cos(\langle \xi, \eta \rangle) \exp\left(-\frac{1}{8}\|\eta\|^2\right) \to 0 \quad \text{as} \quad \eta \to 0.$$

Thus the assertion is proved. □

It follows from this proposition that $(\{U_\eta : \eta \in E\}, (L^2))$ is a unitary representation of the topological group E.

Let G be a transformation group acting on E^*. The concept of G-ergodicity of a G-invariant measure, introduced in the last section, can be generalised to the case where the measure in question is *quasi*-invariant under each member of G. In this case the measure is said to be G-ergodic if any G-invariant measurable set is either the whole space or the empty set, mod 0.

Theorem 5.8 [Y. Umemura (1965)]. *The measure μ of white noise is* **T**-*ergodic.*

PROOF. Let $\varphi(x)$ be a bounded (L^2)-functional and assume that $\varphi(T_\eta x) = \varphi(x)$ a.e. for every $\eta \in E$. Then we must show that $\varphi(x) = 0$, a.e. Take a complete orthonormal system $\{\xi_n\} \subset L^2(\mathbf{R})$ and let $\mathfrak{B}_n = \mathfrak{B}\{\langle x, \xi_j \rangle : 1 \leq j \leq n\}$. Set

$$\varphi_n = E(\varphi | \mathfrak{B}_n), \qquad \varphi_n^\perp = \varphi - \varphi_n.$$

Then $\varphi_n^\perp$ is the component of φ orthogonal to $L^2(E^*, \mathfrak{B}_n, \mu)$. We also have

$$\varphi(T_\eta x) \equiv \varphi(x + \eta)$$
$$= \varphi_n(x + \eta) + \varphi_n^\perp(x + \eta),$$

and we note that $\varphi_n(x+\eta)$ is an $L^2(E^*, \mathfrak{B}_n, \mu)$-functional. For $\varphi_n(x)$ can be expressed in the form $\varphi_n(x) = f(\langle x, \xi_1 \rangle, \langle x, \xi_2 \rangle, \ldots, \langle x, \xi_n \rangle)$ for some function f on $\mathbf{R}^n$, whence $\varphi_n(x+\eta) = f(\langle x, \xi_1 \rangle + \langle \eta, \xi_1 \rangle, \ldots, \langle x, \xi_n \rangle + \langle \eta, \xi_n \rangle)$, still a $\mathfrak{B}_n$-measurable function. We now discuss $\varphi_n^\perp(x+\eta)$. If η is a linear combination $\eta = \sum_1^n b_j \xi_j$ and $\psi(x) = g(\langle x, \xi_1 \rangle, \langle x, \xi_2 \rangle, \ldots, \langle x, \xi_n \rangle)$ belongs to $L^2(E^*, \mathfrak{B}_n, \mu)$,

$$\int_{E^*} \varphi_n^\perp(x+\eta) \overline{\psi(x)} \, d\mu(x) = \int_{E^*} \varphi_n^\perp(x) \overline{g(\langle x, \xi_1 \rangle - b_1, \ldots, \langle x, \xi_n \rangle - b_n)}$$

$$\times \exp\left[\sum_1^n b_j \langle x, \xi_j \rangle - \frac{1}{2} \|\eta\|^2\right] d\mu(x)$$

and this last integral is 0 by definition of $\varphi_n^\perp$, since $\varphi_n^\perp(x+\eta)$ is orthogonal to $L^2(E^*, \mathfrak{B}_n, \mu)$. Thus a rewriting of the assumption

$$\varphi_n(x+\eta) + \varphi_n^\perp(x+\eta) = \varphi_n(x) + \varphi_n^\perp(x)$$

in the form

$$\varphi_n(x+\eta) - \varphi_n(x) = \varphi_n^\perp(x) - \varphi_n^\perp(x+\eta)$$

together with the preceding discussion, shows that both sides of this equation vanish. Now as η runs over the vector spaced spanned by $\xi_1, \xi_2, \ldots, \xi_n$, $\varphi_n(x)$ is invariant under all translations of the space variable when viewed as a function on $\mathbf{R}^n$, and so it must be a constant. Finally, we note that $\varphi_n \to \varphi$ in (L^2) as $n \to \infty$ shows that $\varphi(x) = $ constant, a.e. □

The translations T_η discussed so far have been limited to those for which $\eta \in E$. As suggested in our opening observations, η may vary over a much wider class if only $T_\eta \circ \mu \sim \mu$ (equivalence) is required. Take a general f and consider the mapping

$$T_f: x \to x + f, \qquad x \in E^*. \tag{5.65}$$

Then we find that $T_f \circ \mu$ can be readily defined. The σ-field $\mathfrak{B}_n$ is defined as before, and, noting that $T_f \mathfrak{B}_n = \mathfrak{B}_n$, we can now employ the technique used in the proof of Kakutani's theorem of §1.3. Let μ^n and $T_f \mu^n$ be the restrictions of μ and $T_f \circ \mu$ to $\mathfrak{B}_n$, respectively. Then $\mu^n \sim T_f \mu^n$. We also know that a necessary and sufficient condition for μ and $T_f \circ \mu$ to be equivalent is that $\psi_n = (dT_f \mu^n / d\mu^n)^{1/2}$ forms a Cauchy sequence in (L^2). If this condition is satisfied, then the mean square limit $(\lim_n \psi_n)^2$ turns out to be the density $dT_f \circ \mu/d\mu$. The details can be given as follows: set $\langle f, \xi_n \rangle = a_n, n \geq 1$. Then μ^n is an n-dimensional Gaussian distribution and $T_f \mu^n$ is obtained by translating μ^n by $(a_1, a_2, \ldots, a_n)$. Hence

$$\psi_n(x) = \exp\left[-\frac{1}{2} \sum_1^n a_j \langle x, \xi_j \rangle - \frac{1}{4} \sum_1^n a_j^2\right],$$

5.7 Transformations of White Noise (II): Translations

and it is then easily seen that $\{\psi_n\}$ forms a Cauchy sequence if and only if $\sum_1^\infty a_j^2 < \infty$, i.e. if and only if $f \in L^2(\mathbf{R})$. We thus have

Theorem 5.9. *The measure μ of white noise and the induced measure $T_f \circ \mu$ are equivalent if and only if $f \in L^2(\mathbf{R})$. In the case $f \in L^2(\mathbf{R})$ we have*

$$\frac{dT_f \circ \mu}{d\mu}(x) = \exp\left[-\langle x, f\rangle - \frac{1}{2}\|f\|^2\right]. \tag{5.66}$$

Remark. The term $\langle x, f\rangle$ in the right-hand side of (5.66) denotes the sum $\sum_n a_n \langle x, \xi_n\rangle$ taken in the sense of strong convergence in (L^2). This sum does not depend upon the choice of $\{\xi_n\}$, and agrees with the expression defining $\langle x, f\rangle$ at the beginning of §4.5.

The above result can be extended to the case where f is a linear functional of x, say $f = f(x, a)$, satisfying (a) and (b) below. Let L_u^2 be as in §4.4,

a. $f(\cdot, u) \in \mathcal{H}_1 \cap L_u^2$ for every u;
b. $f(x, \cdot) \in L^2(\mathbf{R})$ for almost all x.

Since $f(\cdot, u)$ is a linear functional of x (and hence a Gaussian random variable), condition (b) is equivalent to $E[\int f(\cdot, u)^2 \, du] < \infty$. With such an f we can define the transformation T_f on E^*:

$$T_f: x \to x + f(x, \cdot), \qquad x \in E^*.$$

Then $T_f \circ \mu$ can be defined and is equivalent to μ.

Conversely, if μ' is a Gaussian measure equivalent to μ, then there exists a unique f satisfying conditions (a) and (b) such that $\mu' = T_f \circ \mu$.

This result is of some interest in its own right, and one with a wide range of applications, but we will have to forego the opportunity to consider greater detail because of the need for further background before doing so. Readers who are interested in this topic are advised to consult L. Shepp (1966), M. Hitsuda (1968) and Yu. A. Rozanov (1971). We also recall that a related topic was discussed in §4.6 (iii)

Now let us return to the group $\mathbf{T}$ to define a larger group involving $O^*(E^*)$. On the product set $O^*(E^*) \times E = \{m = (g^*, \eta): g^* \in O^*(E^*), \eta \in E\}$ we introduce a binary operation as follows: for $m_i = (g_i^*, \eta_i)$, $i = 1, 2$.

$$m_1 m_2 = (g_1^* g_2^*, \eta_1 + g_1^* \eta_2). \tag{5.67}$$

When equipped with this operation the set $O^*(E^*) \times E$ forms a group, and we can also introduce a topology onto the set, namely the product of the μ-topology on $O^*(E^*)$ and the nuclear topology in E. Thus topologised the group is denoted by $M(E^*)$ or M_∞.

Definition 5.3. The topological group $M(E^*)$ ($= M_\infty$) is called the *infinite-dimensional motion group*.

The unit element of $M(E^*)$ is $(e^*, 0)$ where e^* is the unit element of $O^*(E^*)$, and the inverse element of (g^*, η) is $(g^{*-1}, -g^{*-1}\eta)$.
Let the element $m = (g^*, \eta)$ of $M(E^*)$ act on E^* as follows:

$$mx = T_\eta(g^*x) = g^*x + \eta, \qquad x \in E^*. \tag{5.68}$$

Then $M(E^*)$ becomes a transformation group on E^* which behaves rather like the motion group of a finite-dimensional Euclidean space.

Proposition 5.8. *The measure μ of white noise is M_∞-quasi-invariant.*

PROOF. The result is an immediate of the $O^*(E^*)$-invariance and T_η-quasi-invariance of μ. □

For $m \in M(E^*)$ define a unitary operator U_m by

$$U_m = U_g U_\eta, \qquad m = (g^*, \eta), \tag{5.69}$$

when U_η is the unitary operator defined by (5.63) and U_g is the one given in §5.1. The mapping $m \to U_m$, $m \in M(E^*)$, is continuous, and so $(\{U_m : m \in M(E^*)\}, (L^2))$ is a unitary representation of the motion group $M(E^*)$.

We now seek relationships whose forms are similar, although necessarily somewhat modified, to those connecting the Fourier transform and the motion group in the finite-dimensional case. Let $\mathfrak{J}$ be the Fourier-Wiener transform introduced onto (L^2) in §4.7, and set

$$V_m = \mathfrak{J} U_m \mathfrak{J}^{-1},$$
$$V_\eta = \mathfrak{J} U_\eta \mathfrak{J}^{-1},$$
and
$$V_g = \mathfrak{J} U_g \mathfrak{J}^{-1}.$$

Clearly we have $V_m = V_g V_\eta$ when $m = (g^*, \eta)$, and the action of these operators will now be explained.

Proposition 5.9. *For all $\eta \in E$ and $g \in O(E)$ the operators V_η and V_g act on (L^2) as follows:*

$$(V_\eta \varphi)(x) = \exp\left[-\frac{1}{2} i \langle x, \eta \rangle\right] \varphi(x), \tag{5.70}$$

$$(V_g \varphi)(x) = \varphi(g^*x)(= (U_g \varphi)(x)), \qquad \varphi \in (L^2). \tag{5.71}$$

5.7 Transformations of White Noise (II): Translations

PROOF. We first derive the formula (5.70). Let ψ be an element of (L^2). Then we have

$$\mathfrak{J}(U_\eta \psi)(y)$$

$$= \int_{E^*} \psi(\sqrt{2}x + \eta + iy) \exp\left[-\frac{1}{2}\langle \sqrt{2}x + iy, \eta \rangle - \frac{1}{4}\|\eta\|^2\right] d\mu(x)$$

$$= \exp\left[-\frac{1}{2}i\langle y, \eta\rangle - \frac{1}{4}\|\eta\|^2\right] \int_{E^*} \psi(\sqrt{2}x' + iy)$$

$$\times \exp\left[-\frac{1}{2}\langle \sqrt{2}x', \eta\rangle + \frac{1}{2}\|\eta\|^2\right] d\mu(x' - \eta/\sqrt{2})$$

$$= \exp\left[-\frac{1}{2}i\langle y, \eta\rangle\right] \int_{E^*} \varphi(\sqrt{2}x' + iy)\, d\mu(x')$$

$$= \exp\left[-\frac{1}{2}i\langle y, \eta\rangle\right] (\mathfrak{J}\psi)(y).$$

If we set $\varphi = \mathfrak{J}\psi$ then we obtain (5.70).

The equality (5.71) can be proved by imitating the proof of Proposition 4.14, with T_t there replaced by g^*, and for this reason we omit the details. □

There is one important aspect in which the unitary operator U_η differs from U_g and $\mathfrak{J}$, and this is the fact that both U_g and $\mathfrak{J}$ leave invariant all the subspaces $\mathcal{H}_n$ (i.e. $U_g \mathcal{H}_n = \mathcal{H}_n$, $\mathfrak{J}\mathcal{H}_n = \mathcal{H}_n$, $n = 0, 1, 2, \ldots$), whereas U_η is a key operator changing the degree of multiple Wiener integrals. In order to describe the action of U_η, it is convenient to discuss its infinitesimal generator.

Fix a unit vector η (i.e. $\|\eta\| = 1$) of E and observe that in this case both $\{U_{t\eta}: -\infty < t < \infty\}$ and $\{V_{t\eta}: -\infty < t < \infty\}$ are one-parameter unitary groups which are continuous in t. We can then use Stone's theorem to describe their infinitesimal generators. If U_t has a spectral representation

$$U_{t\eta} = \int e^{it\lambda}\, dE_\eta(\lambda),$$

where $\{E_\eta(\lambda)\}$ is a resolution of the identity, then the infinitesimal generator $A_\eta = (d/dt)U_{t\eta}|_{t=0}$ may be expressed as

$$A_\eta = i\int \lambda\, dE_\eta(\lambda).$$

In a similar manner $B_\eta = (d/dt)V_{t\eta}|_{t=0}$ can be obtained, and so we now turn our attention to obtaining explicit expressions for A_η and B_η.

We begin by fixing some notation. As before $\eta \in E$ with $\|\eta\| = 1$ is fixed, and let $\varphi(x)$ be a finite-dimensional (i.e. tame) functional $\varphi(x) = \tilde{f}(\langle x, \tilde{\xi}_n\rangle, \ldots, \langle x, \tilde{\xi}_n\rangle)$. Then we take a complete orthonormal system $\xi_0 = \eta$,

$\xi_1, \ldots, \xi_{n'}$ spanning the vector space generated by $\eta, \tilde{\xi}_1, \ldots, \tilde{\xi}_n$, and re-express $\varphi(x)$ as $f(\langle x, \xi_0 \rangle, \ldots, \langle x, \xi_{n'} \rangle)$. If $(\partial/\partial t_0) f(t_0, \ldots, t_{n'})$ exists, we define $\partial/\partial \eta$ by

$$\frac{\partial}{\partial \eta} \varphi(x) = \frac{\partial}{\partial t_0} f(t_0, \ldots, t_{n'}) \bigg|_{t_j = \langle x, \xi_j \rangle, \, j = 0, 1, \ldots, n'} \tag{5.72}$$

Clearly $\partial/\partial \eta$ is defined for polynomials and exponential functions. We define another operator $\eta \cdot$ by

$$(\eta \cdot \varphi)(x) = \langle x, \eta \rangle \varphi(x). \tag{5.73}$$

Proposition 5.10 a. *The domains of A_η and B_η are both dense in (L^2).*
b. *The operators A_η and B_η are given by*

$$A_\eta = \frac{\partial}{\partial \eta} - \frac{1}{2} \eta \cdot, \quad B_\eta = -\frac{1}{2} i \eta \cdot. \tag{5.74}$$

respectively, when applied to polynomial or exponential functions.

PROOF. The assertions (a) and (b) are both proved by the following computations. If $\varphi(x)$ is a polynomial or an exponential function, then

$$\frac{d}{dt}(U_{t\eta}\varphi)(x)\bigg|_{t=0} = \frac{d}{dt}\left\{\varphi(x + t\eta)\exp\left[-\frac{1}{2}t\langle x, \eta\rangle - \frac{1}{4}t^2\|\eta\|^2\right]\right\}\bigg|_{t=0}$$

$$= \frac{\partial}{\partial \eta}\varphi(x) - \frac{1}{2}(\eta \cdot \varphi)(x),$$

and

$$\frac{d}{dt}(V_{t\eta}\varphi)(x)\bigg|_{t=0} = \frac{d}{dt}\left\{\exp\left[-\frac{1}{2}it\langle x, \eta\rangle\right]\varphi(x)\right\}\bigg|_{t=0}$$

$$= -\frac{1}{2}i(\eta \cdot \varphi)(x).$$

In both cases the (L^2)-norm is used in taking limits, and the results are still in (L^2). □

We can now discuss the important question of how the operators A_η and B_η act on the space $\mathscr{H}_n$. Take a Fourier-Hermite polynomial $\varphi(x)$ of degree n based upon a complete orthonormal system $\{\xi_j\}$ in $L^2(\mathbf{R})$:

$$\varphi(x) = \prod_k H_{n_k}\left(\frac{\langle x, \xi_k \rangle}{\sqrt{2}}\right), \quad \sum n_k = n.$$

Since

$$A_{\xi_j}\varphi(x) = \frac{d}{dt} H_{n_j}\left(\frac{\langle x, \xi_j \rangle + t}{\sqrt{2}}\right)\bigg|_{t=0} \prod_{k \neq j} H_{n_k}\left(\frac{\langle x, \xi_k \rangle}{\sqrt{2}}\right) - \frac{1}{2}\langle x, \xi_j\rangle \varphi(x),$$

we have

$$\begin{cases} A_{\xi_j}\varphi(x) = -\dfrac{1}{\sqrt{8}} \prod_k H_{n_k'}\left(\dfrac{\langle x, \xi_k\rangle}{\sqrt{2}}\right), & \text{if } n_j = 0, \\ & (n_k' = n_k, k \neq j, \text{ and } n_j' = 1) \\ A_{\xi_j}\varphi(x) = \dfrac{n_j}{\sqrt{2}} \prod_k H_{n_k'}\left(\dfrac{\langle x, \xi_k\rangle}{\sqrt{2}}\right) - \dfrac{1}{2\sqrt{2}} \prod_k H_{n_k''}\left(\dfrac{\langle x, \xi_k\rangle}{\sqrt{2}}\right), \\ & \text{if } n_j > 0, \\ & (n_k' = n_k'' = n_k, k \neq j, \text{ and } n_j' = n_j - 1, n_j'' = n_j + 1). \end{cases} \quad (5.75)$$

See the formulae in Appendix §A.5 (i) for details. Similarly, we obtain formulae for B_{ξ_j}:

$$B_{\xi_j}\varphi(x) = -\dfrac{i}{\sqrt{8}} \prod_k H_{n_k'}\left(\dfrac{\langle x, \xi_k\rangle}{\sqrt{2}}\right), \quad \text{for } n_j = 0,$$

$$(n_k' = n_k, k \neq j, \text{ and } n_j' = 1) \quad (5.76)$$

$$B_{\xi_j}\varphi(x) = -\dfrac{i}{\sqrt{2}} n_j \prod_k H_{n_k'}\left(\dfrac{\langle x, \xi_k\rangle}{\sqrt{2}}\right) - \dfrac{i}{2\sqrt{2}} \prod_k H_{n_k''}\left(\dfrac{\langle x, \xi_k\rangle}{\sqrt{2}}\right),$$

$$\text{if } n_j > 0,$$

$$(n_k' = n_k'' = n_k, k \neq j, \text{ and } n_j' = n_j - 1, n_j'' = n_j + 1).$$

Thus we see that both A_ξ and B_ξ carry polynomials in $\mathcal{H}_n$ to polynomials in $\mathcal{H}_{n-1} \oplus \mathcal{H}_{n+1}$. Connections between this result and quantum mechanics can be found in H. Weyl (1928, Chapter 2).

The last part of this section consists of a brief discussion of the infinite-dimensional Laplace-Beltrami operator Δ. Y. Umemura (1966) introduced the operator

$$\Delta = \sum_j \left(\dfrac{\partial^2}{\partial \xi_j^2} - \xi_j \cdot \dfrac{\partial}{\partial \xi_j} \right) \quad (5.77)$$

referred to as the Laplacian, and showed that Δ enjoys properties appropriate to a Laplacian on (L^2). The main characterisation is the following one: Suppose that a symmetric operator H on (L^2) has a sufficiently large domain (e.g. including all exponential functions) and that H commutes with all U_g, $g \in O(E)$. Then H is expressible as a function of Δ, $H = f(\Delta)$.

Using the generator A_η derived from the translation of the variable of (L^2)-functionals, and also the operator $B_\eta = \mathfrak{J} A_\eta \mathfrak{J}^{-1}$, we can express Δ in the form

$$\Delta = \sum_j \left(A_{\xi_j}^2 + B_{\xi_j}^2 + \dfrac{1}{2} \right). \quad (5.78)$$

If we recall the formulae (5.75) and (5.76) describing the action of A_{ξ_j} and B_{ξ_j} on a Fourier-Hermite polynomial $\varphi(x)$ based on the $\{\xi_j\}$, we can immediately prove the relation

$$\Delta \varphi = -n\varphi, \quad n = \text{the degree of } \varphi. \quad (5.79)$$

Since Δ is the sum of infinitely many operators, convergence of the sum needs to be considered when Δ is applied to a general φ. With this in mind we state a few results concerning Δ.

Proposition 5.11. *The operator Δ is Hermitian, and its domain contains $\sum_n \mathcal{H}_n$ (algebraic sum). The subspace $\mathcal{H}_n$ is the eigenspace corresponding to the eigenvalue $-n$, $n \geq 0$.*

PROOF. Since Δ is defined and satisfies (5.79) for all Fourier-Hermite polynomials of degree n, and these form a base for $\mathcal{H}_n$, it is possible to extend Δ to a Hermitian operator defined on all of $\mathcal{H}_n$. The remainder of the statement then follows immediately. □

Corollary. *The definition of the extended operator Δ of the proposition is independent of the choice of the complete orthonormal system $\{\xi_n\}$.*

The next proposition is an immediate consequence of this last corollary.

Proposition 5.12. *The operator Δ commutes with all U_g, $g \in O(E)$, i.e.*

$$\Delta U_g \varphi = U_g \Delta \varphi, \quad \varphi \in \sum_n \mathcal{H}_n. \quad (5.80)$$

Proposition 5.13. *The operator Δ has an extension to the class $\mathbf{A}$ of all exponential functions, which acts as follows:*

$$\Delta \exp[a\langle x, \xi\rangle] = (a^2 \|\xi\|^2 - a\langle x, \xi\rangle)\exp[a\langle x, \xi\rangle], \quad a \in \mathbf{C}. \quad (5.81)$$

PROOF. Choose a system $\{\xi_n\}$ and expand the given ξ as $\xi = \sum_n \alpha_n \xi_n$. Then we have the expression

$$\left(\frac{\partial^2}{\partial \xi_j^2} - \xi \cdot \frac{\partial}{\partial \xi_j}\right)\exp\left[a \sum_n \alpha_n \langle x, \xi_n\rangle\right]$$

$$= (a^2 \alpha_j^2 - a\alpha_j \langle x, \xi_j\rangle)\exp\left[a \sum_n \alpha_n \langle x, \xi_n\rangle\right].$$

Summing over j we obtain the (L^2)-convergent expression

$$\Delta \exp[a\langle x, \xi\rangle] = \left(a^2 \sum_n \alpha_n^2 - a \sum_n \langle x, \alpha_n \xi_n\rangle\right)\exp[a\langle x, \xi\rangle]$$

which leads to (5.81). □

We have seen that the extended operator Δ has a domain including $\sum_n \mathscr{H}_n$ and $\mathbf{A}$, and that it is indeed analogous to the Laplace-Beltrami operator in many respects.

5.8 The Canonical Commutation Relations of Quantum Mechanics

The representation theory of the canonical commutation relations, which was established by von Neumann (1931) in the case of finitely many degrees of freedom, can be extended to the case of a Boson field, and in this case the Hilbert space (L^2) of the measure of white noise arises naturally.

Let us quickly review von Neumann's result for a system of n particles. In quantum mechanics the positions p_j and momenta q_j, $j = 1, 2, \ldots, n$, are understood to be self-adjoint operators on some Hilbert space H, satisfying the commutation relations

$$[q_j, p_k] = i\hbar \delta_{j,k} I,$$
$$[q_j, q_k] = [p_j, p_k] = 0, \qquad 1 \leq j, k \leq n. \tag{5.82}$$

They can be represented as operators on the Hilbert space $L^2(\mathbf{R}^n)$ as

$$p_j \varphi(x) = \left(\frac{\hbar}{i}\right) \frac{\partial}{\partial x_j} \varphi(x), \qquad x = (x_1, \ldots, x_n),$$
$$q_j \varphi(x) = x_j \varphi(x), \qquad \varphi \in L^2(\mathbf{R}^n), \tag{5.83}$$

and this representation is unique up to unitary equivalence. In order to find such a representation von Neumann introduced the system $\{P(\alpha), Q(\beta): \alpha, \beta \in \mathbf{R}^n\}$ of unitary operators derived from the system $\{p_j, q_k: 1 \leq j, k \leq n\}$ by

$$P(\alpha) = \exp\left[\left(\frac{i}{\hbar}\right) \sum_j \alpha_j p_j\right], \quad Q(\beta) = \exp\left[\left(\frac{i}{\hbar}\right) \sum \beta_k q_k\right].$$

These operators satisfy the so-called Weyl commutation relations:

$$\begin{cases} Q(\alpha)P(\beta) = \exp\left[-\left(\frac{i}{\hbar}\right)(\alpha, \beta)\right] P(\beta)Q(\alpha), \\ P(\alpha)P(\beta) = P(\alpha + \beta), \quad Q(\alpha)Q(\beta) = Q(\alpha + \beta). \end{cases} \qquad \alpha, \beta \in \mathbf{R}^n, \tag{5.84}$$

With these relations he was able to determine the type of the representation. More precisely, the problem was to determine a Hilbert space and system $\{P(\alpha), Q(\beta): \alpha, \beta \in \mathbf{R}^n\}$ of unitary operators satisfying the conditions (i) $\{P(\alpha), Q(\beta)\}$ is irreducible, (ii) the commutation relations (5.84) are satisfied, and (iii) $P(\alpha)$ and $Q(\beta)$ are continuous in α and β respectively. For simplicity we take the constant $\hbar$ to be 1 in what follows.

An explicit representation is the following: take H to be $L^2(\mathbf{R}^n)$ and define $P(\alpha)$ and $Q(\beta)$ by

$$(P(\alpha)\varphi)(x) = \varphi(x + \alpha), \quad (Q(\beta)\varphi)(x) = \exp[i(\beta, x)]\varphi(x), \quad (5.85)$$

$\varphi \in L^2(\mathbf{R}^n)$. Obviously conditions (ii) and (iii) are satisfied and irreducibility i.e. (i) above follows from the fact that for any $\varphi \neq 0$ the closed linear subspace spanned by $\{P(\alpha)Q(\beta)\varphi: \alpha, \beta \in \mathbf{R}^n\}$ is the entire space $L^2(\mathbf{R}^n)$.

We now show the uniqueness. Suppose that a representation satisfying (ii) and (iii) is given, and set

$$S(\alpha, \beta) = \exp\left[-\frac{1}{2}i(\alpha, \beta)\right]P(\alpha)Q(\beta), \quad \alpha, \beta \in \mathbf{R}^n.$$

This is a unitary operator and we obtain the relation

$$S(\alpha, \beta)S(\gamma, \delta) = \exp\left[\frac{1}{2}i\{(\alpha, \delta) - (\beta, \gamma)\}\right]S(\alpha + \gamma, \beta + \delta), \quad \alpha, \beta, \gamma, \delta \in \mathbf{R}^n, \quad (5.86)$$

and this is seen to be equivalent to (5.84). It is also easily seen that the adjoint of $S(\alpha, \beta)$ is $S(-\alpha, -\beta)$. Now set

$$A = \int_{\mathbf{R}^{2n}} \exp\left[-\frac{1}{4}(|\alpha|^2 + |\beta|^2)\right]S(\alpha, \beta)\, d\alpha\, d\beta.$$

Then A is a bounded self-adjoint operator on H, and it gives us a relation

$$AS(\gamma, \delta)A = (2\pi)^n \exp\left[-\frac{1}{4}(|\gamma|^2 + |\delta|^2)\right]A,$$

and, in particular,

$$A^2 = (2\pi)^n A.$$

We now take the subspace $M = \{f \in H: Af = (2\pi)^n f\}$ of H. Since A is a bounded operator, M is a closed subspace. An alternative expression for M is as $\{Ag: g \in H\}$, and we also see that $Af = 0$ for any $f \in H \ominus M$. The action of $S(\alpha, \beta)$ may now be illustrated by the following calculation. For $f, g \in M$ we have

$(S(\alpha, \beta)f, S(\gamma, \delta)g)$

$= (2\pi)^{-2n}(S(\alpha, \beta)Af, S(\gamma, \delta)Ag)$

$= (2\pi)^{-2n}(AS(-\gamma, -\delta)S(\alpha, \beta)Af, g)$

$= (2\pi)^{-2n} \exp\left[\frac{1}{2}i\{(\alpha, \delta) - (\beta, \gamma)\}\right](AS(\alpha - \gamma, \beta - \delta)Af, g)$

$= (2\pi)^{-n} \exp\left[-\frac{1}{4}\{|\alpha - \gamma|^2 + |\beta - \delta|^2\} + \frac{1}{2}i\{(\alpha, \delta) - (\beta, \gamma)\}\right](Af, g)$

$= \exp\left[-\frac{1}{4}\{|\alpha - \gamma|^2 + |\beta - \delta|^2\} + \frac{1}{2}i\{(\alpha, \delta) - (\beta, \gamma)\}\right](f, g).$

5.8 The Canonical Commutation Relations of Quantum Mechanics

Thus if $\{f_n\}$ is a complete orthonormal system in M, we have the relations
$(S(\alpha, \beta)f_m, S(\gamma, \delta)f_n)$

$$= \exp\left[-\frac{1}{4}\{|\alpha - \gamma|^2 + |\beta - \delta|^2\} + \frac{1}{2}i\{(\alpha, \delta) - (\beta, \gamma)\}\right]\delta_{m,n}.$$

Once we fix f_n, $\{S(\alpha, \beta)\}$ acts on the cyclic subspace
$$M_n = \{S(\alpha, \beta)f_n : \alpha, \beta \in \mathbf{R}^n\}$$
in the following way:

$$S(\gamma, \delta)f_{\alpha, \beta} = \exp\left[\frac{1}{2}i\{(\beta, \gamma) - (\alpha, \delta)\}\right]f_{\alpha+\gamma, \beta+\delta}, \quad (5.87)$$

where $f_{\alpha, \beta} = S(\alpha, \beta)f_n$. Thus for any γ, δ we have

$$(S(\gamma, \delta)f_{\alpha, \beta}, f_{\alpha, \beta}) = \exp\left[-\frac{1}{4}(|\gamma|^2 + |\delta|^2) + i\{(\beta, \gamma) - (\alpha, \delta)\}\right], \quad (5.88)$$

and the right-hand side is independent of the choice of f_n. But this means that the system $\{S(\alpha, \beta)\}$ is uniquely determined up to unitary equivalence. Indeed if we took another f'_n and used $S(\alpha, \beta)f'_n = f'_{\alpha, \beta}$ to define M'_n, relation (5.88) would continue to hold, so that the mapping

$$U: \sum_k c_k S(\alpha_k, \beta_k)f_n \to \sum_k c_k S(\alpha_k, \beta_k)f'_n$$

can be extended to an isometry of M_n onto M'_n.

It is obvious that $\{S(\alpha, \beta)\}$ is irreducible on each M_n, and we have therefore proved that a system $\{P(\alpha), Q(\beta)\}$ satisfying (i), (ii) and (iii) is uniquely determined and equivalent to the system given by (5.85). One thing to be noted here is the fact that $S(\alpha, \beta)$ reduces to the zero operator on the subspace $H \ominus M$.

The operators P and Q have played a symmetrical role in the discussion so far. From their explicit form (5.85) we see that they are linked by the Fourier transform $\mathscr{F}$:

$$\mathscr{F}^{-1}P(\alpha)\mathscr{F} = Q(\alpha), \quad \text{equivalently,} \quad \mathscr{F}Q(\alpha)\mathscr{F}^{-1} = P(\alpha), \quad \alpha \in \mathbf{R}^n.$$

The triple $P(\alpha)$, $Q(\beta)$ and the multiplication by the scalar $e^{-i\theta}$, which has absolute value unity (and so defines a unitary operator), will be denoted by (α, β, θ), and with this notation the product

$$e^{-i\theta_1}P(\alpha_1)Q(\beta_1)e^{-i\theta_2}P(\alpha_2)Q(\beta_2) = e^{-i(\theta_1+\theta_2)}P(\alpha_1)Q(\beta_1)P(\alpha_2)Q(\beta_2)$$
$$= e^{-i(\theta_1+\theta_2)-i(\beta_1, \alpha_2)}P(\alpha_1 + \alpha_2)Q(\beta_1 + \beta_2)$$

can be expressed in the form

$$(\alpha_1, \beta_1, \theta_1)(\alpha_2, \beta_2, \theta_2) = (\alpha_1 + \alpha_2, \beta_1 + \beta_2, \theta_1 + \theta_2 + (\beta_1, \alpha_2)). \quad (5.89)$$

If we introduce the product (5.89) onto the set
$$G_n = \{(\alpha, \beta, \theta) : \alpha, \beta \in \mathbf{R}^n, \theta \in \mathbf{T}^1\} = \mathbf{R}^n \times \mathbf{R}^n \times \mathbf{T}^1,$$

T^1 denoting the 1-dimensional torus, we obtain a group G_n. Moreover, if we view G_n as the topological product of two copies of $\mathbf{R}^n$ and T^1, then G_n becomes a topological group, indeed a Lie group. And by (5.89) the mapping

$$g = (\alpha, \beta, \theta) \to U_g = e^{-i\theta} P(\alpha) Q(\beta), \qquad g \in G_n \qquad (5.90)$$

defines a unitary representation of G_n. The discussion earlier in this section is now seen to be asserting that there is a unique *irreducible unitary representation* of G_n. This paraphrasing leads us to consider unitary representations of the commutation relations for the Boson field.

The topological group G_n is isomorphic to a factor group of the Heisenberg group that will be introduced in §7.3.

We now generalise the group G_n to the case $n = \infty$, by letting E be a nuclear space and setting $G_B = \{(\xi, \eta, \theta): \xi, \eta \in E, \theta \in T^1\} = E \times E \times T^1$. We introduce a product into G_B: for $(\xi_i, \eta_i, \theta_i) \in G_B$, $i = 1, 2$, write

$$(\xi_1, \eta_1, \theta_1)(\xi_2, \eta_2, \theta_2) = (\xi_1 + \xi_2, \eta_1 + \eta_2, \theta_1 + \theta_2 + (\eta_1, \xi_2)). \quad (5.91)$$

Then G_B forms a group with this product, and when equipped with the product topology (as in the case of G_n), G_B becomes a topological group. In general a topological group is said to be a *nuclear Lie group* if there exists a neighbourhood of the unit element which is isomorphic to a neighbourhood of the zero element of a nuclear space. The group G_B just defined is an example of a nuclear Lie group.

It is easy to see that the nuclear Lie group G_B is generated by $(\xi, 0, 0)$ and $(0, \eta, 0)$, $\xi, \eta \in E$. Suppose that we are given a unitary representation of G_B, and let $P(\xi)$, $Q(\eta)$ be the unitary operators corresponding to the elements $(\xi, 0, 0)$ and $(0, \eta, 0)$ respectively. The centre of G_B consists of the elements $\{(0, 0, \theta): \theta \in T^1\} \simeq T^1$, and if the unitary representation is irreducible we may associate the multiplication operator $e^{-i\theta} \cdot$ with $(0, 0, \theta)$. By using the decomposition $(\xi, \eta, \theta) = (\xi, \eta, 0)(0, 0, \theta)$ and the formulae $(0, \eta, 0)(\xi, 0, 0) = (\xi, \eta, (\eta, \xi))$, $(\xi_1, 0, 0)(\xi_2, 0, 0) = (\xi_1 + \xi_2, 0, 0)$, and $(0, \eta_1, 0)(0, \eta_2, 0) = (0, \eta_1 + \eta_2, 0)$, which are all special cases of (5.91), we obtain the following commutation relations:

$$\begin{aligned} Q(\eta)P(\xi) &= e^{-i(\xi,\eta)}P(\xi)Q(\eta), \\ P(\xi_1)P(\xi_2) &= P(\xi_1 + \xi_2), \qquad Q(\eta_1)Q(\eta_2) = Q(\eta_1 + \eta_2). \end{aligned} \qquad (5.92)$$

This may be viewed as an extension of (5.84).

The above equations describe the Weyl commutation relations in the case of a Boson field (where the degree of freedom is the cardinality of the continuum), and we will illustrate this below. In this case a formal approach involves a system $\{p_t, q_t: t \in \mathbf{R}\}$ of self-adjoint operators satisfying the commutation relations $[q_t, p_{t'}] = i\hbar\delta_{t-t'}$ and $[p_t, p_{t'}] = [q_t, q_{t'}] = 0$. Let us "smear" these operators, forming $p(\xi) = \langle p_\cdot, \xi \rangle$ (more formally, $p(\xi) = \int \xi(t) p_t \, dt$) and $q(\eta) = \langle q_\cdot, \eta \rangle$ from $\{p_t\}$ and $\{q_t\}$ respectively. Then the self-

5.8 The Canonical Commutation Relations of Quantum Mechanics

adjoint operators $p(\xi)$ and $q(\eta)$ satisfy $[q(\eta), p(\xi)] = i(\xi, \eta)\hbar I$, and $[p(\xi), p(\eta)] = [q(\xi), q(\eta)] = 0$. Exponentiating, we are led to the formulae (5.92) satisfied by the unitary operators $P(\xi) = \exp[ip(\xi)]$ and $Q(\eta) = \exp[iq(\eta)]$.

Our aim is now seen to be the search for all possible irreducible representations of the nuclear Lie group G_B. More precisely, we seek a system $\{P(\xi), Q(\eta): \xi, \eta \in E\}$ of unitary operators on H satisfying the conditions (i) $\{P(\xi), Q(\eta)\}$ is irreducible on H; (ii) the commutation relations (5.92) are satisfied; and (iii) $P(\xi)$ and $Q(\eta)$ are continuous in ξ and η respectively.

On a purely formal level our approach to this problem is the same as that adopted in the case of G_n, when there was a finite number of degrees of freedom. A crucial difficulty arises, however, because of the absence of a measure like Lebesgue measure on $\mathbf{R}^n$. With this in mind we present a discussion of the problem which follows I. M. Gelfand and N. Ya Vilenkin (1964) and Y. Umemura (1965).

We begin by forming a cyclic unitary representation of the subgroup $\{(0, \xi, 0): \xi \in E\}$ of G_B. If we suppose that $f \in H$ is a cyclic vector of norm unity, and set

$$C(\xi) = (Q(\xi)f, f),$$

then $C(0) = 1$ and $C(\xi)$ is continuous in ξ. Moreover for $\alpha_j \in \mathbf{C}, \xi_j \in E, j = 1, 2, \ldots, n$,

$$\sum_{j,k=1}^{n} \alpha_j \bar{\alpha}_k C(\xi_j - \xi_k) = \sum_{j,k} \alpha_j \bar{\alpha}_k (Q(\xi_j - \xi_k)f, f)$$

$$= \sum_{j,k} (\alpha_j Q(\xi_j)f, \alpha_k Q(\xi_k)f) \quad [Q(-\xi_k) = Q(\xi_k)^*]$$

$$= \left\| \sum_j \alpha_j Q(\xi_j)f \right\|^2 \geq 0.$$

In other words, C is positive definite. Thus the Bochner-Minlos theorem (Theorem 3.1) implies the existence of a probability measure μ on E^* such that

$$C(\xi) = \int_{E^*} \exp[i\langle x, \xi \rangle] \, d\mu(x),$$

and so the Hilbert space $L^2(E^*, \mu) \equiv (L^2)$ is formed.

Up until now the Hilbert space H has been understood to be an abstract Hilbert space, but it is possible to realise H concretely by introducing the mapping U of H into $L^2(E^*, \mu)$ given by

$$U: g = \sum_j a_j Q(\xi_j)f \to Ug = \sum_j a_j \exp[i\langle x, \xi_j \rangle] \in (L^2).$$

Since we are dealing with a cyclic representation, a discussion similar to that in Theorem 4.1 of §4.2 enables us to extend U to an isometry of H onto (L^2). In addition, for g as above, we have

$$UQ(\xi)g = U\left(\sum_j a_j Q(\xi_j + \xi)f\right)$$

$$= \sum_j a_j \exp[i\langle x, \xi_j + \xi\rangle] = e^{i\langle x, \xi\rangle} \cdot Ug,$$

and hence for any $\varphi \in (L^2)$ we have the relation

$$(UQ(\xi)U^{-1})\varphi(x) = \exp[i\langle x, \xi\rangle] \cdot \varphi(x).$$

From now on we will identify H with (L^2) and accordingly $UQ(\xi)U^{-1}$ will be denoted simply by $Q(\xi)$. Thus we have determined a cyclic unitary representation of the subgroup $\{(0, \xi, 0)\}$ of G_B:

$$Q(\xi)\varphi(x) = \exp[i\langle x, \xi\rangle] \cdot \varphi(x), \qquad \varphi \in (L^2). \tag{5.93}$$

The next step involves the representation $\{P(\xi)\}$ of the subgroup $\{(\xi, 0, 0):$ $\xi \in E\}$ on the space (L^2) on which $Q(\xi)$ has been fixed. Denote by $1(x)$ the (L^2)-functional that is identically equal to 1, and set $(P(\xi)1)(x) = \psi_\xi(x)$. The commutation relations (5.92) imply that

$$(Q(\eta)\psi_\xi, \psi_\xi)_{(L^2)} = \int \exp[i\langle x, \eta\rangle] |\psi_\xi(x)|^2 \, d\mu(x)$$

$$= (\exp[-i(\xi, \eta)]P(\xi)Q(\eta)1, P(\xi)1)$$

$$= \exp[-i(\xi, \eta)] \int \exp[i\langle x, \eta\rangle] \cdot |1(x)|^2 \, d\mu(x)$$

$$= \int \exp[i\langle x - \xi, \eta\rangle] \, d\mu(x)$$

$$= \int \exp[i\langle x, \eta\rangle] \, dT_\xi \circ \mu(x)$$

(cf. the notation of §5.7). Hence it is necessary that μ be T_ξ-quasi-invariant in order for $P(\xi)$ to exist as a unitary operator on (L^2). If this is the case then we have

$$\frac{dT_\xi \circ \mu}{d\mu} = |\psi_\xi|^2. \tag{5.94}$$

This holds because of the one-to-one correspondence between characteristic functionals and probability measures, and thus it is possible to write

$$\psi_\xi(x) = d(\xi, x)\left(\frac{dT_\xi \circ \mu}{d\mu}(x)\right)^{1/2}, \tag{5.95}$$

where d is a functional satisfying $|d(\xi, x)| = 1$, a.e. (μ). At this stage we

5.8 The Canonical Commutation Relations of Quantum Mechanics

re-introduce $Q(\xi)$ by expressing a functional $\varphi(x) = \sum_j a_j \exp[i\langle x, \xi_j\rangle]$ in the form $\sum_j a_j(Q(\xi_j)1)(x)$, so that

$$P(\xi)\varphi(x) = \sum_j a_j \exp[i(\xi, \xi_j)]\exp[i\langle x, \xi_j\rangle](P(\xi)1)(x)$$
$$= \varphi(x + \xi)\psi_\xi(x).$$

This result can be extended to a general (L^2)-functional φ, and combining it with (5.95) we obtain

$$P(\xi)\varphi(x) = \varphi(x + \xi)\, d(\xi, x)\left(\frac{dT_\xi \circ \mu}{d\mu}(x)\right)^{1/2}, \qquad \varphi \in (L^2). \quad (5.96)$$

By considering the relation (5.96) in the light of the identity $P(\xi_1 + \xi_2) = P(\xi_1)P(\xi_2)$, we obtain the further necessary condition on d:

$$d(\xi_1, x)\, d(\xi_2, x + \xi_1) = d(\xi_1 + \xi_2, x). \quad (5.97)$$

On the other hand, if μ is T_ξ-quasi-invariant and d is a functional satisfying $|d| = 1$ and (5.97), then the operators $Q(\xi)$ and $P(\xi)$ defined by (5.93) and (5.96) respectively are continuous, unitary, and satisfy the commutation relations (5.92). We have proved:

Theorem 5.10. *A cyclic unitary representation of the nuclear Lie group G_B:*

$$g = (\xi, \eta; \theta) \to e^{-i\theta}P(\xi)Q(\eta), \qquad g \in G_B, \quad (5.98)$$

is given by (5.93) and (5.96) whenever μ is a T_ξ-quasi-invariant measure on E^* ($\xi \in E$) and d is a functional of absolute value unity which satisfies (5.97).

EXAMPLE 1. Let μ be the measure of white noise, take $d(x, \xi) \equiv 1$, and define operators $P(\xi)$ and $Q(\eta)$ by

$$(P(\xi)\varphi)(x) = \varphi(x + \xi)\exp\left[-\frac{1}{2}\langle x, \xi\rangle - \frac{1}{4}\|\xi\|^2\right],$$

$$(Q(\eta)\varphi)(x) = \exp[i\langle x, \eta\rangle]\varphi(x), \qquad \varphi \in (L^2).$$

It is easily seen that these operators are continuous in ξ and η respectively, and satisfy the commutation relation (5.92).

Our next problem concerns the irreducibility of such representations. A necessary and sufficient condition for a representation $\{P(\xi), Q(\eta)\}$ to be irreducible is that an operator V commuting with all the $P(\xi)$ and all the $Q(\eta)$ has to be of the form cI, c a constant. Let V be such an operator, and set $V1(x) = V(x)$. Since V commutes with $Q(\eta)$ we must have $\exp[i\langle x, \eta\rangle]V(x)$ $(= (Q(\eta)V1)(x) = V\exp[i\langle x, \eta\rangle])$, so that $(V\varphi)(x) = V(x)\varphi(x)$ holds for any $\varphi \in (L^2)$. The further assumption that V commutes with $P(\xi)$ implies that we have $P(\xi)V(x) = V(x)(P(\xi)1)(x) = V(x)\psi_\xi(x)$, i.e. $V(x + \xi)\psi_\xi(x) = V(x)\psi_\xi(x)$ for all $x \in E^*$. Since $\psi_\xi(x) \ne 0$ a.e. (μ), we must have

$$V(x + \xi) = V(x), \qquad \text{a.e. } (\mu), \qquad \text{for all } \xi \in E.$$

In order for this to imply that $V(x) = $ constant, a.e. (μ), it suffices to suppose that μ is T_ξ-ergodic. We have therefore proved

Proposition 5.14. *If μ is T_ξ-ergodic, then the representation given in Theorem 5.10 is irreducible.*

It follows from Theorem 5.8 of §5.7 that Example 1 above is an irreducible representation.

Finally we discuss the unitary equivalence of representations. By Theorem 5.10 we have a certain freedom in the choice of μ and $d(\xi, x)$ when constructing a representation of G_B.

Proposition 5.15. *Suppose that we have two irreducible representations constructed by Theorem 5.10 using the pairs (μ, d) and (μ', d'). These two representations are unitarily equivalent if and only if the following two conditions are satisfied:*

a. *μ and μ' are equivalent;*
b. *there exists a functional $K(x)$ with absolute value unity such that*

$$\frac{d(\xi, x)}{d'(\xi, x)} = K(x + \xi)K(x). \tag{5.99}$$

PROOF. Suppose that two representations $\{P(\xi), Q(\eta)\}$ and $\{P'(\xi), Q'(\eta)\}$ are given, constructed using μ and d and μ' and d', respectively. These representations are unitarily equivalent if and only if there exists a functional $\varphi_0(x)$ in $L^2(E^*, \mu')$, which is a non-zero functional a.e. (μ'), and which satisfies $\int |\varphi_0(x)|^2 \, d\mu'(x) = 1$, such that

$$(P(\xi)Q(\eta)1, 1)_{L^2(E^*, \mu)} = (P'(\xi)Q'(\eta)\varphi_0, \varphi_0)_{L^2(E^*, \mu')} \tag{5.100}$$

for all $\xi, \eta \in E$. In particular if we set $\xi = 0$ we obtain

$$\int \exp[i\langle x, \eta\rangle] \, d\mu(x) = \int \exp[i\langle x, \eta\rangle] |\varphi_0(x)|^2 \, d\mu'(x),$$

which implies that $d\mu = |\varphi_0|^2 \, d\mu'$. Now define $K(x)$ by

$$\varphi_0(x) = K(x)\left(\frac{d\mu}{d\mu'}(x)\right)^{1/2},$$

and the relation (5.99) follows from the fact that (5.100) holds for all η. □

EXAMPLE 2. Take the measure μ_σ of a white noise with variance σ^2. Since μ_σ is T_ξ-ergodic, irreducible representations may be formed by using Proposi-

5.8 The Canonical Commutation Relations of Quantum Mechanics

tion 5.14. If $\sigma_1 \neq \sigma_2$, then the two measures μ_{σ_1} and μ_{σ_2} are mutually singular, so that whatever the choice of d, the unitary representations obtained using them are not unitarily equivalent.

From this example we conclude that there are *uncountably many* mutually inequivalent irreducible unitary representations of G_B.

6 Complex White Noise

Complex Gaussian systems are the most important families of complex-valued random variables, and this chapter begins by presenting the general background to such systems. We then observe that complex white noise, the white noise of Chapter 3 complexified, is a complex Gaussian system. Functionals of complex white noise may also be viewed as functionals of complex Brownian motion and the analysis of such functionals is not only useful in the study of stochastic processes, but is also widely used in applications. Consequently it is an important problem to express these in a concrete form and to develop ways of analysing them (§§6.2–6.3). On the other hand, the infinite dimensional unitary group arises naturally in the study of the probability measure determined on the complex-valued (generalised) function space by complex white noise. This unitary group plays the same role here as the infinite-dimensional rotation group did in describing properties of white noise (§7.1). For added interest, this group is intimately related to aspects of the theory of differential equations and quantum mechanics (§7.5–7.6).

It seems reasonable to ask why complexification of white noise and the rotation group is necessary, and we find that there are several answers. Firstly (i) on the complexified space we can make use of the Fourier transform without any difficulty, and this is one of the basic operations in functional analysis; (ii) the unitary group may be regarded as the complexification of the rotation group and, as in the finite-dimensional case, is more easily dealt with than the latter. In particular our unitary group involves the Fourier transform. (iii) One can develop a theory of holomorphic functionals (of infinitely many variables) on the space of the complex white noise, and (iv) as is seen in many actual examples, complex random variables have important specific meanings in applications of the analysis to mathematical theory of communication and quantum mechanics. The details will be clarified in what follows.

6.1 Complex Gaussian Systems

This section is a preparation for the discussion of the subsequent sections, and its content is perhaps best understood as a generalisation of §1.6. We begin by defining the systems of the title of the section.

Let $Z(\omega)$ be a complex-valued random variable on a probability space $(\Omega, \mathbf{B}, P)$. Denoting the mean of $Z(\omega)$ by $m \in \mathbf{C}$, the real and imaginary part of $Z(\omega) - m$ by $X(\omega)$ and $Y(\omega)$ respectively, we have

$$Z(\omega) = m + X(\omega) + iY(\omega) \qquad i = \sqrt{-1}. \tag{6.1}$$

Definition 6.1. If $X(\omega)$ and $Y(\omega)$ in (6.1) are independent and have the same Gaussian distribution with zero mean, then $Z(\omega)$ is called a *complex Gaussian random variable*, or is simply said to be *complex Gaussian*.

This definition requires a much stronger condition than that (X, Y) simply be real Gaussian (in the sense of §1.6). Indeed such a strengthening is necessary in order to obtain properties similar to those enjoyed by real Gaussian systems (for instance Theorem 1.10).

It follows easily from the definition that the probability distribution of a complex Gaussian random variable $Z(\omega)$ is determined by only its mean m and variance $E\{|Z - m|^2\}$.

Definition 6.2. Let $\mathscr{L} = \{Z_\lambda(\omega): \lambda \in \Lambda\}$ be a system of complex-valued random variables. If any finite linear combination $\sum_j c_j Z_{\lambda_j}, c_j \in \mathbf{C}$, is complex Gaussian, then $\mathscr{L}$ is called a *complex Gaussian system* or just *complex Gaussian*.

If a system $\mathscr{L}$ is complex Gaussian then any sub-system $\mathscr{L}' \subseteq \mathscr{L}$ be again complex Gaussian, and furthermore the system $\bar{\mathscr{L}} = \{\bar{Z}_\lambda(\omega): \lambda \in \Lambda\}$, $\bar{Z}$ the complex conjugate of Z, is also complex Gaussian. In what follows we shall state some easy consequences of the definitions given.

Proposition 6.1. *Suppose that $\mathscr{L} = \{Z_\lambda(\omega): \lambda \in \Lambda\}$ is complex Gaussian with decomposition $Z_\lambda = m_\lambda + X_\lambda + iY_\lambda$ as in (6.1). Then the system $\mathscr{X} = \{X_\lambda: \lambda \in \Lambda\}$ consisting of the real parts, the system $\mathscr{Y} = \{Y_\lambda: \lambda \in \Lambda\}$ consisting of the imaginary parts (of $Z_\lambda - m_\lambda$) and the union $\mathscr{X} \cup \mathscr{Y}$ are all real Gaussian systems.*

PROOF. The fact that the systems $\mathscr{X}$ and $\mathscr{Y}$ are real Gaussian is an immediate consequence of Definitions 6.2 and 1.6. One can prove that the system $\mathscr{X} \cup \mathscr{Y}$ is real Gaussian by viewing a linear combination $\sum c_j X_{\lambda_j} + \sum d_k Y_{\lambda_k}$ as the real part of $\sum c_j Z_{\lambda_j} - \sum i d_k Z_{\lambda_k}$, $c_j, d_k \in \mathbf{R}$. □

Proposition 6.2. *Let $\mathscr{Z} = \{Z_\lambda(\omega): \lambda \in \Lambda\}$ be a system of independent random variables. If each Z_λ is complex Gaussian ($\lambda \in \Lambda$), then the system $\mathscr{Z}$ is complex Gaussian.*

PROOF. It suffices to prove that a linear combination of any two Z_λ's is complex Gaussian. Write $Z_{\lambda_j} = m_j + X_j + iY_j$ as in (6.1), and take $c_j = a_j + ib_j$, $a_j, b_j \in \mathbf{R}$, $j = 1, 2$. Then we have

$$c_1 Z_{\lambda_1} + c_2 Z_{\lambda_2} = c_1 m_1 + c_2 m_2 + (a_1 X_1 - b_1 Y_1 + a_2 X_2 - b_2 Y_2)$$
$$+ i(a_1 Y_1 + b_1 X_1 + a_2 Y_2 + b_2 X_2).$$

Note that the system $\{X_j, Y_j : j = 1, 2\}$ is real Gaussian. Since the variances of the real and imaginary parts of the right hand side (after the means have been subtracted) are equal, and their covariance is zero, we see that $c_1 Z_{\lambda_1} + c_2 Z_{\lambda_2}$ is complex Gaussian. □

Proposition 6.3. *Let $\mathscr{Z} = \{Z_\lambda(\omega): \lambda \in \Lambda\}$ be a complex Gaussian system. Then a necessary and sufficient condition for Z_{λ_1} and $Z_{\lambda_2} (\in \mathscr{Z})$ to be independent is that their covariance $E\{(Z_{\lambda_1} - m_{\lambda_1})(\overline{Z_{\lambda_2} - m_{\lambda_2}})\} = 0$, where the means are $m_{\lambda_j} = E\{Z_{\lambda_j}\}, j = 1, 2$.*

PROOF. The necessity of the condition is obvious and so we turn to the sufficiency.

Writing Z_{λ_1} and Z_{λ_2} in the form used in the previous proposition, the condition for their covariance to be zero is

$$E\{(X_1 + iY_1)(X_2 - iY_2)\} = 0$$

and this gives the following two equalities:

a. $E\{X_1 X_2\} + E\{Y_1 Y_2\} = 0$,
b. $E\{Y_1 X_2\} - E\{X_1 Y_2\} = 0$.

However $\mathscr{Z}$ is complex Gaussian and so $Z_{\lambda_1} + Z_{\lambda_2}$ and $Z_{\lambda_1} + iZ_{\lambda_2}$ are both complex Gaussian random variables. Thus the real and imaginary parts of both are independent and we therefore have the following two equations:

c. $E\{Y_1 X_2\} + E\{X_1 Y_2\} = 0$,
d. $E\{X_1 X_2\} - E\{Y_1 Y_2\} = 0$.

The four equalities (a), (b), (c) and (d) show that $\{X_1, X_2, Y_1, Y_2\}$ is a system of independent real Gaussian random variables and proves that Z_{λ_1} and Z_{λ_2} are independent. □

Let the mean vector of a complex Gaussian system $\mathscr{Z} = \{Z_\lambda(\omega): \lambda \in \Lambda\}$ be denoted by $(m_\lambda: \lambda \in \Lambda)$ and the covariance matrix by $(V_{\lambda, \mu}: \lambda, \mu \in \Lambda)$.

Proposition 6.4. *The covariance matrix $(V_{\lambda, \mu}: \lambda, \mu \in \Lambda)$ of a complex Gaussian system $\mathscr{Z} = \{Z_\lambda(\omega): \lambda \in \Lambda\}$ satisfies the following conditions:*

i. *It is positive definite;*
ii. *If $Z_\lambda = m_\lambda + X_\lambda + iY_\lambda$ as in (6.1), then*

$$\begin{cases} E\{X_\lambda X_\mu\} = E\{Y_\lambda Y_\mu\} = \frac{1}{2} \operatorname{Re} V_{\lambda,\mu}, \\ -E\{X_\lambda Y_\mu\} = E\{Y_\lambda X_\mu\} = \frac{1}{2} \operatorname{Im} V_{\lambda,\mu}. \end{cases} \quad (6.2)$$

(Re *and* Im *denote the real and imaginary part, respectively*).

PROOF. (i) is obvious. For the proof of (ii) it suffices to consider the case $m_\lambda \equiv 0$. Since Z_λ, Z_μ and $Z_\lambda + iZ_\mu$ are all complex Gaussian, the real and imaginary part of these random variables are independent giving the equality $E\{X_\lambda X_\mu\} = E\{Y_\lambda Y_\mu\}$. The equality $E\{X_\lambda Y_\mu\} = -E\{Y_\lambda X_\mu\}$ comes from the fact that $Z_\lambda + Z_\mu$ is complex Gaussian. The remainder of the proof follows from the expression for the covariance of Z_λ and Z_μ and the equalities just obtained. □

Theorem 6.1. *For any vector $(m_\lambda : \lambda \in \Lambda)$ and positive definite matrix $(V_{\lambda,\mu} : \lambda, \mu \in \Lambda)$ on Λ, there exists a complex Gaussian system $\mathscr{Z} = \{Z_\lambda : \lambda \in \Lambda\}$ whose mean vector and covariance matrix coincide with the given vector and matrix. The probability distribution of the system is unique.*

PROOF. Take a set Λ' with the same cardinal number as Λ and a one-to-one correspondence $\lambda \to \lambda'$ between Λ and Λ'. Define a matrix $(v_{\lambda,\mu} : \lambda, \mu \in \Lambda \cup \Lambda')$ on $\Lambda \cup \Lambda'$ in terms of $V_{\lambda,\mu}$ as follows:

$$v_{\lambda,\mu} = \begin{cases} \frac{1}{2} \operatorname{Re} V_{\lambda,\mu} & \text{if } \lambda, \mu \in \Lambda \text{ or } \lambda, \mu \in \Lambda', \\ -\frac{1}{2} \operatorname{Im} V_{\lambda,\mu} & \text{if } \lambda \in \Lambda, \mu \in \Lambda', \\ \frac{1}{2} \operatorname{Im} V_{\lambda,\mu} & \text{if } \lambda \in \Lambda', \mu \in \Lambda. \end{cases}$$

It can be proved that $(v_{\lambda,\mu})$ is positive definite and so by Theorem 1.10 of §1.6 there exists a real Gaussian system $\{U_\lambda : \lambda \in \Lambda \cup \Lambda'\}$ with mean zero and covariance matrix $(v_{\lambda,\mu})$. Put

$$X_\lambda = U_\lambda, \quad \lambda \in \Lambda,$$
$$Y_\lambda = U_{\lambda'}, \quad \lambda' \in \Lambda',$$

and
$$Z_\lambda = m_\lambda + X_\lambda + iY_\lambda, \quad \lambda \in \Lambda.$$

It is easy to see that $\{Z_\lambda : \lambda \in \Lambda\}$ is a complex Gaussian system satisfying the required conditions.

The uniqueness, that two systems satisfying the stated conditions should have the same probability distribution, follows from Proposition 6.4 and the uniqueness result for the system $\{U_\lambda: \lambda \in \Lambda \cup \Lambda'\}$. □

EXAMPLE 1. Let $\{B_1(t, \omega): t \geq 0\}$ and $\{B_2(t, \omega): t \geq 0\}$ be two mutually independent Brownian motions. Define $\mathscr{Z} = \{Z(t, \omega): t \geq 0\}$ by

$$Z(t, \omega) = 2^{-1/2}[B_1(t, \omega) + iB_2(t, \omega)]. \tag{6.3}$$

Then $\mathscr{Z}$ is a complex Gaussian system with mean vector 0 and covariance matrix $(t \wedge s: t, s \geq 0)$:

$$E\{Z(t)\overline{Z(s)}\} = \frac{1}{2} E\{[B_1(t) + iB_2(t)][B_1(s) - iB_2(s)]\} = t \wedge s.$$

This system is nothing but the *complex Brownian motion* defined in §2.3.

We turn now to formulae for the moments of complex Gaussian random variables.

Proposition 6.5. *Let the random variable $Z(\omega)$ be complex Gaussian. Then*

$$E\{\exp[i(tZ + s\bar{Z})]\} = \exp[i(tm + s\bar{m}) - ts\sigma^2] \tag{6.4}$$

where t and s are arbitrary complex numbers, $m = E(Z)$ and $\sigma^2 = E\{|Z - m|^2\}$.

PROOF. Write Z in the form (6.1) and use the integration formula:

$$\frac{1}{\sqrt{\pi\sigma^2}} \int_{-\infty}^{\infty} \exp\left[ax - \frac{x^2}{\sigma^2}\right] dx = \exp\left[\frac{a^2\sigma^2}{4}\right], \quad a \in \mathbf{C}. \quad □$$

Corollary. *Suppose that Z is complex Gaussian and $E\{Z\} = 0$. Then $E\{\exp[tZ]\} \equiv 1, t \in \mathbf{C}$; hence Z has moments of all orders and*

$$\begin{cases} E\{Z^n\} = 0, & n > 0, \\ E\{Z^n \bar{Z}^m\} = \delta_{n,m} n! E\{|Z|^2\}^n. \end{cases} \tag{6.5}$$

Thus it is possible to compute moments of certain functions of Z and $\bar{Z}$, in particular polynomials, but in order to analyse more general functions it is useful to introduce the following complex Hermite polynomials [see K. Itô (1953a)].

Definition 6.3. Let $z \in \mathbf{C}$ be a complex variable. The polynomial given by

$$H_{p,q}(z, \bar{z}) = (-1)^{p+q} e^{|z|^2} \frac{\partial^{p+q}}{\partial \bar{z}^p \, \partial z^q} e^{-|z|^2} \qquad p, q \geq 0 \tag{6.6}$$

is called the *complex Hermite polynomial* of degree (p, q).

In fact $H_{p,q}(z, \bar{z})$ is a polynomial of degree p in z and of degree q in $\bar{z}$. An explicit formula is

$$H_{p,q}(z, \bar{z}) = \sum_{k=0}^{p \wedge q} (-1)^k \frac{p!\, q!}{k!(p-k)!(q-k)!} z^{p-k} \bar{z}^{q-k}. \tag{6.7}$$

Properties of this polynomial and related formulae are listed in Appendix A.5 at the end of this volume; here we pause only to note a few formulae to be used shortly.

$$\overline{H_{p,q}(z, \bar{z})} = H_{q,p}(z, \bar{z}). \tag{6.8}$$

The generating function of the $H_{p,q}$ is

$$\sum_{p,q=0}^{\infty} \frac{s^p t^q}{p!\, q!} H_{p,q}(z, \bar{z}) = \exp[-st + t\bar{z} + sz], \qquad s, t \in \mathbf{C}. \tag{6.9}$$

$$\{(p!\, q!)^{-1/2} H_{p,q}(z, \bar{z}) : p, q \geq 0\} \tag{6.10}$$

is a complete orthonormal system in the space

$$L^2(\mathbf{C}, \pi^{-1} \exp(-|z|^2)\, dx dy), \qquad z = x + iy.$$

Proposition 6.6. *Let $Z(\omega)$ be a complex Gaussian random variable with mean 0 and variance 1 on $(\Omega, \mathbf{B}, P)$. Then the system of functions of $Z(\omega)$:*

$$\{(p!\, q!)^{-1/2} H_{p,q}(Z(\omega), \overline{Z(\omega)}) : p, q \geq 0\}$$

is an orthonormal system in $L^2(\Omega, P)$.

PROOF. This is just a paraphrasing of (6.10). □

6.2 Complexification of White Noise

Let us return now to white noise $(E^*, \mathcal{B}, \mu)$ and, as indicated at the beginning of this chapter, we will consider its complexification.

With E, E^* as in §5.1 we complexify as follows:

$$\begin{aligned} E_c &= E + iE, \qquad i = \sqrt{-1}, \\ E_c^* &= E^* + iE^*. \end{aligned} \tag{6.11}$$

An element ζ of E_c and an element z of E_c^* may be written in the form

$$\begin{aligned} \zeta &= \xi + i\eta & \xi, \eta \in E, \\ z &= x + iy & x, y \in E^*, \end{aligned} \tag{6.12}$$

respectively. The canonical bilinear form $\langle x, \xi \rangle$ connecting E and E^* can be extended naturally to a form $\langle z, \zeta \rangle$, $z \in E_c^*$, $\zeta \in E_c$, connecting E_c and E_c^*. It is readily shown through this extended form that E_c^* becomes the conjugate space of E_c. Indeed if z, ζ are expressed in the form (6.12) we have

$$\langle z, \zeta \rangle = (\langle x, \xi \rangle + \langle y, \eta \rangle) + i(-\langle x, \eta \rangle + \langle y, \xi \rangle). \tag{6.13}$$

Take measures μ_1 and μ_2 on $E^*(\cong iE^*)$ of white noise with variance $\frac{1}{2}$. They are such that (§3.4)

$$\int_{E^*} \exp[i\langle x, \xi \rangle] \, d\mu_j(x) = \exp\left[-\frac{1}{4}\|\xi\|^2\right], \quad j = 1, 2.$$

Introduce on E_c^* the product measure, of course a probability measure:

$$v = \mu_1 \times \mu_2. \tag{6.14}$$

The smallest σ-field generated by the cylinder subsets of E_c^* is denoted by $\mathfrak{B}$, and combining this $\mathfrak{B}$, v given by (6.14), and E_c^*, we obtain a measure space $(E_c^*, \mathfrak{B}, v)$.

Definition 6.4. The measure space $(E_c^*, \mathfrak{B}, v)$ given above is called a *complex white noise*.

Remark. When we wish to discriminate between this and the white noise defined in §5.1 we will call the latter real white noise.

Once an element ζ of E_c is fixed, $\langle z, \zeta \rangle$ is $\mathfrak{B}$-measurable viewed as a function of z, and so is a random variable defined on $(E_c^*, \mathfrak{B}, v)$.

Proposition 6.7. *The collection $\mathcal{H}^0 \equiv \{\langle z, \zeta \rangle : \zeta \in E_c\}$ is a complex Gaussian system of random variables on the probability space $(E_c^*, \mathfrak{B}, v)$ and*

$$\begin{aligned} E\{\langle z, \zeta \rangle\} &= \int_{E_c^*} \langle z, \zeta \rangle \, dv(z) = 0, \\ E\{|\langle z, \zeta \rangle|^2\} &= \|\zeta\|^2. \end{aligned} \tag{6.15}$$

[*Here* $\|\cdot\|$ *denotes the norm on the complex Hilbert space* $L^2(\mathbf{R})$].

PROOF. First we note that $\{\langle x, \xi \rangle : \xi \in E\}$ and $\{\langle y, \eta \rangle : \eta \in E\}$ are independent since v is a product measure. Using the notation (6.12) for z, ζ set $X(z) = \langle x, \xi \rangle + \langle y, \eta \rangle$, $Y(z) = -\langle x, \eta \rangle + \langle y, \xi \rangle$, so that $\langle z, \zeta \rangle = X(z) + iY(z)$. With this notation and the fact that μ_1 and μ_2 are both measures of white noise with variance $\frac{1}{2}$, we see that $X(z)$ is a sum of two independent Gaussian random variables and hence itself Gaussian in distribution with mean 0 and variance $\frac{1}{2}(\|\xi\|^2 + \|\eta\|^2)$. Similarly for $Y(z)$. Noting that the covariance of $X(z)$ and $Y(z)$ is zero, we deduce that they are independent, and have proved that $\langle z, \zeta \rangle$ is complex Gaussian satisfying (6.15). Any finite linear combination of elements of $\mathcal{H}^0$ is again an element of $\mathcal{H}^0$ and so $\mathcal{H}^0$ is complex Gaussian as well. $\square$

6.2 Complexification of White Noise

From the proposition and the evaluation of the variances of $\langle z, \zeta_1 + \zeta_2 \rangle$ and $\langle z, \zeta_1 + i\zeta_2 \rangle$ we have

Corollary. *The covariance matrix of $\mathscr{H}^0$ is given by*

$$E\{\langle z, \zeta_1 \rangle \overline{\langle z, \zeta_2 \rangle}\} = \overline{(\zeta_1, \zeta_2)}, \tag{6.16}$$

where $(\cdot, \cdot)$ denotes the inner product of $L^2(\mathbf{R})$.

Given a complex white noise $(E_c^*, \mathfrak{B}, \nu)$ we can construct the complex Hilbert space $L^2(E_c^*, \mathfrak{B}, \nu)$ which, since it appears frequently, will be written more simply as (L_c^2). The space $\mathscr{H}^0$ above is a linear subspace of (L_c^2) and its closure in (L_c^2) will be denoted by $\mathscr{H}_{(1,0)}$. Then the isometry established by (6.16) between E and $\mathscr{H}^0$ can be extended in a manner similar to that in §4.3 to an isometry between $L^2(\mathbf{R})$ and $\mathscr{H}_{(1,0)}$. Accordingly the notation $\langle z, \zeta \rangle$, $\zeta \in E_c$, is extended in such a way that the element of $\mathscr{H}_{(1,0)}$ corresponding under the isometry to $f \in L^2(\mathbf{R})$ is denoted by $\langle z, f \rangle$. We can also see that as in Corollary 2 to Proposition 1.11 in §1.6 $\mathscr{H}_{(1,0)}$ is a complex Gaussian system.

Proposition 6.8. *In the space $\mathscr{H}_{(1,0)}$ we have*

$$(\langle z, f_1 \rangle, \langle z, f_2 \rangle)_{\mathscr{H}_{(1,0)}} = \overline{(f_1, f_2)}_{L^2(\mathbf{R})}, \tag{6.17}$$

whence $\langle z, f_1 \rangle$ and $\langle z, f_2 \rangle$ in $\mathscr{H}_{(1,0)}$ are independent if and only if f_1 and f_2 are orthogonal in $L^2(\mathbf{R})$.

PROOF. Equation (6.17) is obvious from the definition of $\mathscr{H}_{(1,0)}$ and the remainder follows in a straightforward manner from it using Proposition 6.3. □

Now the system $\bar{\mathscr{H}}^0 = \{\overline{\langle z, \zeta \rangle}: \zeta \in E_c\}$ is also complex Gaussian, and so the closed subspace $\mathscr{H}_{(0,1)}$ of (L_c^2) generated by $\bar{\mathscr{H}}^0$ is complex Gaussian. The correspondence

$$\bar{f} \to \overline{\langle z, f \rangle} \in \mathscr{H}_{(0,1)}, \quad \bar{f} \in L^2(\mathbf{R}),$$

is one-to-one and isometric, but this isometry is somewhat different from that found above in the case of $\mathscr{H}_{(1,0)}$. It is convenient to distinguish one from the other in the following way:

$$\begin{aligned} \mathscr{H}_{(1,0)} &\cong L^2(\mathbf{R}) \\ \mathscr{H}_{(0,1)} &\cong \overline{L^2(\mathbf{R})}. \end{aligned} \tag{6.18}$$

Let $\langle z, f_1 \rangle$ and $\langle z, f_2 \rangle$ be elements of $\mathscr{H}_{(1,0)}$. Then in terms of the decomposition

$$f_2 = \bar{\lambda} f_1 + f_3 \quad \text{where} \quad (f_1, f_3) = 0 \quad \text{and} \quad \lambda = \frac{(f_1, f_2)}{\|f_1\|^2}$$

we have
$$\langle z, f_2 \rangle = \lambda \langle z, f_1 \rangle + \langle z, f_3 \rangle$$
where $\langle z, f_1 \rangle$ and $\langle z, f_3 \rangle$ are independent. Hence we see that
$$\int_{E_c^*} \langle z, f_1 \rangle \langle \overline{z, f_2} \rangle \, dv(z) = \lambda \int_{E_c^*} \langle z, f_1 \rangle^2 \, dv(z) + \int_{E_c^*} \langle z, f_1 \rangle \, dv(z) \int_{E_c^*} \langle z, f_3 \rangle \, dv(z)$$
$$= 0.$$

The left hand side is the inner product of $\langle z, f_1 \rangle$ and $\overline{\langle z, f_2 \rangle}$ in (L_c^2) and so we have proved that

$$\mathcal{H}_{(1, 0)} \perp \mathcal{H}_{(0, 1)} \qquad (\perp \text{ denotes ``orthogonal''}). \tag{6.19}$$

Definition 6.5. Each element of the direct sum
$$\mathcal{H}_1 = \mathcal{H}_{(1, 0)} \oplus \mathcal{H}_{(0, 1)}$$
is called a *complex Wiener integral*.

With the notation of (6.18) we can say that $\mathcal{H}_1$ is isomorphic to the direct sum $L^2(\mathbf{R}) \oplus \overline{L^2(\mathbf{R})}$, which realises elements of $\mathcal{H}_1$ concretely in terms of the well known space $L^2(\mathbf{R})$.

6.3 The Complex Multiple Wiener Integral

An element of the space $\mathcal{H}_{(1, 0)}$ or $\mathcal{H}_{(0, 1)}$ discussed in the previous section may be regarded as a linear functional in z or $\bar{z}$. In this section we consider higher order functionals of z or $\bar{z}$, further discuss general nonlinear functionals, and then proceed to a representation of them using well known concepts from analysis. Our method of representation is simply a generalisation of that developed in §4.3.

As before, polynomials and exponential functions are taken to be the basic functionals.

(i) Polynomials

If a functional $\varphi(z)$ is expressible in the form
$$\varphi(z) = P(\langle z, \zeta_1 \rangle, \ldots, \langle z, \zeta_p \rangle, \overline{\langle z, \zeta_1 \rangle}, \ldots, \overline{\langle z, \zeta_p \rangle}), \qquad \zeta_1, \ldots, \zeta_p \in E_c, \tag{6.20}$$

in terms of a polynomial $P(s_1, \ldots, s_p, t_1, \ldots, t_p)$ in $2p$ variables with complex coefficients, then φ is called a *polynomial* on the space E_c^*. If $\zeta_1, \ldots, \zeta_p$ in the above expression are chosen to be linearly independent, if P is of degree m in $s_1, \ldots, s_p$ and is of degree n in $t_1, \ldots, t_p$, then φ is said to be of degree (m, n). The collection of all polynomials forms an algebra over $\mathbf{C}$ denoted by $\mathbf{P}$. Obviously any polynomial belongs to (L_c^2).

(ii) Exponential functions

A functional of the form

$$\varphi(z) = \exp[i\langle z, \zeta_1 \rangle + i\overline{\langle z, \zeta_2 \rangle}], \qquad \zeta_1, \zeta_2 \in E_c, \tag{6.21}$$

is called an *exponential function* on the space E_c^*.

The algebra over the complex field $\mathbf{C}$ generated by the exponential functions is denoted by $\mathbf{A}'$, and its subalgebra generated by

$$K(z; \zeta) \equiv \exp[i\langle z, \zeta \rangle + i\overline{\langle z, \zeta \rangle}] = \exp[2i \operatorname{Re}\langle z, \zeta \rangle], \qquad \zeta \in E_c, \tag{6.22}$$

is denoted by $\mathbf{A}$. Both $\mathbf{A}'$ and $\mathbf{A}$ are subsets of (L_c^2).

Proposition 6.9. *The algebra $\mathbf{A}$ is dense in (L_c^2).*

PROOF. The space E_c^* can be identified with the product of the spaces E^* and iE^*, and by definition the measure v is a direct product measure [see (6.14)]. Let $(L^2)_x$ and $(L^2)_y$ be the L^2-space over E^* with respect to μ_1 and iE^* with respect to μ_2, respectively. Then (L_c^2) can be regarded as a tensor product

$$(L_c^2) = (L^2)_x \otimes (L^2)_y.$$

Now the algebra $\mathbf{A}_x$ generated by the exponentials $\{\exp[i\langle x, \xi \rangle]: \xi \in E\}$ on E^* is dense in $(L^2)_x$ (§4.2 Theorem 4.1). Defining $\mathbf{A}_y$ similarly we see that $\mathbf{A}_x \otimes \mathbf{A}_y$ is dense in $(L^2)_x \otimes (L^2)_y$.

On the other hand, we have the expression

$$K(z; \zeta) = \exp[2i\langle x, \xi \rangle] \cdot \exp[2i\langle y, \eta \rangle]$$

where $z = x + iy$, $\zeta = \xi + i\eta$, and so it can be proved that $\mathbf{A} = \mathbf{A}_x \otimes \mathbf{A}_y$. Thus $\mathbf{A}$ is dense in $(L^2)_x \otimes (L^2)_y$ and hence in (L_c^2). □

As a consequence of the foregoing discussion we have

$$\overline{\mathbf{A}} = \overline{\mathbf{A}'} = (L_c^2) \qquad \text{(where}^{-} \text{denotes closure).} \tag{6.23}$$

Corollary. *If $\varphi(z) \in (L_c^2)$ satisfies*

$$\int_{E_c^*} K(z; \zeta)\overline{\varphi(z)} \, dv(z) = 0 \tag{6.24}$$

for every $\zeta \in E_c$, then we have $\varphi(z) = 0$, a.e. (v).

PROOF. Any finite product $\prod_j K(z; \zeta_j)$ is again expressible in the form $K(z; \zeta)$ with $\zeta = \sum_j \zeta_j$ and so (6.24) means that $\varphi(z)$ is orthogonal to **A** in (L_c^2). Since **A** is dense in (L_c^2) it follows that φ has to be the zero of (L_c^2). □

We come now to the proof that the algebra **P** generated by polynomials is also dense in (L_c^2). Using (6.5) to evaluate the moments of the random variable $\langle z, \zeta \rangle$ we deduce the convergence in (L_c^2) of the power series expansion

$$K(z; \zeta) = \sum_{n=0}^{\infty} \frac{i^n}{n!} (\langle z, \zeta \rangle + \overline{\langle z, \zeta \rangle})^n$$

for the exponential function. This means that $K(z; \zeta)$ can be approximated by polynomials and the same then applies for finite products of the $K(z; \zeta)$. This together with (6.23) proves that

$$\bar{\mathbf{P}} = (L_c^2). \tag{6.25}$$

Let $\{\omega_n : n \geq 1\}$ be a complete orthonormal system in $L^2(\mathbf{R})$ with each $\omega_n \in E_c$, $n \geq 1$. The expansion $\zeta = \sum_{n=1}^{\infty} a_n \omega_n$ of ζ implies that in (L_c^2)

$$\langle z, \zeta \rangle = \sum_{n=1}^{\infty} \bar{a}_n \langle z, \omega_n \rangle. \tag{6.26}$$

Denote by $\mathbf{P}(\{\omega_n\})$ the algebra over **C** generated by the powers of $\langle z, \omega_n \rangle$ and $\overline{\langle z, \omega_n \rangle}$.

Proposition 6.10. *The algebra $\mathbf{P}(\{\omega_n\})$ is dense in (L_c^2).*

PROOF. The expansion (6.26) implies expansions of powers of $\langle z, \zeta \rangle$ and $\overline{\langle z, \zeta \rangle}$ and by using these it can be proved that any element of **P** is approximable by linear combinations of products of the form

$$\prod_j \langle z, \omega_j \rangle^{p_j} \overline{\langle z, \omega_j \rangle}^{q_j}. \tag{6.27}$$

The vector space spanned by such monomials (6.27) is precisely $\mathbf{P}(\{\omega_n\})$. Thus any element of **P** can be approximated by polynomials in $\mathbf{P}(\{\omega_n\})$ and so, by (6.25), our conclusion follows. □

Let us now recall the complex Hermite polynomials introduced by §6.1.

Theorem 6.2. *The family of all functionals in $\mathbf{P}(\{\omega_n\})$ of the form*

$$\prod_k (p_k! q_k!)^{-1/2} H_{p_k, q_k}(\langle z, \omega_{j_k} \rangle, \overline{\langle z, \omega_{j_k} \rangle}) \tag{6.28}$$

where the product is finite and the j_k's are distinct, is a complete orthonormal system in (L_c^2).

6.3 The Complex Multiple Wiener Integral

PROOF. a. Functionals of the form (6.28) clearly belong to $\mathbf{P}(\{\omega_n\})$. Now the properties of complex Hermite polynomials imply that any monomial $z^p \bar{z}^q$ can be expressed as a linear combination of complex Hermite polynomials of order at most (p, q), and so we see that, conversely, $\mathbf{P}(\{\omega_n\})$ coincides with the collection of all linear combinations of functionals of the form (6.28). Thus completeness follows from Proposition 6.10.

b. Since $\{\omega_n : n \geq 1\}$ is a complete orthonormal system in $L^2(\mathbf{R})$, the family $\{\langle z, \omega_n \rangle : n \geq 1\}$ forms a complex Gaussian system comprising of independent identically distributed (zero mean, unit variance) random variables by Proposition 6.8. Replacing the Z in Proposition 6.6 by $\langle z, \omega_j \rangle$, we see that the family

$$(p!\,q!)^{-1/2} H_{p,q}(\langle z, \omega_j \rangle, \overline{\langle z, \omega_j \rangle}), \qquad p, q \geq 0,$$

forms an orthonormal sequence in (L_c^2) for every $j \geq 1$. Moreover we note that the $\langle z, \omega_j \rangle$ corresponding to distinct ω_j are independent. We are now ready to evaluate the integral

$$\int_{E_c^*} \prod_k H_{p_k, q_k}(\langle z, \omega_{j_k} \rangle, \overline{\langle z, \omega_{j_k} \rangle}) \cdot \prod_l \overline{H_{p_l, q_l}(\langle z, \omega_{j_l} \rangle, \overline{\langle z, \omega_{j_l} \rangle})} \, dv(z)$$

where the $\{\omega_{j_k}\}$ and $\{\omega_{j_l}\}$ are different in each product. The p_k, q_k, p_l, q_l are all permitted to take the value 0 so that the above integral may be written as a finite product

$$\prod_j \int_{E_c^*} H_{p_j, q_j}(\langle z, \omega_j \rangle, \overline{\langle z, \omega_j \rangle}) \overline{H_{p_{j'}, q_{j'}}(\langle z, \omega_j \rangle, \overline{\langle z, \omega_j \rangle})} \, dv(z)$$

where the ω_j appear as many times as they were present in the previous integral. The value of the integral is known to be

$$\prod_j \delta_{p_j, p_{j'}} \delta_{q_j, q_{j'}} p_j!\, q_j!,$$

and so the orthonormality of the system (6.28) is proved. □

Remark. Another proof of (b) goes as follows. Firstly prove that

$$\int_{E_c^*} \exp[i\langle z, \zeta_1 \rangle + i\overline{\langle z, \zeta_2 \rangle}]\, dv(z) = \exp[-(\zeta_2, \zeta_1)] \qquad (6.29)$$

for any $\zeta_1, \zeta_2 \in E_c$. Next take the mean square of the generating function of the $\{H_{p,q}(\langle z, \omega_j \rangle, \overline{\langle z, \omega_j \rangle}) : p, q \geq 0\}$ given by (6.9). Then compare coefficients to obtain the result. Of course this method is essentially the same as the above proof.

The following definition is also a generalisation of that in §4.2.

Definition 6.6. A polynomial on E_c^* expressed (up to a constant multiple) in the form (6.28) is called a *complex Fourier-Hermite polynomial* based on $\{\omega_n : n \geq 1\}$.

Fix $\{\omega_n: n \geq 1\}$ for the moment and denote by $\mathscr{H}_{(m, n)}$ the subspace of (L_c^2) spanned by the Fourier-Hermite polynomials of degree (m, n) based upon $\{\omega_n\}$. In particular $\mathscr{H}_{(0, 0)} = \mathbf{C}$ whilst $\mathscr{H}_{(1, 0)}$ and $\mathscr{H}_{(0, 1)}$ agree with those defined in the previous section. It is easily seen from the definition that the $\mathscr{H}_{(m, n)}$, $m, n \geq 0$, are mutually orthogonal subspaces of (L_c^2).

Proposition 6.11. *The subspace $\mathscr{H}_{(m, n)}$ of (L_c^2) does not depend on the choice of complete orthonormal system $\{\omega_n: n \geq 1\}$ in $L^2(\mathbf{R})$.*

PROOF. Let $\{\omega_n': n \geq 1\}$ be a complete orthonormal system in $L^2(\mathbf{R})$ different from $\{\omega_n: n \geq 1\}$. In order to obtain a relation between the complex Fourier-Hermite polynomials based on $\{\omega_n\}$ and those based on $\{\omega_n'\}$ we present the following formula

$$\int_{E_{c^*}} H_{p, q}(\langle z, \omega_m \rangle, \overline{\langle z, \omega_m \rangle}) \overline{H_{p', q'}(\langle z, \omega_n' \rangle, \overline{\langle z, \omega_n' \rangle})} \, dv(z) \qquad (6.30)$$
$$= \delta_{p, p'} \delta_{q, q'} p! q! (\omega_n', \omega_m)^p \overline{(\omega_n', \omega_m)}^q.$$

This formula can be proved in a manner similar to that explained in the remark following the proof of Theorem 6.2. It tells us that all the complex Fourier-Hermite polynomials of degree (m, n) based on $\{\omega_n'\}$ are orthogonal to $\mathscr{H}_{(m', n')}$ when $(m', n') \neq (m, n)$. However, all the polynomials based upon $\{\omega_n'\}$ span the entire space (L_c^2) and hence the subspace $\mathscr{H}_{(m, n)}'$ formed using $\{\omega_n'\}$ has to coincide with $\mathscr{H}_{(m, n)}$ formed using $\{\omega_n\}$. □

Definition 6.7. An element of $\mathscr{H}_{(m, n)}$ is called a *complex multiple Wiener integral of degree* (m, n) and an element of the direct sum

$$\mathscr{H}_n = \sum_{k=0}^{n} \oplus \mathscr{H}_{(n-k, k)}$$

is called a complex multiple Wiener integral of *degree n*.

We have the following analogue of Theorem 4.3 in §4.2 concerning the decomposition of the space (L_c^2).

Theorem 6.3. *The space (L_c^2) has the direct sum decomposition:*

$$(L_c^2) = \sum_{n=0}^{\infty} \oplus \left(\sum_{k=0}^{n} \oplus \mathscr{H}_{(n-k, k)} \right) = \sum_{n=0}^{\infty} \oplus \mathscr{H}_n. \qquad (6.31)$$

The question now arises of whether a concrete representation of the elements of $\mathscr{H}_{(m, n)}$ can be given in terms of familiar objects, and the answer here is almost the same as that found in §4.3. For this reason we will simply describe the results and make a few remarks upon the differences in the present situation.

6.3 The Complex Multiple Wiener Integral

The exponential function $K(z; \zeta)$ defined at the beginning of this section continues to play a dominant role, for as we see from Proposition 6.9 $\{K(z; \zeta \in E_c\}$ generates a sufficiently rich class of functionals on (L_c^2). Thus it is quite reasonable to consider the following transformation $\mathcal{T}$ on (L_c^2):

$$\varphi \to (\mathcal{T}\varphi)(\zeta) = \int_{E_c^*} K(z; \zeta)\overline{\varphi(z)}\, dv(z), \qquad \zeta \in E_c, \varphi \in (L_c^2), \quad (6.32)$$

as a generalisation of the transformation $\mathcal{T}$ introduced in §4.3. For convenience we continue to use the notation $\mathcal{T}$.

The transformation $\mathcal{T}$ is a linear mapping of (L_c^2) onto a certain space of functionals on E_c. In particular if $\varphi(z) \equiv 1$, call it $1(z)$, then

$$(\mathcal{T}1)(\zeta) = \exp(-\|\zeta\|^2),$$

which is the special case of (6.29) obtained when $\zeta_1 = \zeta_2$. The functional on the right hand side of this last equation will be used frequently in the sequel and is denoted by $C(\zeta)$ (cf. the notation in §3.1).

A definite expression results when φ is taken to be a complex Fourier-Hermite polynomial; if $\varphi(z) = H_{p,q}(\langle z, \omega \rangle, \overline{\langle z, \omega \rangle})$, $\|\omega\| = 1$, then

$$(\mathcal{T}\varphi)(\zeta) = i^{p+q}\overline{(\omega, \zeta)}^p(\zeta, \omega)^q C(\zeta). \quad (6.33)$$

For a complex Fourier-Hermite polynomial $\varphi(z)$ given by (6.28) we have

$$(\mathcal{T}\varphi)(\zeta) = i^{(\sum p_k + \sum q_k)} \prod_k (p_k! q_k!)^{-1/2} \overline{(\omega_{j_k}, \zeta)}^{p_k}(\zeta, \omega_{j_k})^{q_k} C(\zeta), \quad (6.34)$$

by using (6.33) and the fact that $\langle z, \omega_n \rangle$'s are mutually independent for $n \geq 1$. With this result we can argue as in Theorem 4.5 in §4.3 and prove:

Theorem 6.4. *For every $\varphi(z) \in \mathcal{H}_{(m,n)}$ there corresponds a unique function $F_\varphi \in \widehat{L_c^2}(\mathbf{R}^m) \otimes \widehat{L_c^2}(\mathbf{R}^n)$ in such a way that*

$$(\mathcal{T}\varphi)(\zeta) = i^{m+n} C(\zeta) \int_{\mathbf{R}^{m+n}} F_\varphi(u_1, \ldots, u_m; v_1, \ldots, v_n) \quad (6.35)$$
$$\times \overline{\zeta(u_1)} \cdots \overline{\zeta(u_m)} \zeta(v_1) \cdots \zeta(v_n)\, du_1 \cdots du_m\, dv_1 \cdots dv_n,$$

and we also have

$$\|\varphi\|_{(L_c^2)} = \sqrt{m!\, n!}\, \|F_\varphi\|_{L_c^2(\mathbf{R}^{m+n})}. \quad (6.36)$$

The proof is almost the same as that of Theorem 4.5 in §4.3 but in addition we need the following Lemma.

Lemma 6.1. *If* $F \in \widehat{L^2}(\mathbf{R}^m) \otimes \overline{\widehat{L^2}(\mathbf{R}^n)}$ *satisfies*

$$\int_{\mathbf{R}^{m+n}} F_\varphi(u_1, \ldots, u_m; v_1, \ldots, v_n) \times \overline{\zeta(u_1)} \cdots \overline{\zeta(u_m)}\, \zeta(v_1) \cdots \zeta(v_n)\, du_1 \cdots du_m\, dv_1 \cdots dv_n = 0 \quad (6.37)$$

for every $\zeta \in E_c$, *then we have* $F = 0$, *a.e.*

PROOF. We content ourselves with an outline. When m or n is 0 the proof is the same in §4.3 but in the general case it becomes much more complicated. Thus we shall prove the assertion only in the case $m = n = 1$.

Let $\{\omega_n : n \geq 1\}$ be a complete orthonormal system in $L^2(\mathbf{R})$. Then $\{\omega_m \otimes \bar\omega_n : m, n \geq 1\}$ forms a complete orthonormal system in $L^2(\mathbf{R}) \otimes \overline{L^2(\mathbf{R})}$ ($\cong L^2(\mathbf{R}^2)$). With this system we expand $F \in L^2(\mathbf{R}) \otimes \overline{L^2(\mathbf{R})}$ as

$$F = \sum_{m,n} a_{m,n} \omega_m \otimes \bar\omega_n.$$

Now if F satisfies (6.37) we have $a_{j,j} = 0$ and $a_{j,k} = 0, j \neq k$, by taking ζ to be ω_j and $\omega_j + \omega_k$, $\omega_j + i\omega_k$, $j \neq k$, respectively. The conclusion follows. □

Returning to our representation of (L_c^2)-functionals we use Theorem 6.3 to establish the following

Corollary. *The transformation $\mathcal{T}$ induces an isomorphism*

$$(L_c^2) \cong \sum_{n=0}^{\infty} \oplus \sum_{k=0}^{n} \oplus \{(n-k)!\, k!\}^{1/2} \widehat{L^2}(\mathbf{R}^{n-k}) \otimes \overline{\widehat{L^2}(\mathbf{R}^k)}. \quad (6.38)$$

Remark 1. The constant $\{(n-k)!\, k!\}^{1/2}$ is an adjustment to the norm so that the isomorphism (6.38) becomes an isometry as well.

Remark 2. In terms of Physics the expansion (6.38) gives a Fock space for a charged scalar field in quantum mechanics. The space $\widehat{L^2}(\mathbf{R}^{n-k}) \otimes \widehat{L^2}(\mathbf{R}^k)$ corresponds to the state where $n - k$ particles are positively and another k particles are negatively charged.

Any $\varphi \in (L_c^2)$ can be written in the form

$$\varphi = \sum_{m,n} \varphi_{m,n}$$

where $\varphi_{m,n}$ is the projection of φ onto the subspace $\mathcal{H}_{(m,n)}$. Then each $\varphi_{m,n}$ is represented by $F_{\varphi_{m,n}}$ using (6.35).

Definition 6.8. The series of representations of $\varphi \in (L_c^2)$ using $F_{\varphi_{m,n}}$, $m, n \geq 0$, given by (6.35) is called the *integral representation*. The function $F_{\varphi_{m,n}}$ appearing is called the *integral kernel* or simply the *kernel of degree* (m, n).

6.4 Special Functionals in (L_c^2)

The corollary above states that any $\varphi \in (L_c^2)$ can be realised in terms of a series of integral kernals $\{F_{\varphi_{m,n}}: m, n \geq 0\}$, where the norms are appropriately taken into account.

6.4 Special Functionals in (L_c^2)

This section is devoted to an application of the results of the previous section to real-valued functionals over E_c^* and to the presentation of a proposed theory of holomorphic functionals on E_c^*, these being functionals of an infinite dimensional variable z.

(i) Real-valued Functionals

The integral representation of a real-valued functional in (L_c^2) takes a special form so that we can derive some more detailed properties in this case.

The formula $\overline{H_{p,q}(z, \bar{z})} = H_{q,p}(z, \bar{z})$ for complex Hermite polynomials immediately implies that

$$\overline{\mathscr{H}_{(m, n)}} \equiv \{\overline{\varphi(z)}: \varphi \in \mathscr{H}_{(m, n)}\} = \mathscr{H}_{(n, m)}, \qquad m, n \geq 0,$$

but we can state a more precise result for each particular $\varphi \in \mathscr{H}_{(m, n)}$.

Lemma 6.2. *Let F_φ be the kernel of the integral representation of $\varphi(z) \in \mathscr{H}_{(m, n)}$. Then the kernel $F_{\bar{\varphi}}$ associated with $\overline{\varphi(z)} \in \mathscr{H}_{(n, m)}$ is given by*

$$F_{\bar{\varphi}}(v_1, \ldots, v_n; u_1, \ldots, u_m) = (-1)^{m+n} \overline{F_\varphi(u_1, \ldots, u_m, v_1, \ldots, v_n)}. \quad (6.39)$$

PROOF. By using the relation $\overline{K(z; \zeta)} = K(z; -\zeta)$ it follows that

$$\overline{(\mathscr{F}\varphi)(\zeta)} = \overline{\int_{E_c^*} K(z; \zeta)\varphi(z)\, dv(z)} = \int_{E_c^*} K(z; -\zeta)\overline{\varphi(z)}\, dv(z), \qquad \zeta \in E_c.$$

This relation and the expression (6.35) of the previous section easily prove (6.39). □

With this lemma we can prove the following proposition.

Proposition 6.12. *If $\varphi(z) \in \mathscr{H}_{(m, n)}$ is real-valued, then $m = n$ and the kernel $F_\varphi(u, v) \in \hat{L}^2(\mathbf{R}^m) \otimes \hat{L}^2(\mathbf{R}^n)$, $u = (u_1, \ldots, u_m)$, $v = (v_1, \ldots, v_n)$, of the integral representation is Hermitian:*

$$F_\varphi(v, u) = \overline{F_\varphi(u, v)}. \quad (6.40)$$

Since the integral kernel $F_\varphi(u, v)$ for $\varphi \in \mathscr{H}_{(n, n)}$ is in general an L^2-kernel, it defines an integral operator of Hilbert-Schmidt type. If in addition $\varphi(z)$ is assumed to be real-valued, the above result tells us that the kernel defines a Hermitian operator. Thus $F_\varphi(u, v)$ admits an eigenfunction expansion

$$F_\varphi(u, v) = \sum_k \lambda_k^{-1} \psi_k(u)\overline{\psi_k(v)} \tag{6.41}$$

by using the eigenvalues λ_k and normalised eigenfunctions $\{\psi_k\}$ [see F. Smithies (1958)]. It is known that each eigenvalue λ_k is real and that

$$\lambda_k \int_{\mathbf{R}^n} F(u, v)\psi_k(v)\, dv = \psi_k(u),$$

and

$$\sum_k \lambda_k^{-2} = \int_{\mathbf{R}^{2n}} |F_\varphi(u, v)|^2\, du\, dv < \infty.$$

Furthermore we see by the first equation above that each $\psi_k(u)$ belongs to $\widehat{L^2}(\mathbf{R}^n)$ although the system $\{\psi_k\}$ is known to be an orthonormal system in $L^2(\mathbf{R}^n)$.

One might ask what is the counterpart of the expansion (6.41) in (L_c^2). To answer this let us find an element of $\mathcal{H}_{(n, n)}$ with the integral kernel $\psi_k(u)\overline{\psi_k(v)}$. Given a complete orthonormal system $\{\omega_n\}$ in $L^2(\mathbf{R})$ we can take tensor products and form a complete orthonormal system in $L^2(\mathbf{R}^n)$, each element being expressed in the form

$$\bigotimes_j \omega_j^{p_j \otimes}, \quad \sum p_j = n. \tag{6.42}$$

By symmetrisation (see §4.3) we obtain a base for $\widehat{L^2}(\mathbf{R}^n)$. However the space $\mathcal{H}_{(n, n)}$ is spanned by

$$\prod_j H_{p_j, q_j}(\langle z, \omega_j \rangle, \overline{\langle z, \omega_j \rangle}), \quad \sum p_j = \sum q_j = n. \tag{6.43}$$

The associated integral kernels are the tensor products of the symmetrisations of

$$\bigotimes_j \omega_j^{p_j \otimes} \quad \text{and} \quad \bigotimes_j \bar\omega_j^{q_j \otimes}.$$

Now the functional given by (6.43) is obtained from the monomial

$$\prod_j \langle z, \omega_j \rangle^{p_j} \overline{\langle z, \omega_j \rangle}^{q_j}, \quad \sum p_j = \sum q_j = n,$$

of degree (n, n) by subtracting off its projections onto the lower degree spaces $\mathcal{H}_{(m', n')}, m', n' \leq n, m' + n' < 2n$. We note that such an orthogonalisation to lower degree spaces clearly defines a linear operator.

These observations show that the following procedure should give the desired answer. Firstly expand $\psi_k \in \widehat{L^2}(\mathbf{R}^n)$ in a Fourier series using the complete orthonormal system (6.42). Let each term correspond to

$$\prod_j (p_j!)^{-1/2} \langle z, \omega_j \rangle^{p_j}, \quad \sum p_j = n,$$

6.4 Special Functionals in (L_c^2)

from which we find $\tilde{\varphi}_k(z) \in \mathcal{H}_{(n, 0)}$ corresponding to ψ_k. Obviously the integral kernel of $\tilde{\varphi}_k$ is equal to ψ_k. Then take the orthogonalisation $\varphi_k(z)$ of $|\tilde{\varphi}_k(z)|^2$ to lower degree spaces. The functional $\varphi_k(z)$ thus obtained belongs to $\mathcal{H}_{(n, n)}$ and its integral kernel has to be $\psi_k(u)\overline{\psi_k(v)}$. Thus the expansion of $\varphi(z)$ corresponding to (6.41) which we seek is given by

$$\varphi(z) = \sum_k \lambda_k^{-1} \varphi_k(z). \tag{6.44}$$

The actual expressions for each $\varphi_k(z)$ following the above procedures are very complicated. Here we present a simple special example of a functional in the case $n = 1$.

EXAMPLE 2. Let $\varphi(z)$ be a real-valued functional in $\mathcal{H}_{(1, 1)}$. The corresponding integral kernel is expressed in the form

$$F_\varphi(u, v) = \sum_k \lambda_k^{-1} \psi_k(u)\overline{\psi_k(v)} \qquad u, v \in \mathbf{R}$$

by (6.41). Since the system $\{\psi_k\}$ of eigenfunctions is itself an orthonormal system in $L^2(\mathbf{R})$ we may take the $\langle z, \psi_k \rangle$ in $\mathcal{H}_{(1, 0)}$ as the $\tilde{\varphi}_k(z)$. The functional obtained from $|\langle z, \psi_k \rangle|^2$ by subtracting off its projection $\|\psi_k\|^2 = 1$ onto $\mathcal{H}_{(0, 0)}$ is to be $H_{1, 1}(\langle z, \psi_k \rangle, \overline{\langle z, \psi_k \rangle})$, and it is automatically orthogonal to $\mathcal{H}_{(1, 0)}$ and $\mathcal{H}_{(0, 1)}$. Thus the orthogonalisation of $|\langle z, \psi_k \rangle|^2$ has been achieved, and the actual expansion of (6.44) in this case is

$$\varphi(z) = \sum_k \lambda_k^{-1}(|\langle z, \psi_k \rangle|^2 - 1). \tag{6.45}$$

A special remark needs to be made concerning this example. Since $\{\psi_k\}$ is an orthogonal system, $\langle z, \psi_k \rangle$ is a system of independent complex Gaussian random variables and therefore $\{|\langle z, \psi_k \rangle|^2 - 1\}$ is again a system of independent random variables. Thus the expansion (6.45) just obtained expresses $\varphi(z)$ as a sum of not just orthogonal but independent random variables. From this exceptional fact we are able to obtain the probability distribution, other results and applications for a real-valued quadratic functional. For real-valued functionals of higher degree we can get an orthogonal expansion generalising (6.45), but it is not as interesting because in general the terms are not independent.

(ii) Holomorphic Functionals

We wish to present an attempt at defining a holomorphic function (of infinitely many variables) by viewing E_c^* as an infinite-dimensional complex vector space.

Take a complete orthonormal system $\{\omega_n\}$ in $L^2(\mathbf{R})$ and consider polynomials of the form

$$p(z) = \sum_{\substack{(p_1, p_2, \ldots, p_k) \\ k \geq 1}} a_{p_1, p_2, \ldots, p_k} \langle z, \omega_1 \rangle^{p_1} \langle z, \omega_2 \rangle^{p_2} \cdots \langle z, \omega_k \rangle^{p_k} \tag{6.46}$$

based upon $\{\omega_n\}$. The sum is taken as k runs over all positive integers, and $(p_1, p_2, \ldots, p_k)$ run over all k-tuples of positive integers, subject to the restriction that $a_{p_1, p_2, \ldots, p_k}$ is zero except for finitely many subscripts. It seems fitting to call such a $p(z)$ a holomorphic functional. Indeed $\bar{z}$ is not involved; formally

$$\frac{\partial p}{\partial \bar{z}} = 0. \tag{6.47}$$

Recall that $z = x + iy$, $x, y \in E^*$. Then we can express $p(z)$ in the form

$$p(z) = u(x, y) + iv(x, y). \tag{6.48}$$

Set $\langle z, \omega_k \rangle = x_k + i y_k$, $x_k, y_k \in \mathbf{R}$. Then u and v may be expressed as

$$u = u(x_1, x_2, \ldots; y_1, y_2, \ldots), \qquad v = v(x_1, x_2, \ldots; y_1, y_2, \ldots,).$$

Since $p(z)$ is a polynomial, easy computations enable us to prove that equation (6.47) is equivalent to the Cauchy-Riemann differential equations

$$\frac{\partial u}{\partial x_k} = \frac{\partial v}{\partial y_k}, \quad \frac{\partial u}{\partial y_k} = -\frac{\partial v}{\partial x_k}, \qquad k = 1, 2, \ldots,$$

for u and v. In addition u satisfies

$$\frac{\partial^2 u}{\partial x_k \partial x_l} + \frac{\partial^2 u}{\partial y_k \partial y_l} = 0, \quad \frac{\partial^2 u}{\partial x_k \partial y_l} - \frac{\partial^2 u}{\partial x_l \partial y_k} = 0;$$

we have similar equations for v. In other words u and v are *pluriharmonic* on E_c^*. In view of this $p(z)$ could be called a *holomorphic polynomial* on the space E_c^*.

We then compute the (L_c^2)-norm of $p(z)$ by using (6.5) and the fact that $\{\langle z, \omega_n \rangle\}$ is a system of independent complex Gaussian random variables. In fact the functional (6.46) is a sum of mutually orthogonal functionals so that we have

$$\|p\|^2 = \sum |a_{p_1, p_2, \ldots, p_k}|^2 p_1! \, p_2! \cdots p_k!.$$

On the other hand we have

$$\left\| \frac{\partial p}{\partial \omega_j} \right\|^2 = \sum |a_{p_1, p_2, \ldots, p_k}|^2 p_j \cdot p_1! \, p_2! \cdots p_k!,$$

the case $p_j = 0$ inclusive, where $\partial/\partial \omega_j$ is a differential operator defined in a manner similar to that used in §5.7 to define $\partial/\partial \xi_j$. We therefore have

If $p \in \mathcal{H}_{(n, 0)}$, then $\dfrac{\partial p}{\partial \omega_j} \in \mathcal{H}_{(n-1, 0)}$ and $\left\| \dfrac{\partial p}{\partial \omega_j} \right\|^2 \leq n \|p\|^2$.

After expressing it in this way we realise that ω_j may be taken to be an arbitrary unit vector ω in E_c and that p need not be a polynomial based upon $\{\omega_n\}$. In addition the evaluation of the norm $\partial p/\partial \omega$ tells us that p need not

6.4 Special Functionals in (L_c^2)

even be a polynomial, but can be a general functional in $\mathcal{H}_{(n,\,0)}$. Thus we have arrived at the definition of

$$\mathcal{H} = \left\{ \varphi = \sum_n \varphi_n : \varphi_n \in \mathcal{H}_n, \sum_n n^q \|\varphi_n\|^2 < \infty \text{ for every natural number } q \right\}$$

to be the class of *holomorphic functionals*, generalising the class of holomorphic polynomials.

7 The Unitary Group and Its Applications

7.1 The Infinite-Dimensional Unitary Group

As in the last chapter E denotes a nuclear space which is dense in $L^2(\mathbf{R})$. We write E_c and E_c^* for the complexification of E and the dual space of E_c, respectively, and let $(E_c^*, \mathfrak{B}, v)$ be complex white noise.

Let us denote by $U(E_c)$ the collection of all linear transformations g on E_c which satisfy the conditions

1. g is a homeomorphism of E_c onto itself; and
2. g preserves the $L^2(\mathbf{R})$-norm:

$$\|g\zeta\| = \|\zeta\|, \quad \zeta \in E_c.$$

We introduce a product $g_1 g_2$ of elements g_1 and g_2 in $U(E_c)$ by writing

$$(g_1 g_2)\zeta = g_1(g_2 \zeta),$$

and readily see that with this product $U(E_c)$ forms a group. A suitable topology will be introduced shortly but first we give

Definition 7.1. The group $U(E_c)$ is called the *infinite-dimensional unitary group*, or just the *unitary group*.

When there is no need to specify E_c we often write U_∞ instead of $U(E_c)$.

For each $g \in U(E_c)$ a linear transformation g^* on E_c^* is uniquely determined by the following formula:

$$\langle z, g\zeta \rangle = \langle g^*z, \zeta \rangle, \quad z \in E_c^*, \quad \zeta \in E_c.$$

252

7.1 The Infinite-Dimensional Unitary Group

This transformation is an automorphism of E_c^*, and the collection of all such

$$U^*(E_c^*) = \{g^* : g \in U(E_c)\},$$

also forms a group whose product satisfies

$$g_1^* g_2^* = (g_2 g_1)^*, \qquad g_1^*, g_2^* \in U^*(E_c).$$

Now this formula implies that

$$(g^{-1})^* = (g^*)^{-1}, \qquad g \in U(E_c),$$

and so the surjection

$$g \to (g^*)^{-1} \in U^*(E_c^*), \qquad g \in U(E_c), \tag{7.1}$$

defines an isomorphism between the groups $U(E_c)$ and $U^*(E_c^*)$. Thus we have the choice of dealing with either of the two groups, but we will mainly discuss the group $U(E_c)$ in what follows. Before coming to this, however, we state an important proposition concerning $U^*(E_c^*)$.

Proposition 7.1. *For every $g^* \in U^*(E_c^*)$ we have*

$$g^* \circ v = v. \tag{7.2}$$

PROOF. The method of proof is the same as that used in Theorem 5.1 of §5.1 and we begin with some remarks. Firstly, the σ-field $\mathfrak{B}$ is also the one generated by the cylinder sets. Secondly, the two complex Gaussian systems

$$\{\langle z, \zeta \rangle : \zeta \in E_c\}, \quad \text{and} \quad \{\langle z, g\zeta \rangle : \zeta \in E_c\}$$

have the same probability distribution. The former follows directly from the definition, whilst the latter is proved by observing that the two systems have zero mean vectors and covariance matrices

$$(\zeta_1, \zeta_2) \quad \text{and} \quad (g\zeta_1, g\zeta_2) = (\zeta_1, \zeta_2), \qquad \zeta_1, \zeta_2 \in E_c,$$

respectively, that is, the same covariance matrix (cf. Theorem 6.1). If A is a cylinder set then so also is $g^*A = \{g^*z : z \in A\}$, and the above observation concerning covariance matrices shows that

$$v(g^*A) = v(A).$$

This equation can be extended to each set in the σ-field $\mathfrak{B}$ generated by the cylinder sets, and the proof is complete. □

The following corollary is a straightforward consequence of this proposition.

Corollary. *For every $g \in U(E_c)$ the operator U_g defined on (L_c^2) by*

$$U_g \varphi(z) = \varphi(g^*z), \qquad \varphi \in (L_c^2), \tag{7.3}$$

is unitary.

We now return to the group $U(E_c)$. It is clear that the collection $\mathfrak{U} = \{U_g \colon g \in U(E_c)\}$ of operators defined by (7.3) forms a group, and since

$$U_{g_1} U_{g_2} \varphi(z) = U_{g_1} \varphi(g_2^* z) = \varphi(g_2^* g_1^* z) = \varphi((g_1 g_2)^* z), \qquad \varphi \in (L_c^2),$$

we see that $\mathfrak{U}$ is isomorphic to $U(E_c)$. Now $\mathfrak{U}$ can be topologised via the norm in (L_c^2) on which its members act as linear transformations. With this topology on $\mathfrak{U}$ we can then introduce the weakest topology on $U(E_c)$ with respect to which all the mappings

$$g \to U_g \in \mathfrak{U}, \qquad g \in U(E_c),$$

are strongly continuous, in other words, which defines a unitary representation of the group $U(E_c)$. Topologised in this way $U(E_c)$ becomes a topological group. Alternatively, we could introduce the compact-open topology on $U(E_c)$ as it is a transformation group on E_c.

The discussion of $U(E_c)$ so far has been very similar to that of the infinite-dimensional rotation group $O(E)$ given in Chapter 5, §5.1. There is in fact an intimate connection between $O(E)$ and $U(E_c)$ very like that between the orthogonal group $O(n)$ and the unitary group $U(n)$ in the finite-dimensional case. Indeed $O(E)$ may be regarded as a subgroup of $U(E_c)$ in the following sense: for any $\tilde{g} \in O(E)$ and element ζ of E_c, we write $\zeta = \xi + i\eta$ ($\xi, \eta \in E$) and define a transformation g on E_c by

$$g\zeta = \tilde{g}\xi + i\tilde{g}\eta.$$

Then g belongs to $U(E_c)$ and the mapping

$$\tilde{g} \to g \in U(E_c), \qquad \tilde{g} \in O(E),$$

defined in this way is an injection which preserves the group operation. This proves our assertion.

It should be noted, however, that neither $O(E)$ nor $U(E_c)$ is compact or even locally compact, so that the analogy with $O(n)$ and $U(n)$ should be pursued with care.

7.2 The Unitary Group $U(\mathscr{S}_c)$

Until now we have discussed the infinite-dimensional unitary group $U(E_c)$ over an arbitrary nuclear space, E_c, such that $E_c \subset L^2(\mathbf{R}) \subset E_c^*$. In the sections which follow E_c will be taken to be the (complex) Schwartz space $\mathscr{S}_c$, and by making use of the structure of this space some detailed properties of $U(\mathscr{S}_c)$ will be studied.

We begin by explaining why the case $E_c = \mathscr{S}_c$ is emphasized. Firstly (i) the Fourier transform plays an important role. It arises because, as we have seen with the complex white noise $(E_c^*, \mathfrak{B}, \nu)$, each member of E_c^* can be viewed as the derivative of a sample function of a complex Brownian

7.2 The Unitary Group $U(\mathscr{S}_c)$

motion. And in order to carry out a harmonic analysis of the functionals of such sample functions we need the Fourier transform as a basic tool. Viewing z as a generalised function, the Fourier transform of z will be denoted $\mathscr{F} z$ and is defined by the formula

$$\langle \mathscr{F} z, \zeta \rangle = \langle z, \mathscr{F}^* \zeta \rangle, \qquad (\mathscr{F}^* = \mathscr{F}^{-1}),$$

using the inverse Fourier transform on the space E_c. So that $\mathscr{F}$ can be defined as an operator on E_c^*, $\mathscr{F}^*$ must be an element of $U(E_c)$, and such a requirement imposes a restriction on the space E_c. The Schwartz space $\mathscr{S}_c$ certainly satisfies this requirement (see Appendix A.3).

For convenience (ii) we expect that the countable Hilbertian norms $\|\cdot\|_n$, $n \geq 0$, that define the topology of E_c are expressed in the form

$$\|\zeta\|_n = \|D^n \zeta\|, \qquad n \geq 0, \tag{7.4}$$

for a certain differential operator D, the norm $\|\cdot\|$ on the right of (7.4) being the $L^2(\mathbf{R})$-norm.

In addition (iii) the differential operator D defining the norms should commute with the Fourier transform, i.e.

$$\mathscr{F} D = D \mathscr{F}.$$

It is easily seen by elementary computations that an operator D satisfying the last two conditions may be taken, up to a constant multiple, to be

$$D = \frac{d^2}{du^2} - (u^2 + 1). \tag{7.5}$$

This is a well-known differential operator, having as eigenfunctions

$$\xi_n(u) = (2^n n! \sqrt{\pi})^{-1/2} H_n(u) \exp\left(-\frac{1}{2} u^2\right), \qquad n \geq 0, \tag{7.6}$$

with eigenvalues $-2(n+1)$:

$$D \xi_n = -2(n+1) \xi_n, \qquad n \geq 0, \tag{7.7}$$

each of which is simple (i.e. has multiplicity one).

Now the system $\{\xi_n : n \geq 0\}$ of functions constitutes a complete orthonormal system in $L^2(\mathbf{R})$. Let $a_n(\zeta) = (\zeta, \xi_n)$ be the n-th Fourier coefficient of a function $\zeta \in \mathscr{C}^\infty(\mathbf{R}) \cap L^2(\mathbf{R})$. Clearly we have

$$\sum_{n=0}^{\infty} |a_n(\zeta)|^2 < \infty,$$

and the further relation

$$a_n(D\zeta) = -2(n+1) a_n(\zeta), \qquad n \geq 0,$$

tells us that in order to have $\|\zeta\|_p < \infty$ for every p we need the sequence $a_n(\zeta)$ to be rapidly decreasing as $n \to \infty$. But this means that $\zeta \in \mathscr{S}_c$ [see Example 7 in §7 Chapter 7, of L. Schwartz (1950–51)].

In this way we see that $\mathscr{S}_c$ is a suitable nuclear space, satisfying the requirements (i), (ii) and (iii), and in what follows we will always take $E_c = \mathscr{S}_c$. The foregoing discussion leads us to anticipate the next assertion.

Proposition 7.2. *When viewed as linear operator on $\mathscr{S}_c$, both the Fourier transform $\mathscr{F}$ and its inverse $\mathscr{F}^{-1}$ ($= \mathscr{F}^*$) belong to $U(\mathscr{S}_c)$.*

We are now ready to use the structure of the space $\mathscr{S}_c$ to investigate some detailed properties of $U(\mathscr{S}_c)$. As in the case of $O(E)$ we begin by finding several one-parameter subgroups. This topic will be taken up in the next section.

7.3 Subgroups of $U(\mathscr{S}_c)$

This section discusses one-parameter or finite-dimensional subgroups of $U(\mathscr{S}_c)$ which describe probabilistic or dynamical aspects of Brownian motion. The first such subgroup is one locally isomorphic to the Heisenberg group.

(i) The Heisenberg Group

This is a three-dimensional group and we discuss the three associated one-parameter subgroups in turn.

  a. The *Gauge transform* is the simplest one-parameter subgroup given by

$$I_t: \zeta(u) \to (I_t \zeta)(u) = e^{it}\zeta(u), \quad -\infty < t < \infty, \tag{7.8}$$

[see S. S. Schweber (1964), Part Two]. Obviously each I_t is an element of $U(\mathscr{S}_c)$ and $\{I_t: -\infty < t < \infty\}$ forms a one-parameter group which is continuous in t and which has period 2π:

$$I_t I_s = I_{t+s}, \quad -\infty < t, s < \infty;$$
$$I_{t+2\pi} = I_t, \quad -\infty < t < \infty;$$
$$I_t \to I \quad \text{as} \quad t \to 0.$$

The group $\{I_t: -\infty < t < \infty\}$ is called the *Gauge transform*.

We now consider the unitary operator U_t on (L_c^2) derived from I_t by (7.8), i.e.

$$U_t = U_{I_t}.$$

7.3 Subgroups of $U(\mathscr{S}_c)$

Since $U_t\langle z,\zeta\rangle = \langle e^{-it}z, \zeta\rangle = e^{-it}\langle z, \zeta\rangle$ holds we can see that for a complex Fourier–Hermite polynomial $\varphi(z) \in (L_c^2)$ given by (6.28) we have

$$(U_t\varphi)(z) = \prod_k (p_k!\,q_k!)^{-1/2} H_{p_k, q_k}(e^{-it}\langle z, \omega_{j_k}\rangle, \overline{e^{it}\langle z, \omega_{j_k}\rangle})$$

$$= \exp\left[-it\left\{\sum_k (p_k - q_k)\right\}\right]\varphi(z).$$

Hence for any $\varphi(z) \in \mathscr{H}_{(m, n)}$ we have the identity

$$(U_t\varphi)(z) = \exp[-it(m - n)]\varphi(z).$$

In other words, the one-parameter unitary group $\{U_t: -\infty < t < \infty\}$ has only point spectra on the space $\mathscr{H}_n$, consisting of the values

$$-(n - 2k), \qquad k = 0, 1, 2, \ldots, n,$$

and the eigenspace belonging to $-(n - 2k)$ is $\mathscr{H}_{(n-k, k)}$. Thus we can say that $\{U_t\}$ gives us the direct sum decomposition

$$\mathscr{H}_n = \sum_{k=0}^{n} \oplus \mathscr{H}_{(n-k, k)}. \tag{7.9}$$

As for the action of the gauge transform on the measure space $(\mathscr{S}_c^*, \mathfrak{B}, v)$, the fact that $\{I_c\} \subset U(\mathscr{S}_c)$ means that v is invariant under the rotation

$$I_t^* z = (x \cos t + y \sin t) + i(y \cos t - x \sin t), \qquad z = x + iy,$$

$x, y \in \mathscr{S}^*$ where $\mathscr{S}_c^*$ is viewed as the (x, y)-plane.

b. *The shift.* The shift we discuss here is almost the same as the one introduced in §5.2, Chapter 5, the only difference being that it now acts upon the complexification of the earlier space. The one-parameter group $\{S_t: -\infty < t < \infty\}$ is given by

$$S_t: \zeta(u) \to (S_t\zeta)(u) = \zeta(u - t), \qquad -\infty < t < \infty, \tag{7.10}$$

and is clearly a subgroup of $U(\mathscr{S}_c)$.

Setting $S_t^* = T_t$, we see that $\{T_t: -\infty < t < \infty\}$ is a flow on the measure space $(\mathscr{S}_c^*, \mathfrak{B}, v)$, called the *flow of the complex Brownian motion*. It plays an important role in many fields, including probability theory, but as the reasons for this and the interpretations are the same as those in §4.4 Chapter 4, we do not repeat them here.

c. *Multiplication.* Recalling that we have previously emphasised the role of the Fourier transform $\mathscr{F}$, we now introduce a one-parameter group $\{\pi_t: -\infty < t < \infty\}$ which is conjugate to the shift via $\mathscr{F}$,

$$\pi_t = \mathscr{F} S_t \mathscr{F}^{-1}, \qquad -\infty < t < \infty. \tag{7.11}$$

Since $\mathscr{F} \in U(\mathscr{S}_c)$, we see that $\{\pi_t : -\infty < t < \infty\}$ is a subgroup of $U(\mathscr{S}_c)$, called the *multiplication* (subgroup). A simple computation shows that each π_t acts in the following way:

$$(\pi_t \zeta)(u) = e^{iut}\zeta(u), \qquad -\infty < t < \infty. \tag{7.12}$$

The collection $\{\pi_t^* : -\infty < t < \infty\}$ is also a flow on $(\mathscr{S}_c^*, \mathfrak{B}, \nu)$. By (7.2) and (7.11) it is isomorphic to the flow of the complex Brownian motion and so, as a flow, enjoys the same properties as $\{T_t\}$.

We turn now to the unitary operator $U_t = U_{\pi_t}$ given by (7.3). To this end we recall the definition of a generalised stochastic process (see Chapter 1, §1.3, (ii)). Through the bilinear form $\langle z, \zeta \rangle$ each $z \in \mathscr{S}_c^*$ defines a mapping of $\mathscr{S}_c$ into (L_c^2) and so, regarded as a system of random variables, z is a complex generalised stochastic process. Furthermore, the two systems $\{\langle z, \zeta \rangle : \zeta \in \mathscr{S}_c\}$ and $\{\langle z, S_t \zeta \rangle : \zeta \in \mathscr{S}_c\}$ have the same probability distribution, so that we have a stationary generalised stochastic process (see §3.4). It therefore has a spectral representation

$$\langle z, \zeta \rangle = \int_{-\infty}^{\infty} (\mathscr{F}\bar{\zeta})(\lambda) M(d\lambda, z), \tag{7.13}$$

where $\{M(d\lambda)\}$ is a homogeneous $(E|M(d\lambda)|^2 = d\lambda)$ complex Gaussian random measure. Applying U_t to the function $\langle z, \zeta \rangle$, an element of $\mathscr{H}_{(1,0)}$, gives

$$U_t \langle z, \zeta \rangle = \langle \pi_t^* z, \zeta \rangle = \int_{-\infty}^{\infty} (\mathscr{F}\bar{\zeta})(\lambda) M(d\lambda + t).$$

In other words, the operator U_t in this case defines a shift of the random measure M of (7.13):

$$U_t M(d\lambda) = M(d\lambda + t). \tag{7.14}$$

Now a general element of $\mathscr{H}_{(1,0)}$ can be expressed in terms of $M(d\lambda)$ in the form

$$\int_{-\infty}^{\infty} g(\lambda) M(d\lambda), \qquad g \in L_c^2(\mathbf{R}),$$

and so for such an element

$$U_t \int_{-\infty}^{\infty} g(\lambda) M(d\lambda) = \int_{-\infty}^{\infty} g(\lambda - t) M(d\lambda), \tag{7.15}$$

i.e. $\{U_t\}$ defines a shift on the function g.

Turning now to the relationship between U_t and complex conjugation we have

$$U_t \overline{\varphi(z)} = \overline{\varphi(\pi_t^* z)} = \overline{U_t \varphi(z)}.$$

7.3 Subgroups of $U(\mathcal{S}_c)$

We finally note that U_t is multiplicative:

$$U_t(\varphi_1(z)\varphi_2(z)) = \{U_t\varphi_1(z)\}\{U_t\varphi_2(z)\},$$

and we can thus use (7.15) to see clearly how this operator acts on (L_c^2).

Up until now we have studied the three one-parameter subgroups $\{I_t\}, \{S_t\}$ and $\{\pi_t\}$ separately; it is also interesting to observe the relationships between them. Obviously $\{I_t\}$ commutes with both $\{S_t\}$ and $\{\pi_t\}$, whilst $\{S_t\}$ and $\{\pi_t\}$ satisfy the so-called Weyl commutation relations:

$$\pi_t S_s = I_{st} S_s \pi_t, \qquad -\infty < s, t < \infty. \tag{7.16}$$

Dynamically the three one-parameter groups form a three-dimensional subgroup of $U(\mathcal{S}_c)$.

Next let us consider the subgroup H_1 of the linear Lie group $SL(3, \mathbf{R})$ given by

$$H_1 = \left\{ \begin{pmatrix} 1 & a & c \\ 0 & 1 & b \\ 0 & 0 & 1 \end{pmatrix} : a, b, c \in \mathbf{R} \right\}.$$

Any group isomorphic to H_1 is called a (one-dimensional) Heisenberg group, and it is clear that H_1 is generated by the following three one-parameter subgroups:

$$\left\{ \alpha_t = \begin{pmatrix} 1 & t & 0 \\ 0 & 1 & 0 \\ 0 & 0 & 1 \end{pmatrix} \right\}, \quad \left\{ \beta_t = \begin{pmatrix} 1 & 0 & 0 \\ 0 & 1 & t \\ 0 & 0 & 1 \end{pmatrix} \right\} \text{ and } \left\{ \gamma_t = \begin{pmatrix} 1 & 0 & t \\ 0 & 1 & 0 \\ 0 & 0 & 1 \end{pmatrix} \right\}.$$

The centre of H_1 is $\{\gamma_t\}$ and the subgroups $\{\alpha_t\}$ and $\{\beta_t\}$ satisfy the commutation relations:

$$\alpha_t \beta_s = \gamma_{ts} \beta_s \alpha_t.$$

Letting I_t, S_t and π_t correspond to γ_t, β_t and α_t respectively, we see that the three-dimensional Lie group generated by $\{I_t, S_t, \pi_t\}$ is locally isomorphic to the Heisenberg group.

Incidentally, we note that the triple

$$\alpha = \begin{pmatrix} 0 & 1 & 0 \\ 0 & 0 & 0 \\ 0 & 0 & 0 \end{pmatrix}, \quad \beta = \begin{pmatrix} 0 & 0 & 0 \\ 0 & 0 & 1 \\ 0 & 0 & 0 \end{pmatrix}, \quad \gamma = \begin{pmatrix} 0 & 0 & 1 \\ 0 & 0 & 0 \\ 0 & 0 & 0 \end{pmatrix}$$

can be taken as a basis for the Lie algebra associated with H_1, and that their commutation relations are

$$[\alpha, \beta] = \gamma, \qquad [\alpha, \gamma] = [\beta, \gamma] = 0.$$

The n-dimensional Heisenberg group H_n is defined as above, but with a and b in the matrix form of the element of H_1 replaced by n-dimensional vectors. The group

$$Z_n = \left\{ \begin{pmatrix} 1 & 0 & 0 & \cdots & 0 & 2k\pi \\ 0 & 1 & 0 & \cdots & 0 & 0 \\ 0 & 0 & 1 & \cdots & 0 & 0 \\ & \cdots & & & & 1 \end{pmatrix} : k \text{ an integer} \right\}$$

is a normal subgroup of H_n and the group G_n given in §5.8 is isomorphic to H_n/Z_n.

(ii) The Group of Fourier-Mehler Transforms

We now form a one-parameter subgroup of $U(\mathscr{S}_c)$ consisting of powers, not necessarily integral, of the Fourier transform.

The integral kernel

$$K_\theta(u, v) = [\pi(1 - \exp(2i\theta))]^{1/2} \exp\left[-\frac{i(u^2 + v^2)}{2 \tan \theta} + \frac{iuv}{\sin \theta} \right] \quad (7.17)$$

defines an operator $\mathscr{F}_\theta$ by writing

$$(\mathscr{F}_\theta \zeta)(u) = \int_{-\infty}^{\infty} K_\theta(u, v) \zeta(v) \, dv, \qquad \theta \neq \frac{1}{2} k\pi. \quad (7.18)$$

Letting $\{\xi_n(u): n \geq 0\}$ be the system of functions given by (7.6) we will see that

$$\mathscr{F}_\theta \xi_n(u) = e^{in\theta} \xi_n(u), \qquad n \geq 0. \quad (7.19)$$

To prove this relation we first observe that the generating function of $H_n(u)\exp(-\frac{1}{2}u^2)$ is

$$\sum_{n=0}^{\infty} \frac{t^n}{n!} H_n(u) \exp\left(-\frac{1}{2}u^2\right) = \exp\left(2tu - t^2 - \frac{1}{2}u^2\right).$$

Next we apply $\mathscr{F}_\theta$ to the right-hand side obtaining

$$\int_{-\infty}^{\infty} K_\theta(u, v) \exp\left[2tv - t^2 - \frac{1}{2}v^2\right] dv$$

$$= [\pi(1 - \exp(2i\theta))]^{-1/2} \exp\left[-t^2 - \frac{iu^2}{2\tan\theta}\right]$$

$$\times \int_{-\infty}^{\infty} \exp\left[\left(\frac{iu}{\sin\theta} + 2t\right)v - \left(\frac{i + \tan\theta}{2\tan\theta}\right)v^2\right] dv$$

$$= [\pi(1 - \exp(2i\theta))]^{-1/2} [-2i \sin\theta \cdot \exp(i\theta)]^{1/2} \exp\left[-t^2 - \frac{iu^2}{2\tan\theta}\right]$$

$$\times \exp\left[-\frac{1}{2} i \sin\theta \exp(i\theta)\left(\frac{iu}{\sin\theta} + 2t\right)^2\right]$$

(cf. the formula used in the proof of Proposition 6.5)

$$= \exp\left(2ut\exp(i\theta) - [t\exp(i\theta)]^2 - \frac{1}{2}u^2\right)$$

$$= \sum_{n=0}^{\infty} \frac{t^n}{n!} e^{in\theta} H_n(u)\exp\left(-\frac{1}{2}u^2\right).$$

Comparing the coefficients of t^n gives us the result

$$\int_{-\infty}^{\infty} K_\theta(u,v)H_n(v)\exp\left(-\frac{1}{2}v^2\right)dv = e^{in\theta}H_n(u)\exp\left(-\frac{1}{2}u^2\right),$$

which, with the appropriate multiplicative constant, is (7.19).

For the exceptional points $\theta = \frac{1}{2}\pi k$, $k = 0, \pm 1, \pm 2, \ldots$, we proceed as follows: first define $\mathscr{F}_\theta \xi_n(u)$ to be $e^{in\theta}\xi_n(u)$. Then extend $\mathscr{F}_\theta$ linearly to the entire space $\mathscr{S}_c$, this being possible since $\{\xi_n\}$ is a complete orthonormal system in $\mathscr{S}_c$. The linear operators $\mathscr{F}_\theta$ on $\mathscr{S}_c$ so obtained, $-\infty < \theta < \infty$, all belong to $U(\mathscr{S}_c)$ and satisfy:

$$\mathscr{F}_\theta \mathscr{F}_{\theta'} = \mathscr{F}_{\theta+\theta'} = \mathscr{F}_{\theta''}, \quad \theta + \theta' = \theta''(\text{mod } 2\pi);$$

$$\mathscr{F}_\theta \to I \text{ (the identity operator)} \quad \text{as} \quad t \to 0. \tag{7.20}$$

We thus have a one-parameter group $\{\mathscr{F}_\theta: -\infty < \theta < \infty\}$ with period 2π. Particular choices of θ give

$$\mathscr{F}_{\pi/2} = \mathscr{F}, \quad \mathscr{F}_{3\pi/2} = \mathscr{F}^{-1},$$

showing that we have obtained a one-parameter unitary group including the Fourier transform and its inverse. We now recall the basic property of the Fourier transform: that it exchanges the operator consisting of multiplying a function by the independent variable of that function with the differentiation operator (apart from the constant factor i). The following equation involving the kernel $K_\theta(u,v)$ given by (7.17) can be regarded as a generalisation of this property:

$$vK_\theta(u,v) = \left(u\cos\theta - i\frac{\partial}{\partial u}\sin\theta\right)K_\theta(u,v).$$

These remarks lead us to consider the formal identity

$$\mathscr{F}_{\pi r/2} = (\mathscr{F})^r,$$

for any real number r, and to regard $\{\mathscr{F}_\theta\}$ as the subgroup of $U(\mathscr{S}_c)$ consisting of arbitrary powers of the Fourier transform (see Figure 11).

Remark. Arbitrary powers of the Fourier transform were first considered by N. Wiener (1928–1929).

Let us express $\zeta \in \mathscr{S}_c$ in the form

$$\zeta = \sum_{n=0}^{\infty} a_n \zeta_n$$

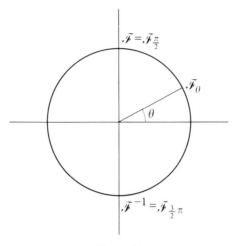

Figure 11

where the $\{\xi_n\}$ are the complete orthonormal system introduced above. Then we see that

$$\mathscr{F}_\theta \zeta = \sum_{n=0}^{\infty} a_n e^{in\theta} \xi_n \tag{7.21}$$

and therefore the unitary operator $U_\theta = U_{\mathscr{F}_\theta}$ given by (7.3) acts in such a way that

$$U_\theta \langle z, \zeta \rangle = \sum_{n=0}^{\infty} a_n e^{-in\theta} \langle z, \xi_n \rangle.$$

In particular, the relation $U_\theta \langle z, \xi_n \rangle = e^{-in\theta} \langle z, \xi_n \rangle$ holds, and U_θ acts multiplicatively, so that for $\varphi(z)$ given by (6.28) (putting $\omega_n = \xi_n$)

$$U_\theta \varphi(z) = \exp\left[i\theta \sum_k (q_k - p_k)\right] \varphi(z),$$

that is, a complex Fourier-Hermite polynomial is an eigenfunction of any U_θ. By Theorem 6.2 the collection of all such polynomials is complete, and so it exhausts the set of eigenfunctions.

Now there are infinitely many choices of triples $(j_k, p_k, q_k: k \geq 1)$ such that $\sum_k j_k(p_k - q_k) = n$, for the sum is over the entire set $\mathbf{Z}$ of integers. Thus the spectrum of the unitary group $\{U_\theta\}$ is $\mathbf{Z}$, and the multiplicity is uniformly countably infinite. We can state this fact in terms of the spectral representation as follows: the support of the spectral measure $dE(\lambda)$ is $\mathbf{Z}$, and for each $n \in \mathbf{N}$, $E(n) - E(n-0)$ is a projection operator onto an infinite-dimensional subspace.

The mapping

$$z \to \{\langle z, \xi_n \rangle : n \geq 0\} \in \mathbf{R}^\infty, \qquad z \in \mathscr{S}_c^*,$$

gives us a coordinate representation of z in the sense that each $\langle z, \xi_n \rangle$ is the projection of $z \in \mathcal{S}_c^*$ along the direction ξ_n. It is known from properties of the Hermite polynomials that ξ_n has n zeros which separate those of ξ_{n-1}, and so as n increases, the number of zeros of ξ_n increases, and thus it oscillates more frequently. This behaviour can be illustrated by considering the L^2-norm of the derivative, for

$$\|\xi_n'\|^2 = \left\| \sqrt{\frac{1}{2}n}\,\xi_{n-1} - \sqrt{\frac{1}{2}(n+1)}\,\xi_{n+1} \right\|^2 = n + \frac{1}{2},$$

which certainly increases with n. The component $\langle z, \xi_n \rangle$ of z in the direction of ξ_n can be taken out of $\langle z, \zeta \rangle$ using U_θ according to the following formula:

$$\frac{1}{2\pi} \int_{-\pi}^{\pi} e^{in\theta} U_\theta \langle z, \zeta \rangle \, d\theta = a_n \langle z, \xi_n \rangle, \qquad a_n = a_n(\zeta) = \langle \zeta, \xi_n \rangle.$$

Thus this method is similar to that involving Fourier coefficients, and in this sense we see a close connection between U_θ and the sample functions of complex white noise.

(iii) The Group of Dilations (Tensions)

The linear transformation τ_t on $\mathcal{S}_c$ which we call the *dilation* is derived from the dilation of the variable u in $\zeta(u)$:

$$\tau_t: \zeta(u) \to (\tau_t \zeta)(u) = \zeta(ue^t)e^{t/2}, \qquad -\infty < t < \infty. \tag{7.22}$$

As in the examples already discussed the family $\{\tau_t: -\infty < t < \infty\}$ forms a one-parameter group and is another important subgroup of $U(\mathcal{S}_c)$. One reason why it is so important is its close connection with the shift, for in the terminology of dynamical systems the shift $\{S_t\}$ is transversal to $\{\tau_t\}$. This fact can be expressed in the formula

$$S_t \tau_s = \tau_s S_{t\exp(s)}, \qquad -\infty < t, s < \infty. \tag{7.23}$$

We can give an illustration of this expression similar to that in (5.15), Chapter 5, and so we do not go into any further details.

We have now discussed five one-parameter subgroups of $U(\mathcal{S}_c)$, with special emphasis on the Fourier transform, and so we turn next to discussing them together. A rough classification is possible using the spectral type of the flows on $(\mathcal{S}_c^*, \mathfrak{B}, \nu)$ given by these subgroups

a. Lebesgue spectrum with countably infinite multiplicity (σ-Lebesgue spectrum): S_t^*, π_t^*, τ_t^*;
b. Discrete spectrum (entire $\mathbf{Z}$) with countably infinite multiplicity: $I_t^*, \mathcal{F}_\theta^*$.

Several other questions also arise, such as the relationships between these one-parameter subgroups [e.g. (7.23)], the nature of the subgroup of $U(\mathcal{S}_c)$

generated by all five, and so on. To answer such questions we must introduce the infinitesimal generators of the one-parameter subgroups and discuss their commutation relations.

7.4 Generators of the Subgroups

The infinitesimal generator α of a one-parameter subgroup $\{g_t\}$ of $U(\mathscr{S}_c)$ is defined to be the derivative of g_t evaluated at $t = 0$:

$$\alpha = \frac{d}{dt} g_t \bigg|_{t=0}. \qquad (7.24)$$

In this chapter we consider only those cases in which the domain of α is the entire space $\mathscr{S}_c$.

In the following table we list the generators of the five one-parameter subgroups discussed in the last section. Certain notation is also introduced there.

Table 1

one parameter subgroup	generator
I_t	iI
S_t	$s \equiv -\dfrac{d}{du}$
π_t	$i\pi \equiv iu$
$\mathscr{F}_\theta$	$if \equiv -\dfrac{1}{2}i\left(\dfrac{d^2}{du^2} - u^2 + 1\right)$
τ_t	$\tau \equiv u\dfrac{d}{du} + \dfrac{1}{2}$

The exact form of each generator except that of $\mathscr{F}_\theta$ can easily be obtained by direct calculation, and in every case the domain is always found to be the entire space $\mathscr{S}_c$. Turning now to $\mathscr{F}_\theta$, the generators given in Table 1 can be derived either from (7.17) and (7.18) by direct calculation, or from the formula

$$\frac{d\mathscr{F}_\theta \xi_n}{d\theta}\bigg|_{\theta=0} = in\xi_n$$

which can be obtained from (7.19), and the fact that the system $\{\xi_n\}$ of eigenfunctions is a complete orthonormal system in $L^2(\mathbf{R})$.

7.4 Generators of the Subgroups

We are now in a position to obtain the commutation relations connecting these infinitesimal generators. The commutator of operators α and β, denoted by $[\alpha, \beta]$, is given by $[\alpha, \beta] = \alpha\beta - \beta\alpha$. After obtaining all the commutators of pairs of generators from our table, we find a new operator $\sigma = \frac{1}{2}[\tau, f]$ that is not in the table. The precise form of σ is given by

$$\sigma = \frac{1}{2}\left(\frac{d^2}{du^2} + u^2\right). \tag{7.25}$$

Formally speaking we might say that there is a one-parameter group $g_t = \exp[i\sigma t]$, $-\infty < t < \infty$, with infinitesimal generator $i\sigma$, but we cannot yet give an interpretation to $\{g_t^*\}$ such as was given to the examples of the previous section. Having introduced the operator σ we now form the following table of *basic commutation relations*.

Table 2 Basic Commutation Relations

$[s, f] = -\pi$	$[\tau, f] = 2\sigma$
$[\pi, f] = -s$	$[\sigma, f] = 2\tau$
$[\pi, s] = I$	$[\tau, \sigma] = 2f + I$
$[\tau, s] = -s$	$[\sigma, s] = \pi$
$[\tau, \pi] = \pi$	$[\sigma, \pi] = -s$

I commutes with every generator

The computation of all these commutators is elementary and will be omitted.

Remark. The table becomes slightly simpler if f is replaced by $f' = f + \frac{1}{2}I$, but we keep f as we wish to emphasise its role as the generator of $\{\mathscr{F}_\theta\}$.

Many things can be obtained from this table of commutation relations. For example, by passing to commutators, f exchanges the generator of the shift with that of the multiplication, and this fact characterises the relationship between the flows $\{T_t\}$ and $\{\pi_t^*\}$. The relationship expressed by (7.23) between $\{S_t\}$ and $\{\tau_t\}$ still holds between $\{\pi_t\}$ and $\{\tau_{-t}\}$, and so on. As a whole the basic commutation relations enable us to describe dynamically the six one-parameter groups.

Viewing the table algebraically enables us to summarise it as follows. Let $\mathfrak{A}$ and $\mathfrak{R}$ be the algebras over $\mathbf{R}$ generated by $\{I, s, \pi, f, \tau, \sigma\}$ and $\{I, s, \pi\}$ respectively.

Proposition 7.3. *If we introduce a product in $\mathfrak{A}$ using the commutator $[\ ,\ ]$, then $\mathfrak{A}$ becomes a Lie algebra and $\mathfrak{R}$ is the radical of $\mathfrak{A}$.*

The proof is an immediate consequence of the basic commutation relations.

Remark. Since the generators forming the base of the vector space have been obtained from the unitary group $U(\mathscr{S}_c)$, it might be thought more reasonable to deal with the complexification $\mathfrak{A}_c = \mathfrak{A} + i\mathfrak{A}$ of $\mathfrak{A}$ rather than the real algebra $\mathfrak{A}$ itself. However, we note that the structure constants of the complex Lie algebra $\mathfrak{A}_c$ using this base are all real, and so it suffices for our purposes to deal with the real form of $\mathfrak{A}_c$, that is, with the algebra $\mathfrak{A}$.

The following proposition rephrases and is a partial summary of what we have just discussed.

Proposition 7.4. *The one-parameter subgroups $\{I_t\}, \{S_t\}, \{\pi_t\}, \{\mathscr{F}_\theta\}, \{\tau_t\}$ generate a six-dimensional subgroup G of $U(\mathscr{S}_c)$. The group G is a Lie group with respect to the topology induced by that of $U(\mathscr{S}_c)$, and its Lie algebra is the complexification $\mathfrak{A}_c$ of $\mathfrak{A}$.*

Remark. The radical $\mathfrak{R}$ of the algebra $\mathfrak{A}$ is the Lie algebra of the Heisenberg group.

As we have seen $\mathfrak{A}$ and $\mathfrak{A}_c$ are neither simple nor semi-simple, but despite this the Lie group G and the Lie algebra $\mathfrak{A}_c$ are of some interest to us. This is because they have a deep connection with the theory of differential equations, in addition to the probabilistic and mechanical properties we have already found.

7.5 The Symmetry Group of the Heat Equation

In this section we illustrate how well the Lie group G introduced in the last section serves us in investigating the symmetry of the solution space of certain partial differential equations, in particular, of the heat equation. The method is that used by W. Miller Jr. (1968); see also his Lecture Notes.

Given the one-dimensional heat equation

$$\frac{\partial}{\partial t} v(x, t) = \frac{1}{2} \frac{\partial^2}{\partial x^2} v(x, t) \qquad x \in \mathbf{R}, \qquad t \in (0, \infty), \qquad (7.26)$$

we clarify the geometric structure of the solution space in the following steps. Firstly we set

$$Q = \frac{\partial}{\partial t} - \frac{1}{2} \frac{\partial}{\partial x^2}, \qquad (7.27)$$

7.5 The Symmetry Group of the Heat Equation

and let $\mathfrak{G}$ be the vector space spanned by all first-order differential operators of the form

$$L = X(x, t)\frac{\partial}{\partial x} + T(x, t)\frac{\partial}{\partial t} + U(x, t) \tag{7.28}$$

where

i. X, T and U are all analytic in x and t; and
ii. $Qv = 0$ implies $QLv = 0$.

The following proposition is an elementary consequence of these definitions.

Proposition 7.5. *A necessary and sufficient condition for a differential operator L of the form* (7.28) *to belong to $\mathfrak{G}$ is that there exists a function $R(x, t)$ analytic in x and t such that*

$$[L, Q] = R(x, t)Q. \tag{7.29}$$

It is also easy to prove

Proposition 7.6. *The vector space $\mathfrak{G}$ forms a Lie algebra when equipped with the commutator product* [,].

For sufficiently small t let us define the exponential map of L by

$$\exp(tL) = \sum_{n=0}^{\infty} \frac{t^n}{n!} L^n.$$

With this notation we have

Proposition 7.7. *The collection G of all operators of the form*

$$\exp(t_1 L_1)\exp(t_2 L_2) \cdots \exp(t_n L_n) \tag{7.30}$$

where $\{L_1, \ldots, L_n\} \subseteq \mathfrak{G}$, $\{t_1, \ldots, t_n\}$ is included in a neighbourhood of 0, and $n \geq 1$, forms a local Lie group.

For the detail of the above results we refer to the Lecture Notes of W. Miller Jr.

Definition 7.2. The group G in Proposition 7.7 is called the *symmetry group* of Q.

Remark. We have only discussed the operator Q associated with the heat equation, but the above results and the definition which follows can easily be generalised to a much wider class of partial differential operators Q having analytic coefficients.

With the above background prepared we can now speak of the symmetry of the heat equation in terms of a Lie algebra.

Theorem 7.1 (W. Miller Jr.). *The Lie algebra $\mathfrak{G}$ is six-dimensional and the following operators can be taken as a base of $\mathfrak{G}$:*

$$I \text{ (the identity)}, \quad L_{-2} = \frac{\partial}{\partial t}, \quad L_{-1} = \frac{\partial}{\partial x}, \quad L_0 = x\frac{\partial}{\partial x} + 2t\frac{\partial}{\partial t},$$

$$L_1 = t\frac{\partial}{\partial x} + x, \quad L_2 = tx\frac{\partial}{\partial x} + t^2\frac{\partial}{\partial t} + \frac{1}{2}(x^2 + t). \tag{7.31}$$

We come now to our main topic, illustrating the role the group G of Proposition 7.7 plays in describing the symmetry of the heat equation. Let $g(t; x)$ be the Gauss kernel

$$g(t; x) = (2\pi t)^{-1/2} \exp\left(-\frac{x^2}{2t}\right),$$

and denote by $\mathfrak{S}$ the class of all complex-valued functions v of the form

$$v \equiv v(x, t; \zeta) = \int_{-\infty}^{\infty} \zeta(u)g(t; x - u)\, du, \qquad \zeta \in \mathcal{S}_c. \tag{7.32}$$

Any function v defined in this way is a solution of the heat equation (7.26) and so $\mathfrak{S}$ is a subspace of the solution space of (7.26). The definition of v gives us a bijection T of $\mathcal{S}_c$ onto $\mathfrak{S}$

$$T: \zeta \to v(x, t; \zeta), \qquad \zeta \in \mathcal{S}_c.$$

and we can thus introduce a topology on the space $\mathfrak{S}$ which ensures that T is a homeomorphism.

Associated with any continuous one-parameter subgroup $\{g_t\}$ of $U(\mathcal{S}_c)$ is a continuous one-parameter group $\{\tilde{g}_t\}$ acting on $\mathfrak{S}$ as follows:

$$\tilde{g}_t = Tg_t, \quad \text{i.e.} \quad \tilde{g}_t v(x, s; \zeta) = v(x, s; g_t\zeta), \qquad -\infty < t < \infty.$$

The infinitesimal generators of $\{\tilde{g}_t\}$ and $\{g_t\}$ are related by the formula

$$\left.\frac{d}{dt}\tilde{g}_t\right|_{t=0} = T\left.\frac{d}{dt}g_t\right|_{t=0}, \tag{7.33}$$

valid on the intersection of the domain of the generator of $\{g_t\}$ with $\mathcal{S}_c$. Now all the infinitesimal generators of the one-parameter subgroups introduced in §7.3 have domains included in $\mathcal{S}_c$, as does the operator σ, and so we can obtain exact formulae for the corresponding differential operators on $\mathfrak{S}$ by using (7.32). These are listed in Table 3.

The computations which give these differential operators are all elementary (integration by parts etc.), it only being necessary to replace $\partial/\partial x^2$ by $2(\partial/\partial t)$, which is a consequence of the fact that $\mathfrak{S}$ is included in the solution space of the heat equation (7.26).

7.5 The Symmetry Group of the Heat Equation

Table 3

$I \to I$ (the identity on $\mathfrak{S}$)

$s \to -\dfrac{\partial}{\partial x}$

$\pi \to t\dfrac{\partial}{\partial x} + x$

$\tau \to 2t\dfrac{\partial}{\partial t} + x\dfrac{\partial}{\partial x} + \dfrac{1}{2}$

$f \to (t^2 - 1)\dfrac{\partial}{\partial t} + tx\dfrac{\partial}{\partial x} + \dfrac{1}{2}(x^2 + t - 1)$

$\sigma \to (t^2 + 1)\dfrac{\partial}{\partial t} + tx\dfrac{\partial}{\partial x} + \dfrac{1}{2}(x^2 + t)$

A consideration of the expressions for the differential operators in Table 3 leads us easily to the following

Proposition 7.8. *Equipped with the product* [,] *the real vector space spanned by the differential operators in Table 3 (including I) forms a Lie algebra isomorohic to the algebra* $\mathfrak{G}$ *of Theorem 7.1.*

Thus we have obtained an alternative description of the symmetry group of Q given by (7.30), this one in terms of a Lie algebra.

We have so far discussed the symmetry under the assumption that the group G is known, and we now show that certain properties of the automorphisms of $\mathfrak{S}$ completely determine this symmetry. In a sense all of the automorphisms of $\mathfrak{S}$ come from G via the mapping T.

The space $\mathfrak{S}$ has been topologised in such a way that it becomes homeomorphic to $\mathscr{S}_c$ under the mapping T. To any smooth curve $\{v_s\}$ in $\mathfrak{S}$ with $v_0 = v$ there exists a smooth curve $\{\zeta_s\}$ in $\mathscr{S}_c$ with $\zeta_0 = \zeta$ such that

$$T\zeta_s = v_s,$$

and if the infinitesimal transformation α on $\mathscr{S}_c$ is given by

$$\alpha\zeta = \dfrac{d}{ds}\zeta_s\bigg|_{s=0},$$

the corresponding infinitesimal transformation L on $\mathfrak{S}$ is related to α by

$$L = T \circ \alpha, \quad \text{where} \quad Lv = \dfrac{d}{ds}v_s\bigg|_{s=0}. \tag{7.34}$$

This expression is nothing but a generalisation of (7.33).

We now take a space **A** of infinitesimal transformations α on $\mathscr{S}_c$ satisfying the following conditions:

i. For all $\alpha \in \mathbf{A}$, $L = T \circ \alpha$ is expressed in the form (7.28) with the properties (i) and (ii) stated there.
ii. The set $\mathbf{A}$ includes the generator f of the Fourier-Mehler transform.
iii. The vector space $\mathbf{A}$ equipped with the commutator product $[\ ,\ ]$ forms a Lie algebra.

The second requirement should seem reasonable in view of our emphasis on the Fourier transform, and our problem now is to determine a maximal Lie algebra $\mathbf{A}$ satisfying these three conditions.

Let L be any differential operator satisfying (i) above. Then for $\zeta \in \mathscr{S}_c$ we have

$$Lv(x, t; \zeta) = \int_{-\infty}^{\infty} \zeta(u) \left[X(x, t)\frac{\partial}{\partial x} + \frac{1}{2} T(x, t)\frac{\partial^2}{\partial x^2} + U(x, t) \right] g(t, x - u)\, du$$

$$= \int_{-\infty}^{\infty} g(t, x - u) \left[-X(x, t)\frac{d}{du} + \frac{1}{2} T(x, t)\frac{d^2}{du^2} + U(x, t) \right] \zeta(u)\, du$$

$$\to \left[-X(x, 0)\frac{d}{dx} + \frac{1}{2} T(x, 0)\frac{d}{dx^2} + U(x, 0) \right] \zeta(x), \qquad t \downarrow 0.$$

These calculations explicitly describe the mapping inverse to T, and assert that, by $\alpha = T^{-1}L$ and condition (i), every $\alpha \in \mathbf{A}$ must be of the same form as the last expression above.

Changing the notation we can express α in the form

$$\alpha = A(u)\frac{d^2}{du^2} + B(u)\frac{d}{du} + C(u), \tag{7.35}$$

$A(u)$, $B(u)$ and $C(u)$ analytic, and thus $\mathbf{A}$ is the set of all such α.

Taking conditions (ii) and (iii) into account now, we see that the commutator of f and any α given by (7.35) is of the form

$$[\alpha, f] = A'(u)\frac{d^3}{du^3} + \frac{1}{2}(A''(u) + 2B'(u))\frac{d^2}{du^2}$$

$$+ \frac{1}{2}(4uA(u) + B''(u) + 2C'(u))\frac{d}{du}$$

$$+ \frac{1}{2}(2A(u) + 2uB(u) + C''(u)).$$

By (iii) the result should still be of the form (7.35), and so we must have $A'(u) \equiv 0$, i.e.

$$A(u) = a \qquad \text{(constant)}.$$

Thus the coefficient of d^2/du^2 in every member of $\mathbf{A}$ has to be constant, and so, returning to the computation above, $A''(u) + 2B'(u) = 2B'(u)$ is a constant, implying that

$$B(u) = bu + c \qquad (b,\ c \text{ constants}).$$

7.5 The Symmetry Group of the Heat Equation

Comparing this with (7.35) once more, we see that $4uA(u) + B''(u) + 2C'(u) = 4au + C'(u)$ has to be a linear function of u, i.e.

$$C(u) = du^2 + eu + k, \quad (d, e, k \text{ constants}).$$

Now it is known that the collection **A** of all α of the form (7.35), with coefficients A, B and C being polynomials of degree 0, 1 and 2 respectively, forms a Lie algebra, and it can be shown that this algebra coincides with the Lie algebra $\mathfrak{A}$ described earlier in the last section.

Thus we have proved the following

Theorem 7.2. *The Lie algebra formed by the differential operators $L = T \circ \alpha$, $\alpha \in \mathbf{A}$, where $\mathbf{A}$ satisfies conditions* (i), (ii) *and* (iii), *coincides with the Lie algebra $\mathfrak{A}$ of Proposition 7.3.*

It can thus be said that the above discussion determines the local Lie group in terms of the algebra **A**, the mapping T and the exponential map. Equivalently, we have obtained the symmetry group of the heat equation using as alternative method based upon a probabilistic idea!

This method of obtaining the symmetry group can easily be extended to the multidimensional case. Let the n-dimensional heat equation be

$$\frac{\partial}{\partial t} v(x, t) = \frac{1}{2} \Delta_n v(x, t), \quad \Delta_n = \sum_{j=1}^{n} \frac{\partial^2}{\partial x_j^2}, \qquad (7.36)$$

where $t > 0$ and $x = (x_1, \ldots, x_n) \in \mathbf{R}^n$. As in the case when $n = 1$ the class of initial functions is taken to be the Schwartz space, this time $\mathscr{S}_c^n$, the space of rapidly decreasing $\mathscr{C}^\infty$-functions on $\mathbf{R}^n$. The Green function of (7.36) is again a Gauss kernel

$$g(t; x) = (2\pi t)^{-n/2} \exp\left(-\frac{\|x\|^2}{2t}\right), \quad t > 0, \quad x \in \mathbf{R}^n,$$

where $\|x\|^2 = \sum_{j=1}^{n} x_j^2$, and we define the mapping T as above where (7.32) was used, namely

$$T: \zeta \to v = v(x, t; \zeta) = \int_{\mathbf{R}^n} \zeta(u) g(t; x - u) \, du, \quad \zeta \in \mathscr{S}_c^n. \qquad (7.37)$$

The function $v(x, t; \zeta)$ is obviously a solution to (7.36) with initial function ζ, so that T is an injection of $\mathscr{S}_c^n$ into its solution space. As before we set

$$\mathfrak{S}^n = \{T\zeta : \zeta \in \mathscr{S}_c^n\}$$

and topologise $\mathfrak{S}^n$ in such a way as to make T a homeomorphism. With the help of the structure of $\mathfrak{S}^n$ we can now obtain the symmetry group of the differential operator

$$Q_n = \frac{\partial}{\partial t} - \frac{1}{2}\Delta_n. \qquad (7.38)$$

The proof is similar to the case $n = 1$ already discussed, so that we shall only mention the points of difference. Having obtained a space $\mathbf{A}_n$ of differential operators α on $\mathscr{S}_c^n$ from the tangent vectors of curves in $\mathfrak{S}^n$ as before, we change the conditions (i), and (ii) on $\mathbf{A}$ to (i_n) and (ii_n) respectively on $\mathbf{A}_n$, with (iii) being left unchanged, where

i_n. For all $\alpha \in \mathbf{A}_n$ every $L = T \circ \alpha$ is expressible in the form

$$\sum_{j=1}^{n} X_j(x, t) \frac{\partial}{\partial x_j} + T(x, t) \frac{\partial}{\partial t} + U(x, t), \qquad (7.39)$$

where

a. $X_j(x, t)$, $1 \leq j \leq n$, $T(x, t)$ and $U(x, t)$ are all analytic in x, t and
b. $[L, Q_n] = R(x, t)Q_n$ for a function $R(x, t)$ which is analytic in x, t.

ii_n. $\mathbf{A}_n$ contains the differential operator

$$f_n = -\frac{1}{2} \sum_{j=1}^{n} \left(\frac{\partial^2}{\partial u_j^2} - u_j^2 + 1 \right), \qquad \text{(a generalisation of } f\text{).} \qquad (7.40)$$

Under the conditions (i_n), (ii_n) and (iii) every $\alpha \in \mathbf{A}_n$ can be written as

$$\alpha = a_0 + \sum_k a_k u_k + \sum_k b_k \frac{\partial}{\partial u_k} + \sum_k c_k u_k^2 + \sum_k d_k \frac{\partial^2}{\partial u_k^2} + \sum_{k,j} e_{k,j} u_k \frac{\partial}{\partial u_j}. \qquad (7.41)$$

Naturally some restrictions must be placed upon the coefficients $a_0, a_k, \ldots, e_{k,j}$ and it is convenient to state these in terms of the Fourier transform. Denoting by $\hat{\zeta}$ the Fourier transform of ζ we have

$$\widehat{\alpha\zeta}(\lambda) = \Big| a_0 - i \sum_k a_k \frac{\partial}{\partial \lambda_k} - i \sum_k b_k \lambda_k - \sum_k c_k \frac{\partial^2}{\partial \lambda_k^2} \\ - \sum_k d_k \lambda_k^2 - \sum_{k,j} e_{k,j} \frac{\partial}{\partial \lambda_k} \lambda_j \Big| \hat{\zeta}(\lambda), \qquad (7.42)$$

$\lambda = (\lambda_1, \ldots, \lambda_n)$. The function $v(x, t; \zeta)$ can be rewritten as

$$v(x, t; \zeta) = (2\pi)^{-n} \int_{\mathbf{R}^n} \hat{\zeta}(\lambda) \exp\left(-\frac{1}{2} t \sum_k \lambda_k^2 - i \sum_k x_k \lambda_k \right) d\lambda.$$

Therefore we have

$$v(x, t, \alpha\zeta) = (2\pi)^{-n} \int_{\mathbf{R}^n} \widehat{\alpha\zeta}(\lambda) \exp\left(-\frac{1}{2} t \sum_k \lambda_k^2 - i \sum_k x_k \lambda_k \right) d\lambda$$

$$= \Big[a_0 - i \sum_k a_k \left(ix_k + it \frac{\partial}{\partial x_k} \right) - i \sum_k b_k \left(i \frac{\partial}{\partial x_k} \right)$$

$$- \sum_k c_k \left(-t - x_k^2 - 2tx_k \frac{\partial}{\partial x_k} \right) + \sum_{k,j} e_{k,j} x_k \frac{\partial}{\partial x_j} \Big] v(x, t; \zeta)$$

$$+ (2\pi)^{-n} \int_{\mathbf{R}^n} \left(-t^2 \sum_k c_k \lambda_k^2 - \sum_k d_k \lambda_k^2 - t \sum_{k,j} e_{k,j} \lambda_k \lambda_j \right) \hat{\zeta}(\lambda)$$

$$\times \exp\left(-\frac{1}{2} t \sum_k \lambda_k^2 - i \sum_k x_k \lambda_k \right) d\lambda.$$

A necessary and sufficient condition for the last expression to be of the form $Lv(x, t; \zeta)$ and to satisfy (i_n) is that

$$c_k = c, \quad d_k = d, \quad e_{k,k} = e, \quad k = 1, \ldots, n;$$
$$e_{k,j} = -e_{j,k}, \quad j \neq k, \quad j, k = 1, \ldots, n. \tag{7.43}$$

The other constants a_0, a_k, b_k, $1 \leq k \leq n$, may, of course, be chosen arbitrarily. The above computation easily shows that the requirement (7.43) is necessary in order that (i_n) holds, and a direct computation easily proves that the collection $\mathbf{A}_n$ of all α of the form (7.41), subject to (7.43), satisfies the conditions (i_n), (ii_n) and (iii). Thus we have proved

Theorem 7.3. *The collection $\mathbf{A}_n$ of all differential operators of the form (7.41) which satisfy the restrictions (7.43) forms a Lie algebra isomorphic to the Lie algebra $\mathfrak{A}_n$ associated with the symmetry group of the operator Q_n given by (7.38). The isomorphism is described by the mapping T introduced in (7.37):*

$$T: \alpha \to L = T \circ \alpha \in \mathfrak{A}_n, \quad \alpha \in \mathbf{A}_n.$$

Corollary. *The symmetry group of Q_n is $(\tfrac{1}{2}n(n+3) + 4)$-dimensional.*

PROOF. The dimension in question is equal to that of $\mathfrak{A}_n$ and so of $\mathbf{A}_n$. It must therefore be the sum of $2n + 1$, the number of arbitrary constants $(a_0, a_k, b_k, 1 \leq k \leq m)$, and $3 + \tfrac{1}{2}n(n-1)$, the number of constants $(c_k, d_k, e_{k,j}, 1 \leq j, k \leq n)$ subject to (7.43), appearing in the expression (7.41) for $\alpha \in \mathbf{A}_n$. But this is just

$$2n + 1 + 3 + \frac{1}{2}n(n-1) = \frac{1}{2}n(n+3) + 4. \qquad \square$$

Remark. For a reason similar to that mentioned just after Proposition 7.3 of the previous section, it suffices to discuss only the real form $\mathfrak{A}_n$ of the Lie algebra.

EXAMPLE. In the special case $n = 2$ we will indicate the actual forms of $\mathfrak{A}_2$ and $\mathbf{A}_2$ as well as describing the mapping T between them. A base for $\mathbf{A}_2$ is:

$$I, \quad \frac{\partial}{\partial u_1}, \quad \frac{\partial}{\partial u_2}, \quad u_1, \quad u_2,$$
$$\frac{\partial^2}{\partial u_1^2} + \frac{\partial^2}{\partial u_2^2}, \quad u_2 \frac{\partial}{\partial u_1} - u_1 \frac{\partial}{\partial u_2}, \tag{7.44}$$
$$u_1 \frac{\partial}{\partial u_1} + u_2 \frac{\partial}{\partial u_2}, \quad u_1^2 + u_2^2.$$

Each of these is transformed by the mapping T into

$$I, \quad \frac{\partial}{\partial x_1}, \quad \frac{\partial}{\partial x_2}, \quad t\frac{\partial}{\partial x_1} + x_1, \quad t\frac{\partial}{\partial x_2} + x_2,$$

$$2\frac{\partial}{\partial t}, \quad x_2\frac{\partial}{\partial x_1} - x_1\frac{\partial}{\partial x_2}, \tag{7.45}$$

$$2t\frac{\partial}{\partial t} + x_1\frac{\partial}{\partial x_1} + x_2\frac{\partial}{\partial x_2}, \quad 2t^2\frac{\partial}{\partial t} + 2t\left(x_1\frac{\partial}{\partial x_1} + x_2\frac{\partial}{\partial x_2}\right) + (x_1^2 + x_2^2 + 2t),$$

in order, and the collection of images forms a base for $\mathfrak{A}_2$. Needless to say, the dimension of $\mathfrak{A}_2$ is $9(=\frac{1}{2}2(2+3)+4)$.

Let us note the commutation relations between the differential operators in (7.45) and $Q_2 = (\partial/\partial t) - \frac{1}{2}([\partial^2/\partial x_1^2] + [\partial^2/\partial x_2^2])$. Each of the seven operators on the first two lines of (7.45) commutes with Q_2, whilst

$$\left[2t\frac{\partial}{\partial t} + x_1\frac{\partial}{\partial x_1} + x_2\frac{\partial}{\partial x_2}, Q_2\right] = -2Q_2,$$

and

$$\left[2t^2\frac{\partial}{\partial t} + 2t\left(x_1\frac{\partial}{\partial x_1} + x_2\frac{\partial}{\partial x_2}\right) + (x_1^2 + x_2^2 + 2t), Q_2\right] = -4tQ_2.$$

Remark. We note the presence in this example of a generator of the rotation in the plane:

$$x_2\frac{\partial}{\partial x_1} - x_1\frac{\partial}{\partial x_2}.$$

This is as one would expect, and is in contrast with the one-dimensional case.

7.6 Applications to the Schrödinger Equation

It is known from the theory of wave mechanics that the wave functions which describe the states of a mechanical system are solutions of the Schrödinger equation

$$i\hbar\frac{\partial \psi}{\partial t} = H\psi. \tag{7.46}$$

The operator H in this equation is the Hamiltonian operator, obtained from the Hamiltonian function $H(p, q)$ of classical mechanics by replacing the variables q, p with the operators $q \cdot$ (multiplication by the variable q), $(\hbar/i)(\partial/\partial q)$ respectively. When p and q are vectors these are understood to be componentwise operations.

7.6 Applications to the Schrödinger Equation

The collection of all solutions of equation (7.46) forms a vector space, the space of all wave functions, and we will investigate its structure. It is not possible to deal with such structure in full generality and so we only consider the symmetry group in a few cases where H takes a simple form.

In what follows we set the Planck constant $\hbar$ equal to 1 as is usual, and also put $m = 1$.

(i) The Free Particle

To be consistent with our previous notation the space variable will be denoted by x, u, etc. When there are n free particles the wave function $\psi(x, t)$, $x \in \mathbf{R}^n$, $t > 0$, is a solution of the Schrödinger equation

$$\frac{1}{i}\frac{\partial}{\partial t}\psi(x, t) = \frac{1}{2}\Delta_n \psi(x, t), \qquad \Delta_n = \sum_1^n \frac{\partial^2}{\partial x_j^2}. \qquad (7.47)$$

For the moment we take initial states from the Schwartz space $\mathscr{S}_c^n$:

$$\psi(x, 0) \in \mathscr{S}_c^n, \qquad (7.48)$$

This is convenient as the class $\mathscr{S}_c^n$ consists of complex-valued functions.

The Green function K of equation (7.47) is

$$K(t, x - u) = (2\pi i t)^{-n/2} \exp\left(-\frac{\|x - u\|^2}{2it}\right) \qquad (7.49)$$

and is of the same form as the Gauss kernel, the Green function of the heat equation, if we replace it in K by t. Thus the symmetry can be formally obtained by adopting arguments similar to those in the last section. However it should be noted that the absolute value of $K(t, x - u)$ is $(2\pi t)^{-n/2}$, and so K is not integrable in u. Nonetheless, if $\zeta \in \mathscr{S}_c^n$, then the integral

$$\psi(x, t; \zeta) = (2\pi i t)^{-n/2} \int_{\mathbf{R}^n} \zeta(u)\exp\left(-\frac{\|x - u\|^2}{2it}\right) du \qquad (7.50)$$

certainly exists, and the partial derivatives of ψ in t and x also exist, as those differential operators interchange with the integral sign without any difficulty.

With this remark we now topologise the space

$$\mathfrak{S}^n = \{\psi(x, t; \zeta): \zeta \in \mathscr{S}_c^n\}$$

of wave functions in such a way that $\mathfrak{S}^n$ is homeomorphic to $\mathscr{S}_c^n$, and then prove that the Lie algebra of the automorphism group of $\mathfrak{S}^n$ is isomorphic to that of the heat equation, whence it is $(\frac{1}{2}n(n + 3) + 4)$-dimensional.

For concreteness and simplicity we will discuss explicit formulae for the case of one-dimensional wave functions $\psi(x, t)$, $x \in \mathbf{R}$, $t \geq 0$. As before let T be the mapping

$$T: \zeta \to \psi(x, t; \zeta) \in \mathfrak{S}.$$

Then (cf. §7.4) a one-parameter subgroup $\{g_t\}$ of the Lie group G defines an infinitesimal generator α, which in turn defines an infinitesimal generator L on $\mathfrak{S}$ using T:

$$\psi(x, t; \alpha\zeta) = L\psi(x, t; \zeta).$$

This correspondence is illustrated by the following table, similar to Table 3 above.

Table 4

$$I \to I$$

$$s \to L_s = -\frac{\partial}{\partial x}$$

$$\pi \to L_x = it\frac{\partial}{\partial x} + x^\dagger$$

$$\tau \to L_\tau = 2t\frac{\partial}{\partial t} + x\frac{\partial}{\partial x} + \frac{1}{2}$$

$$f \to L_f = i(t^2 + 1)\frac{\partial}{\partial t} + itx\frac{\partial}{\partial x} + \frac{1}{2}(x^2 + it - 1)$$

$$\sigma \to L_\sigma = i(t^2 - 1)\frac{\partial}{\partial t} + itx\frac{\partial}{\partial x} + \frac{1}{2}(x^2 + it)$$

† This operator is connected with the Galilean transformation.

It can be seen from this table that each member is obtained from the corresponding member of Table 3 by replacing t with it. This is of course consistent with the relation between (7.47) and (7.36). Similarly, speaking formally, if we replace $\partial/\partial t$ with $\frac{1}{2}i\,\partial^2/\partial x^2$, then the differential operators on the right side of Table 4 approach the corresponding ones on the left side as $t \downarrow 0$.

There are many other similarities with the heat equation, but there is also one difference which deserves mention. This is the fact that we are required by the form of the operators L in the table to form a complex Lie algebra. On the complex vector space spanned by these operators we introduce a commutator product as follows:

$$[A + iB, A' + iB'] = [A, A'] - [B, B'] + i([A, B'] + [B, A']),$$

where A, A', B, B' are elements of the obvious Lie algebra over the reals (cf. $\partial/\partial x, t\,\partial/\partial x$). With this product the vector space $\mathfrak{L}$ spanned by the operators L of Table 4 forms a Lie algebra and these operators (together with I) constitute a base for the algebra. Their commutation relations are exactly the same as the basic commutation relations, that is $\mathfrak{L} \cong \mathfrak{A}$.

Remark. The last statement implies that L has real form, see for example S. Helgason (1962) Chapter 3. However we still keep the above notation for the generators L as it draws attention to their significance.

7.6 Applications to the Schrödinger Equation

We have now clarified the structure of the Lie algebra $\mathfrak{L}$, and so the symmetry group of $(1/i)(\partial/\partial t) - \frac{1}{2}(\partial^2/\partial x^2)$, which describes the structure of the space $\mathfrak{S}$, is generated by the one-parameter groups obtained by applying the exponential map to L_s, L_τ, iI, iL_π, iL_f and iL_σ.

(ii) Particles in a Constant External Field

We turn now to a simple generalisation of (i) in which a single particle is influenced by a constant external field F. The one-dimensional Schrödinger equation in this case is

$$\frac{1}{i}\frac{\partial}{\partial t}\psi = \frac{1}{2}\frac{\partial^2}{\partial x^2}\psi + Fx\psi, \qquad (7.51)$$

and the Green function K is given by

$$K(t, x, u) = (2\pi it)^{-1/2} \exp\left[\frac{1}{2t}i(x-u)^2 + \frac{1}{2}iFt(x+\mu) - \frac{iF^2 t^3}{24}\right]. \qquad (7.52)$$

We can proceed as in §7.5 in this case too, forming the space $\mathfrak{S}$ of wave functions of the form

$$\psi(x, t; \zeta) = \int_{-\infty}^{\infty} \zeta(u) K(t, x, u)\, du, \qquad \zeta \in \mathscr{S}_c.$$

Then we can compute the infinitesimal transformations such as I, L_s and L_π on $\mathfrak{S}$ by applying elements of the Lie algebra $\mathfrak{A}$ to elements $\zeta \in \mathscr{S}_c$. For example we can obtain

$$s \to L_s = -\frac{\partial}{\partial x} + \frac{1}{2}iFt; \qquad \pi \to L_\pi = it\frac{\partial}{\partial x} + x + \frac{1}{2}Ft^2.$$

The other correspondences can be obtained in the same manner, but we omit them to avoid unnecessary repetition.

(iii) The Harmonic Oscillator

This is a simple but important example and we will obtain the symmetry explicitly. In classical mechanics the one-dimensional harmonic oscillator with frequency ν is described by the equation

$$\ddot{q}(t) = -\omega^2 q(t), \qquad \omega = 2\pi\nu,$$

where the equilibrium point is taken to be zero. Setting $m\dot{q}(t) = p(t)$ we see that the Hamiltonian function is

$$H(p, q) = \frac{1}{2}mp(t)^2 + \frac{1}{2}m\omega^2 q(t)^2.$$

Returning to our method we set $\hbar = m = 1$ as before and, after replacing q by x, the wave function $\psi(x, t)$, $x \in \mathbf{R}$, $t > 0$, is a solution of the Schrödinger equation

$$\frac{1}{i}\frac{\partial}{\partial t}\psi = \frac{1}{2}\frac{\partial^2}{\partial x^2}\psi - \frac{1}{2}\omega^2 x^2 \psi. \tag{7.53}$$

The Green function for this case is also known to be of the form

$$K(t, x, u) = \left(\frac{\omega}{2\pi i \sin \omega t}\right)^{1/2} \exp\left[\frac{i\omega(x^2 + u^2)}{2 \tan \omega t} - \frac{i\omega x u}{\sin \omega t}\right]. \tag{7.54}$$

Computing the operators L as we did in (i) gives the following:

Table 5

T
$I \to I$
$s \to L_s = -\cos \omega t \dfrac{\partial}{\partial x} - i\omega x \sin \omega t$
$\pi \to L_\pi = i\dfrac{\sin \omega t}{\omega}\dfrac{\partial}{\partial x} + x \cos \omega t$
$\tau \to L_\tau = x \cos 2\omega t \dfrac{\partial}{\partial x} + \dfrac{\sin 2\omega t}{\omega}\dfrac{\partial}{\partial t} + i\omega x^2 \sin 2\omega t + \cos^2 \omega t - \dfrac{1}{2}$
$f \to L_f = \left[\dfrac{ix \sin 2\omega t}{\omega}\dfrac{\partial}{\partial x} + \dfrac{2i \sin^2 \omega t}{\omega^2}\dfrac{\partial}{\partial t} + x^2 \cos 2\omega t + \dfrac{i}{2\omega}\sin 2\omega t\right]$ $\times \dfrac{(1-\omega^2)}{2} + i\dfrac{\partial}{\partial t} - \dfrac{1}{2}$
$\sigma \to L_\sigma = \left[\dfrac{ix \sin 2\omega t}{\omega}\dfrac{\partial}{\partial x} + \dfrac{2i \sin^2 \omega t}{\omega^2}\dfrac{\partial}{\partial t} + x^2 \cos 2\omega t + \dfrac{i}{2\omega}\sin 2\omega t\right]$ $\times \dfrac{(1+\omega^2)}{2} - i\dfrac{\partial}{\partial t}$

Remark. Notice that L_f has a simple expression. It is also of interest to point out the intimate connections between such disparate concepts as the harmonic oscillator, the Fourier-Mehler transform, the system ξ_n, $n \geq 0$, of functions defined by (7.6), the Schwartz space $\mathscr{S}_c$, and so on. However we omit further discussion of this point.

Let us recall the approach of W. Miller Jr. discussed at the beginning of §7.5. Setting

$$Q = i\frac{\partial}{\partial t} + \frac{1}{2}\frac{\partial^2}{\partial x^2} - \frac{1}{2}\omega^2 x^2,$$

7.6 Applications to the Schrödinger Equation

we consider the commutation relations between the operators of Table 5. They are found to be

$$[Q, L_s] = 0, \quad [Q, L_\pi] = 0,$$

$$[Q, L_\tau] = (2\cos 2\omega t)Q, \quad [Q, L_f] = \frac{i(1-\omega^2)\sin 2\omega t}{\omega} Q,$$

$$[Q, L_\sigma] = \frac{i(1+\omega^2)\sin 2\omega t}{\omega} Q,$$

and show that each such L certainly satisfies the relation (7.29).

The method used to obtain the symmetry groups for the examples (i), (ii) and (iii) discussed so far has been exactly the same as that used in the case of the heat equation. In particular the Lie algebra L is always isomorphic to $\mathfrak{A}$, and we therefore have mutually locally isomorphic Lie groups. Thus we can prove

Theorem 7.4. *If the Green function K of the Schrödinger equation can be expressed in the form*

$$K(t, x, u) = f(t)\exp[ig(t, x, u)] \tag{7.55}$$

with $g(t, x, u)$ a polynomial of degree 2 in x, u involving the term xu, then the symmetry group is a six-dimensional Lie group locally isomorphic to the symmetry group of a free particle.

This result can easily be generalised to include higher dimensional cases.

8 Causal Calculus in Terms of Brownian Motion

In this chapter a new viewpoint of the analysis of non-linear functionals is presented, one which leads naturally to a generalisation of these functionals. Our approach is along the lines of the so-called *causal analysis*, where the passage of time is taken into account in the analysis of functionals of Brownian motion, and we begin the discussion by setting out our general point of view.

Functionals of Brownian motion, which we call simply *Brownian functionals*, can be expressed in terms of white noise by viewing the latter as the time derivative of Brownian motion. Let us start with the probability measure μ of white noise on the space $\mathscr{S}^*$ of generalised functions (tempered distributions). It is uniquely determined by the characteristic functional

$$C(\xi) = \exp[-\tfrac{1}{2}\|\xi\|^2]. \qquad \xi \in \mathscr{S}, \tag{8.1}$$

in such a way that

$$C(\xi) = \int_{\mathscr{S}^*} \exp[i\langle x, \xi \rangle]\, d\mu(x). \tag{8.2}$$

With this measure μ each $x \in \mathscr{S}^*$ may be viewed as a sample function of white noise $\dot{B}(t) = dB(t)/dt$[1], and the Hilbert space $(L^2) = L^2(\mathscr{S}^*, \mu)$ is therefore the collection of all complex-valued Brownian functionals with finite variance.

Our main interest is in discussing differential and integral calculus on the space (L^2), and our approach involves the following three steps. (i) The first

---

[1] Since the topics in this chapter were inspired by Mechanics and Engineering, we prefer $\dot{B}(t)$ rather than the earlier notation $B'(t)$ to express the time derivative of a Brownian motion $B(t)$.

thing which should be done is to use standard tools from analysis to help visualize members of (L^2). For this purpose the transformation $\mathscr{T}$ given by

$$(\mathscr{T}\varphi)(\xi) = \int \exp[i\langle x, \xi\rangle]\varphi(x)\,d\mu(x), \qquad \varphi \in (L^2), \tag{8.3}$$

again (see §4.3) plays a basic role. As we saw in §4.3, the collection $\mathbf{F} = \{\mathscr{T}\varphi : \varphi \in (L^2)\}$ forms a reproducing kernel Hilbert space such that $\mathbf{F}$ is isomorphic to (L^2) under $\mathscr{T}$. The Fock space expression for $\mathbf{F}$ follows, in a manner different from that of §4.2, and we then get the integral representation of multiple Wiener integrals. (ii) The next step is new to this chapter. We introduce a suitable coordinate system in the space $\mathscr{S}^*$ equipped with the measure μ. This allows us to carry out the causal analysis, that is, the analysis in which the passage of time is taken into account. (iii) With this system we shall be able to introduce a certain class of *generalised* Brownian functionals. This generalisation uses the integral representation of (L^2)-functionals, the idea coming from P. Lévy's approach to functional analysis (P. Lévy, 1951).

We pause now to add a few words on the motivation of the work which follows. Inspiration has come from several actual problems arising in quantum mechanics, in particular the theory of the Feynman path integral in field theory, and also from problems in the areas of stochastic control theory and the stochastic evolution equations from population biology. These problems require a non-linear causal analysis of noise or fluctuations, a mathematical expression of which is to be the white noise $\{\dot{B}(t)\}$.

8.1 Summary of Known Results

We begin with a quick review of known results concerning Brownian functionals and their integral representations. For details we refer to Chapter 4.

Let $(\mathscr{S}^*, \mu)$ be the measure space of *white noise*, where μ is given by (8.1) and (8.2), and set $(L^2) = L^2(\mathscr{S}^*, \mu)$. The functional $C(\xi - \eta)$, $(\xi, \eta) \in \mathscr{S} \times \mathscr{S}$, where $C(\xi)$ is given by (8.1), is positive definite, and so defines a reproducing kernel Hilbert space which is denoted by $\mathbf{F}$. Letting $\mathscr{T}$ be given by (8.3) we have an alternative method of obtaining an integral representation of (L^2)-functionals.

Proposition 8.1. *The reproducing kernel Hilbert space $\mathbf{F}$ with kernel $C(\cdot)$ is isomorphic under the transformation $\mathscr{T}$ to the Hilbert space (L^2).*

Now consider the Taylor expansion of the kernel $C(\cdot)$, (ξ, η) denoting the inner product in $L^2(\mathbf{R})$,

$$C(\xi - \eta) = \sum_{n=0}^{\infty} C(\xi) \frac{(\xi, \eta)^n}{n!} C(\eta). \tag{8.4}$$

If we set

$$C_n(\xi, \eta) = C(\xi) \frac{(\xi, \eta)^n}{n!} C(\eta), \tag{8.5}$$

then C_n is positive definite, and so defines a reproducing kernel Hilbert space which we call $\mathbf{F}_n$. The space $\mathbf{F}_n$ turns out to be a subspace of $\mathbf{F}$, and we can in fact prove that

$$(C_n(\cdot, \eta_1), C_m(\cdot, \eta_2))_\mathbf{F} = 0 \quad \text{for} \quad \eta_1, \eta_2 \in \mathscr{S}, m \neq n,$$

showing that the subspaces $\{\mathbf{F}_n : n \geq 0\}$ are mutually orthogonal. A direct sum decomposition is thus obtained:

$$\mathbf{F} = \sum_{n=0}^{\infty} \oplus \mathbf{F}_n \quad \text{(Fock space)}. \tag{8.6}$$

Now let us set

$$\mathscr{H}_n = \mathscr{T}^{-1} \mathbf{F}_n, \quad n \geq 0.$$

The space $\mathscr{H}_n$ is nothing but the *multiple Wiener integral* of degree n introduced in §4.2, and corresponding to the decomposition (8.6) of $\mathbf{F}$ is one of (L^2):

$$(L^2) = \sum_{n=0}^{\infty} \oplus \mathscr{H}_n. \tag{8.7}$$

Proposition 8.2. i. *For $\varphi(x) \in \mathscr{H}_n$, we have*

$$(\mathscr{T}\varphi)(\xi) = i^n C(\xi) \int \cdots \int_{\mathbf{R}^n} F(u_1, \ldots, u_n) \xi(u_1) \cdots \xi(u_n) \, du_1 \cdots du_n, \tag{8.8}$$

where $F \in \hat{L}^2(\mathbf{R}^n)$, the class of symmetric $L^2(\mathbf{R}^n)$-functions, and the map

$$\varphi \to F \in \hat{L}^2(\mathbf{R}^n), \quad \varphi \in \mathscr{H}_n,$$

is one-to-one.

ii. *Under the relationship established in* (i) *we have*

$$\|\varphi\|_{(L^2)} = \sqrt{n!} \, \|F\|_{L^2(\mathbf{R}^n)}. \tag{8.9}$$

Definition 8.1. The expression (8.8) for φ is called the *integral representation* of $\varphi \in \mathscr{H}_n$, and F is called the *kernel* of the representation.

For a general $\varphi \in (L^2)$, we use the expansion

$$\varphi = \sum_{n=0}^{\infty} \varphi_n, \quad \varphi_n \in \mathscr{H}_n,$$

and have a series of integral representations and of kernels $\{F_n\}$.

We recall from §4.6 (i) that special interest is attached to the case in which $\varphi \in \mathcal{H}_2$. Associated with each such φ is a symmetric function $F(u, v) \in \hat{L}^2(\mathbf{R}^2)$, and if, in addition, we assume that φ is real-valued, then F determines a self-adjoint Hilbert-Schmidt operator acting on $L^2(\mathbf{R})$. We can therefore appeal to the theory of Hilbert-Schmidt operators to get the eigenfunction expansion of F

$$F(u, v) = \sum_{n=1}^{\infty} \lambda_n^{-1} \eta_n(u) \eta_n(v),$$

where $\{\lambda_n, \eta_n : n \geq 1\}$ is the system of eigenvalues and eigenvectors, respectively. This gives an expansion of φ as the sum of a sequence of independent random variables:

$$\varphi(x) = \sum_{n=1}^{\infty} \lambda_n^{-1}(\langle x, \eta_n \rangle^2 - 1).$$

Once this expansion is obtained, the probability distribution of φ is easily obtained by computing its characteristic function or its cumulants; it then leads us to generalized quadratic functionals [cf. (8.16) below].

8.2 Coordinate Systems in $(\mathcal{S}^*, \mu)$

Next we discuss coordinate systems in $\mathcal{S}^*$ and bases for the Hilbert space (L^2), but before coming to the main topics in this area, we recall some elementary concepts from classical analysis on $L^2([0, 1])$ [cf. P. Lévy (1951)].

Let $\{\xi_n\}$ be a complete orthonormal system in $L^2([0, 1])$ and define Ψ_n, Φ_n as follows:

$$\Psi_n(u, v) = \frac{1}{n} \sum_{m=1}^{n} \xi_m(u) \xi_m(v),$$

$$\Phi_n(u) = \frac{1}{n} \sum_{m=1}^{n} \xi_m(u)^2.$$

We see immediately that

$$\int_0^1 \int_0^1 \Psi_n(u, v)^2 \, du \, dv = \frac{1}{n} \to 0, \quad \text{as } n \to \infty, \tag{8.10}$$

i.e. Ψ_n converges to 0 in $L^2([0, 1] \times [0, 1])$. As for $\Phi_n(u)$, we have

$$\int_0^1 \Phi_n(u) \, du = 1, \quad n \geq 1,$$

but the convergence

$$\Phi_n(u) \to 1 \quad \text{in} \quad L^2([0, 1]) \tag{8.11}$$

is not always true. When (8.11) does hold, we say that the complete orthonormal system $\{\xi_n\}$ is *equally dense*.

EXAMPLE 1. The following are examples of equally dense systems.
i. $\{1, \sqrt{2}\sin 2k\pi t, \sqrt{2}\cos 2k\pi t: k \geq 1\}$.
ii. Walsh functions.

Remark. We could also consider the notion of an equally dense complete orthonormal system in $L^2(\mathbf{R})$. The analogue of (8.11) would be the property that Φ_n should, in an appropriate sense, behave like the density of a uniform probability measure on $\mathbf{R}$. An example of such a system is given by

$$\xi_n(u) = (2^n n! \sqrt{\pi})^{-1/2} H_n(u) \exp(-\tfrac{1}{2}u^2), \qquad n \geq 0, \tag{8.12}$$

where $H_n(u)$ is the Hermite polynomial of degree n.

We come now to a coordinate system for $\mathscr{S}^*$.

I. Let $\{\xi_n\}$ be a complete orthonormal system in $L^2(\mathbf{R})$. Each $x \in \mathscr{S}^*$ has a coordinate representation in the form

$$x \sim \{x_n : n \geq 1\}, \qquad x_n = \langle x, \xi_n \rangle.$$

Furthermore, if the measure μ is introduced onto $\mathscr{S}^*$, $\{\langle x, \xi_n \rangle\}$ forms an independent system of standard Gaussian random variables on the probability space $(\mathscr{S}^*, \mu)$. Hence we have (by the strong law of large numbers):

$$\frac{1}{N} \sum_{n=1}^{N} \langle x, \xi_n \rangle \langle y, \xi_n \rangle \to 0, \qquad \text{a.e. on } (\mathscr{S}^* \times \mathscr{S}^*, \mu \times \mu),$$

$$\frac{1}{N} \sum_{n=1}^{N} \langle x, \xi_n \rangle^2 \to 1, \qquad \text{a.e. on } (\mathscr{S}^*, \mu), \text{ as } N \to \infty.$$

These observations show that the complete orthonormal system $\{\langle x, \xi_n \rangle\}$ enjoys the property we have termed "equally dense" in the Hilbert space $\mathscr{H}_1$.

As before, we are naturally led to the complete orthonormal system in $\mathscr{H}_n$ defined by the Fourier-Hermite polynomials by

$$\prod_j (n_j! \, 2^{n_j})^{-1/2} H_{n_j}\!\left(\frac{\langle x, \xi_j \rangle}{\sqrt{2}}\right), \qquad \sum_j n_j = n. \tag{8.13}$$

Unfortunately this system is not suited to the *causal calculus*, because it does not permit a particularly easy understanding of the passage of time.

II. To fix our ideas, let us now take the unit interval $[0, 1]$, and let $\{\pi_n : n \geq 1\}$ be a sequence of partitions of $[0, 1]$ with the property that π_{n+1} is finer than π_n for each n, denoted $\pi_n < \pi_{n+1}$. If $\pi = \{\Delta_i\}$ with $\Delta_i = [t_i, t_{i+1}]$, then as in §2.2 (ii) we can define the quadratic variation of a function $f(t)$, $0 \leq t \leq 1$, over π:

$$\pi^2 f = \sum_{j=0}^{n-1} |f(t_{i+1}) - f(t_i)|^2.$$

8.2 Coordinate Systems in $(\mathscr{S}^*, \mu)$

Let $B(t, x)$ be a version of Brownian motion given by

$$B(t, x) = \langle x, \chi_{[0, t]} \rangle, \qquad 0 \le t \le 1.$$

Then we have

Lemma. *If* $\delta\pi_n = \max |t_{j+1} - t_j| \to 0$, *then*

$$\lim_{n \to \infty} \pi_n^2 B(\cdot, x) = 1, \qquad \text{a.e. } (\mu).$$

For a proof we refer to the Corollary to Theorem 2.3 given in §2.2. In particular if π_n is the uniform partition with $|\Delta_i| = 2^{-n}$, then

$$\sum_{k=1}^{2^n} (\Delta_k B)^2 = 2^{-n} \sum_{k=1}^{2^n} (2^{n/2} \Delta_k B)^2 \to 1 \qquad \text{a.e. } (\mu).$$

Speaking formally, this asserts that $\{(dt)^{-1/2} dB(t)\}$ is an equally dense complete orthonormal system in $\mathscr{H}_1$. More precisely, the projective system

$$\{|\Delta_k|^{-1/2} \Delta_k B : \Delta_k \in \pi_n\}, \qquad \pi_n \text{ uniform}, n = 1, 2, \ldots,$$

defines a complete orthonormal system which is equally dense.

We next proceed to a complete orthonormal system in $\mathscr{H}_2$. By using Fourier-Hermite polynomials, all members of the system are classified as follows:

$$(dt)^{-1/2} dB(t) \cdot (ds)^{-1/2} dB(s) \qquad (t \ne s) \qquad \text{class (1)},$$

$$2^{-1/2}[((dt)^{-1/2} dB(t))^2 - 1] \qquad \text{class (2)}.$$

If we use the following Hermite polynomials with parameter given by

$$H_n(x; \sigma^2) = \left[\frac{(-\sigma^2)^n}{n!}\right] \exp\left[\frac{x^2}{2\sigma^2}\right] \frac{d^n}{dx^n} \exp\left[\frac{-x^2}{2\sigma^2}\right], \qquad n \ge 0,$$

where $\sigma > 0$, then a complete orthonormal system is easily obtained for $\mathscr{H}_n$, $n \ge 0$, in the form

$$\prod_{j=1}^n \left\{ H_1\left(\dot{B}(t_j); \frac{1}{dt_j}\right) \sqrt{dt_j} \right\}, \qquad (t_j \text{ different}) \qquad \text{class (1)},$$

$$\prod_j \left\{ c_{n_j} H_{n_j}\left(\dot{B}(t_j); \frac{1}{dt_j}\right) (\sqrt{dt_j})^{n_j} \right\}, \qquad \sum_j n_j = n, \qquad \text{class (2)}, \tag{8.14}$$

where the t_j are different in the second expression, and $c_{n_j} = (n_j!)^{-1/2}$. These expressions are still purely formal, but we note that they are consistent with the partitions π_n via the addition formula for $H_n(x; \sigma^2)$:

$$\sum_{k=0}^n H_{n-k}(x; \sigma^2) H_k(y; \tau^2) = H_n(x + y, \sigma^2 + \tau^2);$$

see formula (A.36) of §A.5.

8.3 Generalised Brownian Functionals

We first consider an example which suggests to us the idea of defining a class of generalised Brownian functionals. If we set

$$B(t) = B(t, x) = \langle x, \chi_{[t \wedge 0, t \vee 0]} \rangle,$$

then $\{B(t)\}$ is a version of a Brownian motion. Bearing in mind the class (2) base of (8.14) (in the case $n = 2$), we are led to consider functionals in $\mathscr{H}_2$ of the form

$$\varphi_\pi = \sum_j a_j 2^{-1/2} \{(|\Delta_j|^{-1/2} \Delta_j B)^2 - 1\}, \quad \pi = \{\Delta_j\} \text{ a partition of } \mathbf{R}.$$

Its integral representation is

$$(\mathscr{T}\varphi_\pi)(\xi) = i^2 C(\xi) 2^{-1/2} \iint \sum_j |\Delta_j|^{-1} a_j \chi_{\Delta_j^2}(u, v) \xi(u) \xi(v) \, du \, dv.$$

If the partition $\pi = \{\Delta_j\}$ is increasingly refined, so that $\delta\pi \to 0$, and $\sum_j a_j \chi_{\Delta_j}(u)$ approaches a function f, then the integral representation approaches

$$i^2 C(\xi) 2^{-1/2} \int f(u) \xi(u)^2 \, du. \tag{8.15}$$

Such an expression can never belong to $\mathscr{F}_2$, but if it is written as

$$i^2 C(\xi) 2^{-1/2} \iint f(\tfrac{1}{2}(u + v)) \, \delta(u - v) \xi(u) \xi(v) \, du \, dv,$$

then we are led to think of a much wider class of functionals than $\mathscr{F}_2$. At the same time we consider a limit of the φ_π as $\delta\pi \to 0$, in some sense, but not in the (L^2)-sense. Formally speaking, this limit would be expressible in the form

$$\sqrt{2} \int f(u) H_2\left(\dot{B}(u); \frac{1}{du}\right) du, \tag{8.16}$$

and its integral representation would be given by (8.15).

We are now in a position to define generalised Brownian functionals. Let $H^m(\mathbf{R}^n)$ be the Sobolev space of order m on $\mathbf{R}^n$, and set $\hat{H}^m(\mathbf{R}^n) = H^m(\mathbf{R}^n) \cap \hat{L}^2(\mathbf{R}^n)$. Then define $\mathscr{F}_n^{(n)}$ to be

$$\left\{ U(\xi) = \int_{\mathbf{R}^n} \cdots \int F(u_1, \cdots, u_n) \xi(u_1) \cdots \xi(u_n) \, du_1 \cdots du_n : F \in \hat{H}^{(n+1)/2}(\mathbf{R}^n) \right\},$$

and

$$\mathbf{F}_n^{(n)} = \{i^n C(\xi) U(\xi) : U \in \mathscr{F}_n^{(n)}\},$$

and introduce the topology on $\mathscr{F}_n^{(n)}$ as a subset of $H^{(n+1)/2}(\mathbf{R}^n)$; i.e. if $U(\xi)$ is in $\mathscr{F}_n^{(n)}$ with kernel F, then the norm of the functional is defined to be $(n!)^{-1/2}$ times the $H^{(n+1)/2}(\mathbf{R}^n)$-norm of F [cf. Proposition 8.2, (ii)]. Set

$$\mathscr{H}_n^{(n)} = \mathscr{T}^{-1}(\mathbf{F}_n^{(n)}),$$

and topologise $\mathscr{H}_n^{(n)}$ in such a way that $\mathscr{T}$ becomes an isometry. Writing $\mathscr{H}_n^{(-n)}$ and $\mathscr{F}_n^{(-n)}$ for the dual space of $\mathscr{H}_n^{(n)}$ and $\mathscr{F}_n^{(n)}$, respectively, we have the following diagram:

$$\begin{array}{ccccc} \mathscr{F}_n^{(n)} & \hookrightarrow & \mathscr{F}_n & \hookrightarrow & \mathscr{F}_n^{(-n)} \\ \updownarrow & & \updownarrow & & \updownarrow \\ \mathscr{H}_n^{(n)} & \hookrightarrow & \mathscr{H}_n & \hookrightarrow & \mathscr{H}_n^{(-n)}, \end{array}$$

where $\hookrightarrow$ denotes a continuous injection, and the vertical double-headed arrows denote an isomorphism under $\mathscr{T}$ (with $\mathscr{T}$ naturally extended to $\mathscr{H}_n^{(-n)}$).

Definition 8.2. A member of $\mathscr{H}_n^{(-n)}$ is called a *generalised Brownian functional* of degree n.

8.4 Generalised Random Measures

It is easily seen that the following expression is a functional in $\mathscr{F}_n^{(-n)}$:

$$\int \cdots \int f(u_1, \ldots, u_k)\xi(u_1)^{n_1} \cdots \xi(u_k)^{n_k} du_1 \cdots du_k. \qquad \sum_j n_j = n. \tag{8.17}$$

We would expect that the relationship between (8.15) and (8.16) will extend to the case in which (8.15) is replaced by (8.17). Now start with an $\mathscr{H}_n$-functional

$$\varphi_\pi = \sum_{i_1, \ldots, i_k} a_{i_1, \ldots, i_k} \prod_j \left\{ n_j! |\Delta_{i_j}| H_{n_j}\left(|\Delta_{i_j}|^{-1} \Delta_{i_j} B; \frac{1}{|\Delta_{i_j}|}\right) \right\}.$$

where $i_1, \ldots, i_k$ are all distinct, and $\pi = \{\Delta_i\}$ is a finite partition of $\mathbf{R}$. If the function $f(u_1, \ldots, u_k)$ in (8.17) is approximated by

$$\sum_{i_1, \ldots, i_k} a_{i_1, \ldots, i_k} \prod_j \chi\Delta_{i_j}^{n_j}/|\Delta_{i_j}|^{n_j-1},$$

then it can be shown that the integral representation of φ_π approaches the functional (8.17) in the space $\mathscr{F}_n^{(-n)}$.

Bearing this last remark in mind, we introduce some further notation. Write

$$M_n(dt) = H_n\left(\dot{B}(t); \frac{1}{dt}\right) dt,$$

and also

$$\prod_{j=1}^k \cdot M_{n_j}(dt_j) \equiv M_{n_1}(dt_1) \cdots M_{n_j}(dt_k)$$

$$= \begin{cases} \prod_{j=1}^k M_{n_j}(dt_j), & \text{if } t_1, \ldots, t_k \text{ are all distinct}, \\ 0 & \text{otherwise}. \end{cases} \quad (8.18)$$

Of course a rigorous definition of the product $\prod \cdot$ should be given as in $\prod$, §8.2, using a sequence $\{\pi_n\}$ of partitions of **R** with $\delta\pi_n \to 0$. In this notation, the foregoing may be summarised as

Theorem 8.1. *The product* $\prod_{j=1}^k \cdot M_{n_j}(dt_j)$, $\sum_j n_j = n$, *is a random measure, and integrals with respect to it belong to* $\mathscr{H}_n^{(n)}$. *Further,*

$$\mathscr{T}\left(\prod_{j=1}^k \cdot M_{n_j}(dt_j)\right) = \prod_{j=1}^k (n_j!)^{-1} \delta_u^{n_j} du^{n_j}, \quad (8.19)$$

where $\delta_u^n du^n$ *is the measure given by*

$$\int_{\mathbf{R}^n} \cdots \int f(u_1, \ldots, u_n) \delta_u^n du^n = \int f(u, \ldots, u) du.$$

Definition 8.3. The measure defined by (8.18) is called a *generalised random measure*.

Although symbolic, the integral with respect to a generalised random measure of the form (8.18) may be written as

$$\int_{\mathbf{R}^k} \cdots \int f(u_1, \ldots, u_k) M_{n_1}(du_1) \cdots M_{n_k}(du_k).$$

At the moment, multiplication of generalised random measures can be defined only for special cases. Multiplication by $M_1(dt)$ is possible as:

$$M_n(dt) \cdot M_1(ds) = M_n(dt) \cdot M_1(ds) + \delta_{t-s} M_{n-1}(dt),$$

$$\left(\prod_{j=1}^k \cdot M_{n_j}(dt_j)\right) \cdot M_1(ds) = \sum_{j=1}^k \left\{M_{n_j}(dt_j) M_1(ds) \prod_{i \neq j} \cdot M_{n_i}(dt_i)\right\}.$$

For example,

$$\int f(u)M_n(du) \cdot \int g(v)M_1(dv)$$
$$= \iint (f \otimes g)(u,v)M_n(du) \cdot M_1(dv) + \int f(u)g(u)M_{n-1}(du),$$

where $\otimes$ denotes the tensor product. Three remarks are now in order.

Remark 1. Integrals with respect to a generalised random measure should *not* be thought of as a definite integral, but rather they should be viewed as continuous analogues of polynomials in finitely many variables. Such an interpretation is based on the discussion of §8.2.

Remark 2. The space $\mathcal{H}_n$ coincides with the class of all integrals with respect to the random measure $M_1(du_1) \cdot \cdots \cdot M_1(du_n)$.

Remark 3. The space $\mathcal{H}_n^{(-n)}$ cannot be exhausted by integrals with respect to generalised random measures of the form (8.18) with $\sum_j n_j = n$. An example of a member of $\mathcal{H}_n^{(-n)}$ not expressible as such an integral will be given in the next section. In the terms used by P. Lévy, a functional in $\mathcal{F}_n$ is said to be *regular*, and a $U(\xi)$ such that $i^n C(\xi)U(\xi)$ can be expressed in the form (8.17) is termed a *normal* functional.

8.5 Causal Calculus

By "causal calculus" we mean the differential and integral calculi, as well as related operations, in which time is explicitly involved in the computations and the formulae. A complete orthonormal system for the basic space which is most suitable for this calculus was introduced in §8.2, and with this base and the notion of generalised random measures, we have already discussed:

I. Integration

The next topic has to be

II. Differentiation

We are interested in the derivative $\partial/\partial \dot{B}(t)$, where an exact meaning for this expression will be given in what follows.

By analogy with $\mathscr{F}_n$, a functional space $\mathscr{F}_n^{(n)}$ can be defined and topologised in such a way that $\mathscr{F}_n^{(n)}$ is isomorphic to $\mathbf{F}_n^{(n)}$, and the same can be done for $\mathscr{F}_n^{(n)} \cong \mathbf{F}_n^{(-n)}$. If $\varphi \in \mathscr{H}_n^{(n)}$, we can find $U(\xi)$ in $\mathscr{F}_n^{(n)}$ such that $(\mathscr{T}\varphi)(\xi) = i^n C(\xi) U(\xi)$. The functional derivative (in the sense of Fréchet) of $U(\xi)$ always exists and belongs to $\mathscr{F}_{n-1}^{(n-1)}$ for every t, and we denote it by $U'_\xi(\xi; t)$. By applying the transformation $\mathscr{T}^{-1}$ we obtain an $\mathscr{H}_{n-1}^{(n-1)}$-functional which we denote by $\varphi'(t, x)$:

$$\mathscr{T}(i^{n-1} C(\xi) U'_\xi(\xi; t))(x) = \varphi'(t, x),$$

and the mapping $\varphi(x) \to \varphi'(t, x)$ is denoted by

$$\left|\frac{\partial}{\partial \dot{B}(t)}\right| \varphi(x) = \varphi'(t, x). \tag{8.20}$$

This is a derivative, indeed the $\dot{B}(t)$-derivative of a Brownian functional φ.

Higher order derivatives such as $\partial^2/\partial \dot{B}(t)^2$, $\partial^2/d\dot{B}(t)\partial \dot{B}(s)$ can also be defined by using second-order functional derivatives. They are simply illustrated by the following:

$$\mathscr{H}_n^{(n)} \leftrightarrow \mathscr{F}_n^{(n)}$$

$$\frac{\partial^2}{\partial \dot{B}(t)^2} \leftrightarrow U''_{\xi^2}(\xi; t) \tag{8.21}$$

$$\frac{\partial^2}{\partial \dot{B}(t)\,\partial \dot{B}(s)} \leftrightarrow U''_{\xi\eta}(\xi; t, a)$$

The second-order derivatives U''_{ξ^2} and $U''_{\xi\eta}$ of a functional U are determined by the procedures (i) and (ii) below where:

  i. The variation δU when ξ varies by $\delta\xi$ is a functional which is

a. linear in $\delta\xi$, and such that
b. $U(\xi + \delta\xi) - U(\xi) = \delta U + o(\delta\xi)$.

The functional derivative is determined by the formula

$$\delta U = \int U'_\xi(\xi; t)\, \delta\xi(t)\, dt.$$

  ii. The second variation $\delta^2 U$ is a function which is

a. quadratic in $\delta\xi$, such that

b. $U(\xi + \delta\xi) - U(\xi) = \delta U + \tfrac{1}{2}\delta^2 U + o(\delta\xi)^2$.

The functional derivatives $U''_{\xi^2}(\xi; t)$ and $U''_{\xi\eta}(\xi; t, s)$ are determined by the formula

$$\delta U'_\xi(t) = U''_{\xi^2}(\xi; t)\, \delta\xi(t) + \int U''_{\xi\eta}(\xi; t, s)\, \delta\xi(s)\, ds.$$

8.5 Causal Calculus

An infinite-dimensional *Laplacian operator* Δ on $\mathcal{H}_n^{(-n)}$ can now be defined by

domain $\mathcal{D}(\Delta) = \mathcal{T}^{-1}\{i^n C(\xi)U(\xi): U''_{\xi^2}, U''_{\xi\eta}$ exist, $U''_{\xi^2}(\xi; t)$ is t-integrable$\}$

$$\Delta = \int \left[\frac{\partial^2}{\partial \dot{B}(t)^2} \right] dt. \tag{8.22}$$

It is straightforward to extend Δ to a suitable subset of $\sum_n \mathcal{H}_n^{(-n)}$.

EXAMPLE 2.

i. Putting $\varphi_t(x) = \exp[B(t) - \tfrac{1}{2}t]$, we find that

$$\frac{\partial}{\partial \dot{B}(s)} \varphi_t = \begin{cases} \varphi_s & 0 < s < t; \\ 0 & s > t. \end{cases}$$

ii. $\{\partial/\partial \dot{B}(s)\} H_n(B(t); t) = H_{n-1}(B(t); t), \quad s \le t.$

iii. For $\varphi \equiv \int f(u) M_2(du)$, we have

$$\frac{\partial}{\partial \dot{B}(t)} \varphi = f(t)\dot{B}(t), \quad \frac{\partial^2}{\partial \dot{B}(t)^2} \varphi = f(t);$$

$$\frac{\partial^2}{\partial \dot{B}(t) \partial \dot{B}(s)} \varphi = 0, \quad \Delta\varphi = \int f(t)\, dt.$$

Definition 8.4. A functional φ is said to be *harmonic* if $\varphi \in \mathcal{D}(\Delta)$ and

$$\Delta\varphi = 0.$$

Theorem 8.2. *Every functional in $\mathcal{H}_n$, $n \ge 0$, is harmonic.*

III. Multiplication by $\dot{B}(t)$

Since the system $\{e^{i\langle x, \eta \rangle}: \eta \in \mathcal{S}\}$ generates the whole of (L^2), we start with $\dot{B}(t) e^{i\langle x, \eta \rangle}$, which is approximated by

$$\langle x, |\Delta|^{-1} \chi_\Delta \rangle e^{i\langle x, \eta \rangle}.$$

Applying $\mathcal{T}$ to this approximation yields the result

$$iC(\xi)(\xi + \eta, |\Delta|^{-1}\chi_\Delta) \exp[-\tfrac{1}{2}\|\eta\|^2 - (\xi, \eta)],$$

and as $\Delta \to \{t\}$, this tends to

$$iC(\xi)(\xi(t) + \eta(t)) \exp[-\tfrac{1}{2}\|\eta\|^2 - (\xi, \eta)].$$

This enables us to prove that if $\|\eta\| = 1$,

$$\mathcal{T}\left(\dot{B}(t) H_n\left(\frac{\langle x, \eta \rangle}{\sqrt{2}} \right) \right)(\xi) = i^{n+1} 2^{n/2} [\xi(t)(\xi, \eta)^n + n\eta(t)(\xi, \eta)^{n-1}] C(\xi). \tag{8.23}$$

This means that when $\|\eta\| = 1$, $\dot{B}(t)H_n(\langle x, \eta\rangle/\sqrt{2})$ belongs to $\mathscr{H}_{n+1}^{(-n-1)}$ $+ \mathscr{H}_{n-1}$, and that its associated kernel is

$$2^{n/2}\widehat{\{\eta^{n\otimes} \otimes \delta_t} - n\eta(t)\eta^{(n-1)\otimes}\}, \wedge \text{ denoting symmetrisation}.$$

Thus we can prove the following theorem.

Theorem 8.3. *Functionals in $\mathscr{H}_n^{(n)}$ can be multiplied by $\dot{B}(t)$, and the resulting product lies in $\mathscr{H}_{n+1}^{(-n-1)} + \mathscr{H}_{n-1}$.*

IV. Fourier Transforms

In order to indicate the need for a Fourier transform acting on the space of generalised Brownian functionals, we examine three transformations which have been met to date, each looking in some respect like a Fourier transform. Although neither of them is satisfactory for the purposes of the causal calculus, their study may be suggestive in the search for an *ideal* such transform.

a. The transformation $\mathscr{T}$. This was given in §4.3 by

$$(\mathscr{T}\varphi)(\xi) = \int \exp[i\langle x, \xi\rangle] \, \varphi(x) \, d\mu(x).$$

It looks like a Fourier transform, but transforms $\varphi(x)$ to a functional on E rather than E^*. Despite this there is, of course, no need to emphasise further its important role in our theory.

b. The Fourier-Wiener transform. This was defined in §4.7 by

$$\mathfrak{I}(y) = \int \varphi(\sqrt{2}x + iy) \, d\mu(x), \qquad \varphi \in (L^2),$$

and satisfies

$$\mathfrak{I}(y) = i^n\varphi(y), \qquad \varphi \in \mathscr{H}_n.$$

This is a beautiful transformation in (L^2), but does not seem appropriate for the causal calculus.

c. Another Fourier transform. Start with $\varphi \in \mathscr{H}_n$, and let $F \in \hat{L}^2(\mathbf{R}^n)$ be the kernel associated with φ. Denoting the ordinary Fourier transform of F by $\hat{F}$, we can find an element $\hat{\varphi}$ of $\mathscr{H}_n$ for which $\hat{F}$ is the associated kernel, and we see that the transformation $\varphi \to \hat{\varphi}$ is unitary. However we cannot be satisfied by this transformation either.

What is needed for the causal calculus is a "Fourier transform" which carries the operation $\partial/\partial\dot{B}(t)$ into multiplication by $i\dot{B}(t)$, and conversely. This search for a Fourier transform with which to establish the appropriate harmonic analysis is a very interesting open problem. It is worth noting that the generalized functional $\exp[i \int \dot{B}(t)^2 \, dt]$ would, with suitable modification, play a basic role in our analysis. (See L. Streit and T. Hida (1979).)

Appendix

A.1 Martingales

In this section we will explain the basic inequalities and limit theorems for martingales that are frequently used in the book. An heuristic approach to such results was made by P. Lévy (1937, Chapter VIII) and a systematic investigation of them was made by J. L. Doob, particularly in his book which appeared in 1953.

(i) Discrete Parameter Martingales

Although this notion was introduced in §2.2, we repeat the definition for convenience. Let $\{X_n(\omega): n \in I\}$ be a stochastic process defined on a probability space $(\Omega, \mathbf{B}, P)$, and take the parameter set I to be one of the following: the whole set N of integers; the set N^+ of all positive integers; the set N^- of all negative integers (where we sometimes add $+\infty$ and/or $-\infty$ to these sets); or any subset of N consisting of consecutive integers. We take as given an increasing family $\{\mathbf{B}_n, n \in I\}$ of sub-σ-fields on $\mathbf{B}$:

$$\mathbf{B}_n \subseteq \mathbf{B}, \qquad n \in I,$$
$$\mathbf{B}_n \subseteq \mathbf{B}_{n+1}, \qquad n, n+1 \in I.$$

Definition A.1. A system $\{X_n, \mathbf{B}_n: n \in I\}$ is called a *martingale* if for every $n \in I$ the following three conditions are satisfied:

a. X_n is $\mathbf{B}_n$-measurable;
b. $E(|X_n|) < \infty$;
c. $E(X_m | \mathbf{B}_n) = X_n$, a.e., for every $m > n$.

If (a) and (b) above are satisfied, but in place of (c) we have

c′ $E(X_m|\mathbf{B}_n) \geq X_n$, a.e., for every $m > n$,
then $\{X_n, \mathbf{B}_n: n \in I\}$ is called a *submartingale*.

If $\{-X_n, \mathbf{B}_n: n \in I\}$ is a submartingale, then $\{X_n, \mathbf{B}_n: n \in I\}$ is referred to as a *supermartingale*.

For notational convenience we often write $\{X_n, \mathbf{B}_n: n \in I\}$ as $\{X_n: n \in I\}$ or, more simply, as $\{X_n\}$, when $\{\mathbf{B}_n\}$ and I are clearly known.

If $\{X_n, \mathbf{B}_n: n \in I\}$ is a martingale, then for $m, n \in I$ with $m > n$, we have $E(X_m) = E(E(X_m|\mathbf{B}_n)) = E(X_n)$, and so $E(X_n)$ is constant. Similarly, if $\{X_n\}$ is a submartingale, then $E(X_n)$ is non-decreasing in n.

Proposition A.1. *Let $\Phi(x)$ be a continuous convex function on $\mathbf{R}$. If $\{X_n, \mathbf{B}_n: n \in I\}$ is a martingale, and if $\Phi(X_{n_0})$ is integrable for some $n_0 \in I$, then $\{\Phi(X_n), \mathbf{B}_n: n \in I, n \leq n_0\}$ is a submartingale.*

PROOF. It is clear that $\{\Phi(X_n)\}$ satisfies condition (a) of Definition A.1, whilst for $n \leq n_0$, we may use Jensen's inequality to deduce

$$\Phi(X_n) = \Phi(E(X_{n_0}|\mathbf{B}_n)) \leq E(\Phi(X_{n_0})|\mathbf{B}_n), \quad \text{a.e.} \quad (A.1)$$

Now $\Phi(X_n)$ is $\mathbf{B}_n$-measurable, and we have just shown it to be bounded above by an integrable $\mathbf{B}_n$-measurable function. On the other hand, it is well-known that for a convex function Φ, there exists a positive constant c such that $\Phi(x) \geq cx + a$ for all negative x, and so we see that $\Phi(X_n)$ is bounded below by the integrable function $cX_n + a$. Thus $\Phi(X_n)$ is integrable and so condition (b) of Definition A.1 is seen to be satisfied.

Finally we remark that the result $\Phi(X_n) \leq E(\Phi(X_m)|\mathbf{B}_n)$ for $n < m \leq n_0$ can be proved in the same way as (A.1) above, thus showing that (c′) and hence the whole of Definition A.1 is satisfied. $\square$

EXAMPLE 1. If $\{X_n\}$ is a martingale, and if $\alpha \geq 1$, then $\{|X_n|^\alpha\}$ is a submartingale.

EXAMPLE 2. If we write $X_n^+ = X_n \vee 0$, then $\{X_n^+\}$ is a submartingale whenever $\{X_n\}$ is a martingale.

Corollary. *For a martingale $\{X_n, \mathbf{B}_n: n \in I\}$ we have the inequality*

$$E(|X_n|) \leq E(|X_{n+1}|), \quad n, n+1 \in I.$$

We now turn to Doob's fundamental inequality, which is an extension to submartingales of the Kolmogorov inequality for sequences of independent random variables (§1.5, Lemma 1.3).

A.1 Martingales

Theorem A.1. *Let $\{X_n, \mathbf{B}_n : n \in I\}$ be a submartingale. Then for any $\lambda > 0$ we have the inequality*

$$P\left(\max_{k \leq n} X_k \geq \lambda\right) \leq \lambda^{-1} E(X_n^+). \tag{A.2}$$

PROOF. Let B_k be the event that the first time X_j attains λ is at k:

$$B_k = \bigcap_{j < k} (X_j < \lambda) \cap (X_k \geq \lambda).$$

Each B_k is an element of the corresponding σ-field $\mathbf{B}_k$, $k \leq n$, and they form a mutually exclusive set of events. Setting $B = \bigcup_{k \leq n} B_k$, we note that $P(B) = P(\max_{k \leq n} X_k \geq \lambda)$, and

$$E(X_n^+) \geq \int_B X_n^+ \, dP = \sum_{k \leq n} \int_{B_k} E(X_n^+ | \mathbf{B}_k) \, dP$$

$$\geq \sum_{k \leq n} \int_{B_k} X_k^+ \, dP$$

$$\geq \lambda \sum_{k \leq n} P(B_k) = \lambda P(B).$$

Thus the inequality (A.2) is proved. □

From this theorem and Proposition A.1, the assertion below follows.

Corollary. *Let $\{X_n : 1 \leq n \leq N\}$ be a martingale with $E(X_N^2) < \infty$. Then $\{X_n^2 : 1 \leq n \leq N\}$ is a submartingale, and for $\lambda > 0$ we have the inequality*

$$P\left(\max_{1 \leq n \leq N} X_n^2 \geq \lambda\right) \leq \lambda^{-1} E(X_N^2).$$

Let $a_1, a_2, \ldots, a_N$ be real numbers and let $[r_1, r_2]$ be an interval. The upcrossing number of the interval $[r_1, r_2]$ by the sequence $a_1, a_2, \ldots, a_N$ [see J. L. Doob (1953) Chapter VII] is defined to be the number of times the sequence a_i, $1 \leq i \leq N$, passes from below r_1 to above r_2.

Proposition A.2. *Let $\{X_n(\omega): 1 \leq n \leq N\}$ be a submartingale, and let $\beta(\omega)$ be the upcrossing number of $[r_1, r_2]$ by a sample sequence $X_1(\omega), X_2(\omega), \ldots, X_N(\omega)$. Then we have the inequality*

$$E(\beta) \leq (r_2 - r_1)^{-1}[E(|X_N|) + |r_1|]. \tag{A.3}$$

PROOF. Assume initially that all $X_k \geq 0$ and that $r_1 = 0$, and let $\sigma_1(\omega)$ be the smallest k (if one exists), for which $X_k(\omega) \leq 0$. In general let $\sigma_j(\omega)$ be the first i (if one exists) after $\sigma_{j-1}(\omega)$ for which

$$X_i(\omega) \geq r_2 \quad (j \text{ even}),$$

$$X_i(\omega) \leq 0 \quad (j \text{ odd}).$$

If the above definitions leave σ_j undefined, then, for convenience, we set $\sigma_j(\omega) = N + 1$. Now denote by $U_j(\omega)$, $2 \leq j \leq N$, the indicator function of the event $\bigcup_i (\sigma_{2i} < j \leq \sigma_{2i+1}) \cup (j \leq \sigma_1)$, and set

$$X = X_1 + \sum_{j=2}^{N} U_j(X_j - X_{j-1}).$$

Since U_j is $\mathbf{B}(X_1, X_2, \ldots, X_{j-1})$-measurable, the submartingale property implies that

$$\int_{(U_j=1)} (X_j - X_{j-1}) \, dP \geq 0.$$

We therefore have

$$E(X) = E(X_1) + \sum_{j=2}^{N} \int_{(U_j=1)} (X_j - X_{j-1}) \, dP \geq 0.$$

On the other hand

$$X(\omega) \leq X_N^+(\omega) - r_2 \beta(\omega),$$

so that

$$0 \leq E(X) \leq E(X_N^+) - r_2 E(\beta),$$

which implies (A.3) in the case $r_1 = 0$. For the case r_1 general we set

$$X_n'(\omega) = (X_n(\omega) - r_1) \vee 0.$$

Noting that $\{X_n'\}$ is also a submartingale, the proof being similar to that of Proposition A.1, we can apply the above special case to $\{X_n'\}$. This gives us

$$E(\beta) \leq (r_2 - r_1)^{-1} E(X_N') = (r_2 - r_1)^{-1} \int_{\{X_N \geq r_1\}} (X_N - r_1) \, dP,$$

which implies (A.3). □

The above proposition is a key result in the proof of the following:

Theorem A.2. Let $\{X_n, \mathbf{B}_n : n \in \mathbf{N}^+\}$ be a martingale. If $\lim_{n \to \infty} E(|X_n|) = K < \infty$, then the limit

$$\lim_{n \to \infty} X_n = X_\infty$$

exists almost surely, and $E(|X_\infty|) \leq K$ holds.

PROOF. Set $\limsup_{n \to \infty} X_n(\omega) = X^*(\omega)$, and $\liminf_{n \to \infty} X_n(\omega) = X_*(\omega)$. Clearly

$$(X^* - X_* > 0) = \bigcup_{\substack{r_1, r_2 \\ \text{rational}}} (X^* > r_2 > r_1 > X_*).$$

A.1 Martingales

Take two rationals r_1 and $r_2 > r_1$ and let $\beta_n(\omega)$ denote the upcrossing number of the interval $[r_1, r_2]$ by $X_1(\omega), X_2(\omega), \ldots, X_n(\omega)$. Then for an ω for which $X^*(\omega) > r_2 > r_1 > X_*(\omega)$, we must have $\beta_n(\omega) \to \infty$ as $n \to \infty$. However Proposition A.2 implies that

$$E(\beta_n) \leq (r_2 - r_1)^{-1}[K + |r_1|] < \infty,$$

and so $\beta_n \to \infty$ on a set of zero P-measure. Thus $P(X^* > r_2 > r_1 > X_*) = 0$ and, since r_1, r_2 are an abitrary pair of rationals, we conclude that $X_*(\omega) = X^*(\omega)$, a.e. This equality could be $\infty = \infty$, but our assumption allows the use of Fatou's lemma to show that $X_\infty (= X^* = X_*)$ is finite with probability 1, and that

$$E(|X_\infty|) \leq \lim_{n \to \infty} E(|X_n|) = K. \qquad \square$$

We now state another result which can be similarly proved.

Theorem A.3. *If $\{X_n, \mathbf{B}_n : n \leq -1\}$ is a martingale, then*

$$\lim_{n \to -\infty} X_n = X_{-\infty}$$

exists, and is finite, almost surely.

Remark. The assumption in Theorem A.3 is much weaker than that of Theorem A.2.

(ii) Continuous Parameter Martingales

Let $\{X(t) : t \in T\}$ be a stochastic process with a continuous parameter, the set T being either the entire real line $\mathbf{R}$ or an interval subset of the form $[0, \infty)$, $(-\infty, 0]$, or $[a, b]$. Further, let $\{\mathbf{B}_t\}$ be in family of σ-fields which are increasing in t. Then a pair $\{X(t), \mathbf{B}_t : t \in T\}$ is defined to be a martingale (submartingale, or supermartingale) in a manner similar to Definition A.1. Theorems and propositions established in (i) can usually be carried over directly to the continuous parameter case, but we often need the additional assumption that the process $\{X(t)\}$ is separable, in the following sense:

Definition A.2. A stochastic process $\{X(t) : t \in T\}$ is said to be *separable* if there exists a countable subset S of T such that for any subinterval $(a, b) \subseteq T$:

$$\sup_{t \in (a,b)} X(t) = \sup_{t \in (a,b) \cap S} X(t), \quad \inf_{t \in (a,b)} X(t) = \inf_{t \in (a,b) \cap S} X(t), \quad \text{a.e.} \qquad (A.4)$$

In what follow we will always suppose $\{X(t)\}$ to be separable. We now state without proof two theorems which are valid under the separability assumption.

Theorem A.4. *Let $\{X(t), \mathbf{B}_t: t \in T\}$ be a martingale. Then for any $\lambda > 0$ and $t \in T$ we have the inequality:*

$$P\left(\sup_{u \leq t, u \in T} X(u) \geq \lambda \right) \leq \lambda^{-1} E(X(t)^+). \tag{A.5}$$

In particular, if $E(X(t)^2) \leq \infty$, then we also have

$$P\left(\sup_{u \leq t, u \in T} |X(u)| \geq \lambda \right) \leq \lambda^{-2} E(X(t)^2). \tag{A.6}$$

Theorem A.5. *For a martingale $\{X(t), \mathbf{B}_t: t \in [0, \infty)\}$ we have the following limits existing, and being finite, almost surely:*

$$\lim_{u \to t-0} X(u), \quad \lim_{u \to t+0} X(u).$$

A.2 Brownian Motion with a Multidimensional Parameter

(i) Definition and Some Basic Properties

Definition A.3. A system $\{B(A, \omega): A \in \mathbf{R}^n\}$ of real-valued random variables $B(A, \omega)$ with parameter space $\mathbf{R}^n$ is called a *Brownian motion with n-dimensional parameter* if it satisfies the following two conditions:

a. $B(O, \omega) = 0$, a.e., where $O \in \mathbf{R}^n$ denotes the origin of $\mathbf{R}^n$;
b. The system $\{B(A, \omega): A \in \mathbf{R}^n\}$ is Gaussian and $B(A) - B(A')$ has zero mean and variance $d(A, A')$, d denoting the usual metric on $\mathbf{R}^n$.

In the particular case $n = 1$, this definition agrees with that given in §2.1, although there the parameter only varied over a half-space. In cases where the dimension n does not specifically enter, $B(A)$ is simply called a *Brownian motion with multidimensional parameter*.

Set $\mathbf{R}^n_+ = \{A = (a_1, a_2, \ldots, a_n) \in \mathbf{R}^n: \text{for every } i, a_i \geq 0\}$. A *Wiener process with multi-dimensional parameter* is a system $\{B(A, \omega): A \in \mathbf{R}^n_+\}$ of random variables $B(A, \omega)$ which satisfy condition (a) of Definition A.3 and also

c. the system $\{B(A, \omega): A \in \mathbf{R}^n_+\}$ is Gaussian, $B(A)$ has zero mean, and

$$E(B(A)B(A')) = \prod_{j=1}^{n} (a_j \wedge a'_j),$$

where $A = (a_1, \ldots, a_n)$, $A' = (a'_1, \ldots, a'_n)$ are arbitrary elements of $\mathbf{R}^n_+$.

A.2 Brownian Motion with a Multidimensional Parameter

In order to avoid confusion a process $\{B(A)\}$ satisfying Definition A.3 is often called a *Lévy Brownian motion*. Note that in the case $n = 1$ the Wiener process and Lévy Brownian motion are the same concept, when the parameter set of the latter restricted to $[0, \infty)$.

The Brownian motion just defined has little connection with the main topics of this book and so a detailed discussion of it will be omitted. We will, however, note that as a Gaussian process, $\{B(A)\}$ possesses interesting and complicated probabilistic properties, a consequence of the fact that the parameter runs over a multidimensional space. This Brownian motion $\{B(A)\}$ is certainly an object of investigation for the future.

A glimpse of what we call complicated properties may be obtained from the following remarks. The Brownian motion $\{B(A): A \in \mathbf{R}^n\}$ is invariant under rotations of the parameter space $\mathbf{R}^n$ around the origin, and as for translations of the parameter, we can see that the new system $\{B(A) - B(A_0): A \in \mathbf{R}^n\}$ is again a Brownian motion, where A_0 fixed plays the role of the origin. If A is constrained to run over a hyperplane passing through A_0, then we obtain a Brownian motion with lower dimensional parameter. In particular, if A runs over a straight line l ($A_0 \in l$), then $\{B(A) - B(A_0): A \in l\}$ is a Brownian motion in the ordinary sense.

From the definition, we easily obtain the result:

$$E\{B(A)B(A')\} = \frac{1}{2}\{d(A, O) + d(A', O) - d(A, A')\},$$

and with this equation we find:

$$E\{(B(A_1) - B(A'_1))(B(A_0) - B(A'_0))\}$$
$$= \frac{1}{2}\{d(A'_0, A_1) + d(A_0, A'_1) - d(A_0, A_1) - d(A'_0, A'_1)\}.$$

Once A_0 and A'_0 are fixed, we can state an interesting property, namely, that a necessary and sufficient condition for a difference $B(A_1) - B(A'_1)$ to be independent of the fixed difference $B(A_0) - B(A'_0)$, is that both A_1 and A'_1 lie on the same sheet of a two-sheeted hyperboloid of revolution with foci A_0 and A'_0. This fact indicates how difficult it is to obtain a property equivalent to the additivity of ordinary Brownian motion.

The projective invariance (see §§2.1, 5.4) of a Brownian motion $\{B(t): t > 0\}$ is indeed remarkable. In the multiparameter case, we find a generalization in Takenaka (1977). Here we note the following. Let A^* be the inverse or reciprocal point of A relative to the unit sphere in $\mathbf{R}^n$, and set

$$B^*(A) = \begin{cases} d(O, A)B(A^*) & A \neq O, \\ 0 & A = O. \end{cases}$$

Then $\{B^*(A): A \in \mathbf{R}^n\}$ is again a Brownian motion with n-dimensional parameter.

(ii) Sample Path Properties

We can discuss the continuity of sample paths of the Brownian motion with n-dimensional parameter exactly as in the one-dimensional case; in particular we can show that for almost all ω, $B(A, \omega)$ is a continuous function of $A \in \mathbf{R}^n$. The next two results imply local and uniform continuity as well.

$$P\left(\lim_{A \to 0} \sup\{2\, d(O, A)\log\log d(O, A)^{-1}\}^{-1/2} B(A) = 1\right) = 1; \quad (A.7)$$

$$P\left(\lim_{\substack{d(A, A') \to 0 \\ A, A' \in S}} \sup\{2n\, d(A, A')\log d(A, A')^{-1}\}^{-1/2} |B(A) - B(A')| = 1\right) = 1, \quad (A.8)$$

where S is the unit ball.

Since P. Lévy (1937) there have been many approaches to problems concerning the continuity of sample paths, and a superb, final result was given by T. Sirao in 1960.

(iii) Markov Properties

As the last topic, we now discuss Markov properties of Lévy's Brownian motion. In the course of the investigating the Brownian motion $\{B(A): A \in \mathbf{R}^n\}$, P. Lévy conjectured that it possessed a Markov property when n is odd, but had no such property when n is even. Provoking his conjecture were many results about the process $M_n(t)$, which is the average of $B(A)$ over the sphere of radius t in $\mathbf{R}^n$, and, in connection with $M_n(t)$, results concerning the Dirichlet problem relative to spheres. What interests us at the moment is the definition of the *Markov property*: what sort of probabilistic property should we term Markov in the case of a multidimensional parameter stochastic process? Lévy's original idea was the following: take a domain D whose boundary ∂D is smooth and separates $\mathbf{R}^n$ into two parts, and suppose that $\{B(A): A \in D\}$ and $\{B(A'): A' \in \text{interior of } D^c\}$ become independent when we know the values of $B(A)$, A belonging to some neighbourhood of ∂D.

H. P. McKean (1963) introduced the concept of a splitting field, and using it he defined the Markov property and gave an affirmative answer to Lévy's conjecture. He also gave other interesting results which indicate how to investigate general Gaussian processes with a multidimensional parameter. For details we refer to his paper, here mentioning only the relationship between a Brownian motion and a white noise. In the case $n = 1$ we have often used the fact that a white noise can be viewed as the time-derivative of a Brownian motion, but this simple relationship cannot be generalised so easily to the case $n > 1$. Indeed there are many open problems in this line, although in the paper of McKean noted above, some results of other authors

(Lévy, Chentsov) are cited. Just recently S. Takenaka (1977) discovered important relationships between Brownian motion with a multidimensional parameter, the associated white noise, the group of projective transformations, as well as the related Radon transform.

A.3 Examples of Nuclear Spaces

This section begins with the definition of nuclear space, postponed from §3.1. A vector space E is said to be a *countably Hilbert* space if E is topologised by countably many compatible Hilbertian norms $\|\cdot\|_n$, $n \geq 0$, with respect to which E is complete. A Hilbertian norm is one derived from an inner product. Let E_n be the completion of E with respect to the n-th norm $\|\cdot\|_n$. Then by definition we have

$$E = \bigcap_n E_n.$$

Since the norms $\|\cdot\|_n$ are compatible, i.e., if a sequence approaches 0 in $\|\cdot\|_m$ and is a Cauchy one in $\|\cdot\|_n$, then it approaches 0 in $\|\cdot\|_n$, we may assume that they are arranged in increasing order:

$$\|\cdot\|_0 \leq \|\cdot\|_1 \leq \cdots \leq \|\cdot\|_n \leq \cdots,$$

and this implies the inclusions

$$E_0 \supset E_1 \supset \cdots \supset E_n \supset \cdots.$$

Taking $\|\cdot\|_0$ to be the basic norm, we form the dual space E_n^* of E_n, and have

$$E_0 = E_0^* \subset E_1^* \subset \cdots \subset E_n^* \subset \cdots.$$

Letting $\|\cdot\|_{-n}$ be the norm of the Hilbert space E_n^*, we see that $\{\|\cdot\|_n : -\infty < n < \infty\}$ is an increasing family of Hilbertian norms. Further, the dual space E^* of E is expressible as

$$E^* = \bigcup_n E_n^*.$$

In what follows a countably Hilbert space is always assumed to have a structure such that the countably many norms, the E_n, and the E_n^*, are all arranged in linear order. (If necessary we may rearrange the $\|\cdot\|_n$ without changing the topology on E.)

Definition A.4. Let E be a countably Hilbert space. If for any m there exists $n > m$ such that the injection mapping

$$T_m^n : E_n \to E_m$$

is nuclear, then E is called a *countably Hilbert nuclear space* or simply a *nuclear space*.

Remark. The definition of nuclear space can, of course, be generalised. We have taken this restricted but more concrete one as it fits in better with the purpose of this book.

We are now ready to give several important examples of nuclear spaces, and to discuss some topics related to each of these spaces.

EXAMPLE 1. *The space* $\hat{\mathscr{D}}(\pi)$. Let $\hat{\mathscr{D}}(\pi)$ be the collection of all $\mathscr{C}^\infty$-functions on the unit circle (radius 1), and let $\xi(\theta)$, $-\pi \le \theta \le \pi$, denote an arbitrary member of it.

We now introduce countably many norms $\|\cdot\|_n$. For $\xi \in \mathscr{D}(\pi)$ define

$$\|\xi\|_n^2 = \sum_{k=0}^n \int_{-\pi}^\pi |\xi^{(k)}(\theta)|^2 \, d\theta, \qquad n \ge 0, \tag{A.9}$$

where $\xi^{(k)}$ is the k-th derivative of ξ, and we understand that $\xi^{(0)} = \xi$. Clearly each $\|\cdot\|_n$ is a Hilbertian norm, and we have

$$\|\cdot\|_n \le \|\cdot\|_{n+1}, \qquad n \ge 0. \tag{A.10}$$

In particular $\|\xi\|_0$ is the usual $L^2([-\pi, \pi])$-norm.

Upon completing $\hat{\mathscr{D}}(\pi)$ with respect to the norm $\|\cdot\|_n$, we obtain a Hilbert space denoted by $\hat{H}_n$, and the inequality (A.10) implies that

$$\hat{H}_n \supset \hat{H}_{n+1},$$

where $\hat{H}_0$ coincides with $L^2([-\pi, \pi])$. It can be proved that the space $\hat{H}_n$ coincides with the space of $(n-1)$-times continuously differentiable functions, whose $(n-1)$st derivatives are absolutely continuous and have Radon-Nikodym derivatives belonging to $L^2([-\pi, \pi])$. As a result we have

$$\bigcap_n \hat{H}_n = \hat{\mathscr{D}}(\pi). \tag{A.11}$$

We are now in a position to indicate a most useful aspect of the sequence $\hat{H}_n$, $n \ge 0$, of Hilbert spaces. Namely, there exists a common complete orthogonal system (normalisation depends on $\hat{H}_n$!), given by

$$\{(2\pi)^{-1/2}, \pi^{-1/2} \sin k\theta, \pi^{-1/2} \cos k\theta : k \ge 1\}. \tag{A.12}$$

It is an orthonormal system only in $\hat{H}_0$. In $\hat{H}_n$ for $n \ge 1$ we see that

$$\|(2\pi)^{-1/2}\|_n^2 = 1$$
$$\|\pi^{-1/2} \sin k\theta\|_n^2 = \|\pi^{-1/2} \cos k\theta\|_n^2 = 1 + k^2 + \cdots + k^{2n}. \tag{A.13}$$

It is a straightforward consequence of the definitions of $\|\cdot\|_n$ and $\hat{H}_n$ to show that for $m < n$ the injection T_m^n from $\hat{H}_n$ into $\hat{H}_m$ is continuous. Indeed we have the stronger assertion:

Proposition A.3. *For any $n \ge 0$ the injection T_n^{n+1} is of Hilbert-Schmidt type.*

A.3 Examples of Nuclear Spaces

PROOF. It suffices for us to prove that for some complete orthonormal system $\{\xi_k\}$ in $\hat{H}_{n+1}$ we have

$$\sum_k \|T_n^{n+1}\xi_k\|_n^2 < \infty,$$

for this sum is nothing but the square of the Hilbert-Schmidt norm $\|T_n^{n+1}\|_2$. We take the complete orthogonal system (A.12), and normalise it using (A.13), obtaining a complete orthonormal system in $\hat{H}_{n+1}$:

$$\left\{(2\pi)^{-1/2}, \left(\pi \sum_{j=0}^{n+1} k^{2j}\right)^{-1/2} \sin k\theta, \left(\pi \sum_{j=0}^{n+1} k^{2j}\right)^{-1/2} \cos k\theta: k \geq 1\right\} \quad (A.12')$$

This system will be taken as $\{\xi_k\}$ in the above discussion. Noting that T_n^{n+1} simply carries a function of $\hat{H}_{n+1}$ into the same function viewed as a member of $\hat{H}_n$, the sum $\sum_k \|T_n^{n+1}\xi_k\|_n^2$ is easily computed using (A.13). Its value is

$$1 + 2\sum_{k=1}^{\infty} \frac{\sum_{j=0}^{n} k^{2j}}{\sum_{j=0}^{n+1} k^{2j}}$$

which is finite, proving that T_n^{n+1} is a Hilbert-Schmidt operator. □

We now see that T_n^{n+2} is nuclear, and so the space $\hat{D}(\pi)$ is topologised by the norms $\|\cdot\|_n$, $n \geq 1$, in such a way as to be a nuclear space.

EXAMPLE 2. The space $\mathscr{D}(K)$, K a finite interval. Let $\mathscr{D}(K)$ be the space of $\mathscr{C}^\infty$-functions defined on K such that the functions and their derivatives of all orders vanish at the boundary of K.

Let us take $K = [-\pi, \pi]$. Then $\mathscr{D}([-\pi, \pi]) \equiv \mathscr{D}(\pi)$ may be viewed as a subset of $\hat{\mathscr{D}}(\pi)$, and we introduce the relative topology derived from $\hat{\mathscr{D}}(\pi)$ onto the space $\mathscr{D}(\pi)$, i.e. we take the norms $\|\cdot\|_n$, $n \geq 0$, given by (A.9). The completion of $\mathscr{D}(\pi)$ with respect to $\|\cdot\|_n$ will be denoted by H_n and although the space H_0 coincides with $L^2([-\pi, \pi])$ as before, for $n \geq 1$ we have

$$H_n \subsetneq \hat{H}_n.$$

The subspace $\hat{H}_n \ominus H_n$, the orthogonal complement of H_n in $\hat{H}_n$, is finite-dimensional; indeed

$$H_n = \{\xi \in \hat{H}_n: \xi^{(k)}(-\pi) = \xi^{(k)}(\pi) = 0, k = 0, 1, \ldots, n-1\}.$$

With this result and Proposition A.3 of Example 1, we can prove that the injection

$$H_{n+1} \to H_n$$

is of Hilbert-Schmidt type.

It can also be shown that

$$\bigcap_n H_n = \mathscr{D}(\pi),$$

so that $\mathscr{D}(\pi)$ is proved to be a nuclear space.

For the case in which K is a general (finite) interval, we can reduce the discussion to that concerning $[-\pi, \pi]$ by applying a linear transformation to the variable u in $\xi(u)$; such a transformation involves no essential change in the structure of the nuclear space.

The next example is also obtained from $\hat{\mathcal{D}}(\pi)$ by a transformation of u which leaves the nuclear structure invariant, but this time the transformation is non-linear.

EXAMPLE 3. The space D_0. This space is given by

$$D_0 = \{f(u) \in \mathscr{C}^\infty : |u|^{-1} f(u^{-1}) \in \mathscr{C}^\infty\}$$

A topology will be introduced onto D_0 which gives a homeomorphism with $\hat{\mathcal{D}}(\pi)$. Firstly, we consider the transformation γ that maps $\xi(\theta)$, $-\pi < \theta < \pi$, to a function $f(u)$, $-\infty < u < \infty$, as follows:

$$\gamma: \xi(\theta) \to f(u) \equiv (\gamma\xi)(u) = \sqrt{2}(1 + u^2)^{-1/2} \xi(2 \tan^{-1} u). \quad \text{(A.14)}$$

If $\xi \in L^2(S^1)$, S^1 denoting the unit circle, then $f = \gamma\xi \in L^2(\mathbf{R})$, and the converse is also true. Further,

$$\|\xi\|_{L^2(S^1)} = \|f\|_{L^2(\mathbf{R})},$$

and we can thus prove that γ is a linear isomorphism between $L^2(S^1)$ and $L^2(\mathbf{R})$.

If ξ is differentiable, then so is $f = \gamma\xi$, and conversely. To see this converse, we write the inverse transformation γ^{-1} in the form:

$$(\gamma^{-1} f)(\theta) \equiv \xi(\theta) = \left(\sqrt{2} \left|\cos \frac{1}{2}\theta\right|\right)^{-1} f\left(\frac{\sin \theta/2}{\cos \theta/2}\right)$$

We can then see that $\gamma^{-1} f$ is in $\hat{D}(\pi)$ only if $f(u)$ and $|u|^{-1} f(u^{-1})$ are both in $\mathscr{C}^\infty$, and the equality

$$\gamma(\hat{D}(\pi)) = D_0$$

is proved.

Having established the correspondence between $\hat{\mathcal{D}}(\pi)$ and D_0, it is quite easy to determine the Hilbert spaces H_n^0 and the norms $\|\cdot\|_n$ explicitly. We use the following diagram, in which γ is always an isomorphism.

$$\begin{array}{ccccc} \hat{\mathcal{D}}(\pi) & \subset \cdots \subset & \hat{H}_n & \subset \cdots \subset & L^2(S^1) \\ \downarrow \gamma & & \downarrow \gamma & & \downarrow \gamma \\ D_0 & \subset \cdots \subset & H_n^0 & \subset \cdots \subset & L^2(\mathbf{R}). \end{array}$$

The Hilbert space H_n^0 is the completion of D_0 with respect to the norm $\|\cdot\|_n$ given by

$$\|f\|_n^2 = \sum_{j=0}^n \int_{-\infty}^\infty 2^{-2j} \left|\left\{(1 + u^2)\frac{d}{du} + u\right\}^j f(u)\right|^2 du,$$

A.3 Examples of Nuclear Spaces

this norm corresponding to the n-th norm given by (A.9) for the nuclear space $\hat{\mathscr{D}}(\pi)$. We note in passing that the complete orthonormal system for H_n^0 corresponding to the system (A.12) for $\hat{H}_n$ can be expressed in the form:

$$\pi^{-1/2}(1+u^2)^{-1/2};$$

$$\sqrt{2}\left\{\pi\sum_{j=0}^{n}k^{2j}\right\}^{-1/2}(1+u^2)^{-k-1/2}\sum_{l=0}^{k-1}\binom{2k}{2l+1}(-1)^l u^{2l+1}, \qquad k\geq 1;$$

$$\sqrt{2}\left\{\pi\sum_{j=0}^{n}k^{2j}\right\}^{-1/2}(1+u^2)^{-k-1/2}\sum_{l=0}^{k-1}\binom{2k}{2l}(-1)^l u^{2l}, \qquad k\geq 1.$$

It can be proved that

$$\bigcap_n H_n^0 = D_0,$$

and that D_0 is topologised by the norms $\|\cdot\|_n$, $n \geq 0$ given above. The injection from H_{n+1}^0 into H_n^0 is also of Hilbert-Schmidt type, so that D_0 is a nuclear space. Finally, we note that further detailed structure of D_0 may be obtained from that of $\hat{\mathscr{D}}(\pi)$ via the transformation γ.

EXAMPLE 4. The Schwartz space $\mathscr{S}$. This space is the family of functions on **R** given by

$$\mathscr{S} = \left\{\xi \in \mathscr{C}^\infty(\mathbf{R}): \||\xi\||_n = \max_{0\leq k\leq n}\sup_u |(1+u^2)^n \xi^{(k)}(u)| < \infty, \right.$$

$$\left. n = 0, 1, 2, \ldots\right\},$$

and it is topologised by the norms $\||\cdot\||_n$, $n \geq 0$. Equivalently, we can take the topology for which the sets

$$U_{n,k}(\varepsilon) = \left\{\xi: \sup_u |(1+u^2)^n \xi^{(k)}(u)| < \varepsilon\right\}, \qquad n\geq 0, 0\leq k\leq n, \varepsilon > 0,$$

are a fundamental system of neighbourhoods of 0.

We first show that $\mathscr{S}$ is a countably Hilbert space, and to do this we introduce another system $\|\cdot\|_n$, $n \geq 0$ of norms:

$$\|\xi\|_n^2 = \sum_{k=0}^{n}\int_{-\infty}^{\infty}(1+u^2)^n |\xi^{(k)}(u)|^2\, du.$$

These obviously constitute an increasing system of Hilbertian norms, and the two systems $\{\||\cdot\||_n: n\geq 0\}$ and $\{\|\cdot\|_n: n\geq 0\}$ of norms define the same topology on $\mathscr{S}$. This is an immediate consequence of the following lemma.

Lemma A.1. *For any $n \geq 1$ there exist constants C_n and D_n such that for every $\xi \in \mathscr{S}$ we have*

$$C_n\||\xi\||_{n-1} \leq \|\xi\|_n \leq D_n\||\xi\||_{n+1}. \tag{A.15}$$

PROOF. It is easy to obtain the right-hand inequality since for $k \leq n$,

$$\int_{-\infty}^{\infty} (1+u^2)^n |\xi^{(k)}(u)|^2 \, du \leq \sup_u [(1+u^2)^{n+1}|\xi^{(k)}(u)|^2] \int_{-\infty}^{\infty} \frac{(1+u^2)^n}{(1+u^2)^{n+1}} \, du.$$

On the other hand, if $k < n$

$$|(1+u^2)^{n-1}\xi^{(k)}(u)|$$

$$= \left| \int_{-\infty}^{u} \{(1+u^2)^{n-1}\xi^{(k)}(u)\}' \, du \right|$$

$$\leq \int_{-\infty}^{\infty} |(1+u^2)^{n-1}\xi^{(k+1)}(u)| \, du + n \int_{-\infty}^{\infty} |(1+u^2)^{n-1}\xi^{(k)}(u)| \, du$$

$$\leq \left\{ \int_{-\infty}^{\infty} (1+u^2)^{-2} \, du \right\}^{1/2} \left[\left\{ \int_{-\infty}^{\infty} (1+u^2)^{2n} |\xi^{(k+1)}(u)|^2 \, du \right\}^{1/2} \right.$$

$$\left. + n \left\{ \int_{-\infty}^{\infty} (1+u^2)^{2n} |\xi^{(k)}(u)|^2 \, du \right\}^{1/2} \right]$$

$$\leq C_n^{-1} \|\xi_n\|, \qquad C_n \text{ a constant.}$$

Taking the supremum over u in the first expression, we obtain the left-hand side of the inequality (A.15). □

Let $\mathscr{S}_n$ be the Hilbert space obtained by completing $\mathscr{S}$ with respect to the Hilbertian norm $\|\cdot\|$. It is easy to show that

$$\bigcap_n \mathscr{S}_n = \mathscr{S},$$

and so $\mathscr{S}$ is proved to be a countably Hilbert space.

The next step is to prove that $\mathscr{S}$ is nuclear. For this purpose we first establish an isomorphism between $\mathscr{S}$ and a space s to be defined below. Let us introduce a family $\|\cdot\|_p$, $p \geq 0$, of Hilbertian norms into the vector space of all sequences $x = \{x_n : n \geq 0\}$, $x_n \in \mathbf{R}$, by

$$\|x\|_p^2 = \sum_{n=0}^{\infty} (n+1)^p x_n^2. \tag{A.16}$$

Denoting by s this vector space topologised by the norms $\|\cdot\|_p$, it can be shown that s is complete, and that s is dense in l^2. We now prove

Lemma A.2. *The Schwartz space $\mathscr{S}$ is isomorphic to the topological vector space s defined above.*

PROOF. We use the complete orthonormal system $\{\xi_n\}$ in $L^2(\mathbf{R})$:

$$\xi_n(u) = (2^n n! \sqrt{\pi})^{-1/2} H_n(u) \exp(-\tfrac{1}{2} u^2), \qquad n \geq 0,$$

see formula (A.25) in §A.5. Expanding an element $\xi \in \mathscr{S}$ in a Fourier series

$$\xi = \sum_{n=0}^{\infty} x_n \xi_n,$$

we obtain a mapping from $\mathscr{S}$ into l^2:

$$\xi \to x = \{x_n : n \geq 0\} \in l^2, \qquad \xi \in \mathscr{S}.$$

We can then use the formulae (A.26) in §A.5 for $\xi_n'(u)$ and $u\xi_n(u)$ to prove that the above mapping is one-to-one and onto, and further, an elementary computation proves that the mapping is a homeomorphism. Note, however, that under the above mapping, the norm $\|\xi\|_n$ does not necessarily correspond to $\|x\|_n$. □

It is quite easy to prove that s is a nuclear space. Indeed the proof is similar in spirit to that used for $\hat{\mathscr{D}}(\pi)$, but even simpler. This fact, together with Lemma A.2 above, proves that $\mathscr{S}$ is a nuclear space.

EXAMPLE 5. The sequence space s. As indicated in the above example, s is a nuclear space. Since it is a space of sequences of real numbers, it is somewhat easier to deal with than other nuclear spaces. If we wish to transfer a result to a function space such as $\mathscr{S}$, we need to use properties of s and of a complete orthonormal system such as $\{\xi_n\}$ above.

EXAMPLE 6. The space $\mathscr{D}$. Let $\{K_n : n \geq 0\}$ be an increasing sequence of compact intervals such that $\bigcup_n K_n = \mathbf{R}$. The space $\mathscr{D}$ is defined to be the inductive limit of the spaces $\mathscr{D}(K_n)$:

$$\mathscr{D} = \varprojlim_n \mathscr{D}(K_n).$$

It is easy to show that this limit does not depend upon the particular sequence $\{K_n\}$ used. Clearly $\mathscr{D}$ is a vector space consisting of $\mathscr{C}^\infty$-functions having compact support, and the topology can be described as follows: a sequence $\{\xi_j\} \subseteq \mathscr{D}$ converges to 0 if and only if the supports of all of the ξ_j are contained in some compact set, and, for any fixed k, the derivatives of the ξ_j order k converge uniformly to zero.

A continuous linear functional on $\mathscr{D}$ is a generalised function in the sense of L. Schwartz (1950-1), and is referred to as a distribution.

The space $\mathscr{D}$ is not a countably Hilbert nuclear space as we have defined it in this book, but the Bochner-Minlos theorem (§3.2, Theorem 3.2) still holds for this space, the proof using the fact that $\mathscr{D}$ is a limit of nuclear spaces.

A.4 Wiener's Non-linear Circuit Theory

In his 1958 book, N. Wiener described a practical method of constructing electrical circuits, first proposed jointly by Wiener and Y. W. Lee, which allows the formation of any (L^2)-functional from a given white noise input. We will illustrate this Lee-Wiener circuit theory, and also indicate its connection with the Wiener-Itô decomposition of (L^2).

(i) The Lee-Wiener Electrical Circuits

First take a four-terminal circuit as indicated in Fig. A.1. Let the inductance of the coil and the capacitance of the condenser be L henry and C farad, respectively. The resistance should be $(L/C)^{1/2}$ ohm.

By the first and second network laws of Kirchhoff, we can compute the current and voltage at each part of the circuit when a voltage $E_\omega \equiv E_\omega(t) = \exp[i\omega t]$, $\omega = 2\pi f$, of frequency f is supplied. In fact the input impedance is a pure resistance equal to $(L/C)^{1/2}$, and the transfer function is given by

$$\frac{V_\omega}{E_\omega} = \frac{1 - i\omega(LC)^{1/2}}{1 + i\omega(LC)^{1/2}}.$$

The phase delay is $\sin^{-1}[2\omega(LC)^{1/2}]$. The most interesting fact is that these circuits can be dovetailed into one another, for if n circuits of this kind are put together in series, then the transfer function turns out to be

$$\left| \frac{1 - i\omega(LC)^{1/2}}{1 + i\omega(LC)^{1/2}} \right|^n.$$

As a result $|V|$ is always equals to $|E|$. By placing an inductance of L henry between the source E and the first circuit, indicated by $*$ in Fig. A.2 below, it is possible to multiply the above transfer function by an additional factor $(L/C)^{1/2}\{1 + i\omega(LC)^{1/2}\}^{-1}$.

We are now able to construct mechanically a circuit which has transfer function

$$\frac{V_{n,\omega}}{E_\omega} = \left(\frac{L}{C}\right)^{1/2} \frac{\{1 - i\omega(LC)^{1/2}\}^n}{\{1 + i\omega(LC)^{1/2}\}^{n+1}}, \qquad n \geq 0.$$

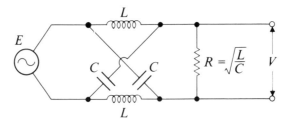

Figure A.1.

A.4 Wiener's Non-linear Circuit Theory

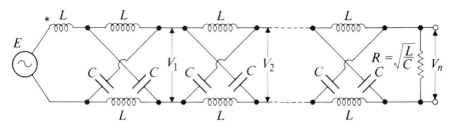

Figure A.2.

Hereafter we put $LC = 1$ (which may be thought of as simply changing the time-scale). Then we have a family of functions

$$L \cdot \frac{(1 - i\omega)^n}{(1 + i\omega)^{n+1}}, \qquad n \geq 0,$$

and we will see that they are obtained by taking the Fourier transforms of Laguerre functions.

(ii) Laguerre Functions

The system $\{L_n^{(\alpha)}(x): x \geq 0, n \geq 0\}$ of Laguerre functions is given by

$$L_n^{(\alpha)}(x) = (n!)^{-1} e^x x^{-\alpha} \frac{d^n}{dx^n} (e^{-x} x^{n+\alpha}), \qquad \alpha > -1.$$

In the case $\alpha = 0$ we simply write $L_n(x)$. These are polynomials in x expressible in the form

$$L_n(x) = \sum_{k=0}^{n} \binom{n}{k} \frac{(-x)^k}{k!}.$$

If we set

$$l_n(x) = 2^{1/2} e^x L_n(-2x), \qquad x \leq 0,$$

then $\{l_n(x): n \geq 0\}$ is, as is well-known, a complete orthonormal system in $L^2((-\infty, 0])$. We extend $l_n(x)$ to a function on $(-\infty, \infty)$ by putting $l_n(x) = 0$ if $x > 0$, and continue to use the same notation. The Fourier transform $\hat{l}_n(\omega)$, where the variable is here denoted by ω to match the notation in (i) above, is given by

$$\hat{l}_n(\omega) = (2\pi)^{-1/2} \int_{-\infty}^{0} e^{i\omega x} l_n(x) \, dx = (2\pi)^{-1/2} \frac{(1 - i\omega)^n}{(1 + i\omega)^{n+1}}.$$

This agrees (up to a constant multiplier) with the transfer function of the circuit shown in Fig. A.2 above.

(iii) Outputs for White Noise Input

Let us take a white noise $(\mathscr{S}^*, \mathfrak{B}, \mu)$. As illustrated in the last paragraph of §3.4, the spectrum of white noise is flat and extends over the entire real line. In other words, "all frequencies are equally represented" in white noise, and so a white noise input may be thought of as the superposition of components $e^{i\omega t}M(d\omega)$, where $M(d\omega)$ is a uniform random measure. Consequently, a smeared output can be expressed in the form [see formula (3.32) in §3.4 with λ replaced by ω]:

$$\langle x, \xi \rangle = \int \hat{\xi}(\omega) M(d\omega), \qquad \xi \in \mathscr{S}.$$

In particular, if ξ is taken to be l_n, that is, if white noise is passed through a circuit with transfer function $\hat{l}_n$, then we obtain

$$\langle x, l_n \rangle = (2\pi)^{-1/2} \int \frac{(1-i\omega)^n}{(1+i\omega)^{n+1}} M(d\omega).$$

An important conclusion from this is the following: since the support of l_n is $(-\infty, 0)$, if $t = 0$ is viewed as the present instant, then $\langle x, l_n \rangle$ may be viewed as information obtainable from only the past inputs. In other words, the system is *causal*.

We also have another observation. Since $\{l_n : n \geq 0\}$ is a complete orthonormal system in $L^2((-\infty, 0])$, Fourier-Hermite polynomials can be obtained as in §4.2, and are a basis of L_0^2 (see the notation of Corollary 1 in §4.4). Moreover, the collection of all such polynomials of degree n forms a base of $L_0^2 \cap \mathscr{H}_n$.

The last question asks about a practical method of forming non-linear circuits corresponding to Fourier-Hermite polynomials. We already know that the Lee-Wiener circuits correspond to $\langle x, l_n \rangle$, $n \geq 0$, so that we now only need multiplication, and it suffices for us to consider multiplication of elements of the form $\hat{l}_n(\omega) e^{i\omega t} M(d\omega)$. Such multiplications can eventually be reduced to the formation of squares, and for this there are many known methods.

We have thus seen how to construct non-linear circuits corresponding to the Fourier-Hermite polynomials, which constitute a base for L_0^2. This validates the assertion that the results of §4.6, (iv), and §5.5 can be effectively applied to practical problems.

A.5 Formulae for Hermite Polynomials

(i) Hermite Polynomials $H_n(x)$

Definition.
$$H_n(x) = (-1)^n e^{x^2} \frac{d^n}{dx^n} e^{-x^2}, \qquad n \geq 0.$$

A.5 Formulae for Hermite Polynomials

Generating function

$$\sum_{n=0}^{\infty} \frac{t^n}{n!} H_n(x) = e^{-t^2 + 2tx}$$

$$H_n(x) = n! \sum_{k=0}^{[n/2]} \frac{(-1)^k}{k!} \frac{(2x)^{n-2k}}{(n-2k)!} \tag{A.17}$$

EXAMPLES.

$$H_0(x) \equiv 1, \quad H_1(x) = 2x, \quad H_2(x) = 4x^2 - 2,$$
$$H_3(x) = 8x^3 - 12x, \quad H_4(x) = 16x^4 - 48x^2 + 12,$$
$$H_5(x) = 32x^5 - 160x^3 + 120x$$

$$H_n''(x) - 2xH_n'(x) + 2nH_n(x) = 0 \tag{A.18}$$

$$H_n'(x) = 2nH_{n-1}(x) \tag{A.19}$$

$$H_{n+1}(x) - 2xH_n(x) + 2nH_{n-1}(x) = 0 \tag{A.20}$$

$$H_n(ax + \sqrt{1-a^2}\, y) = \sum_{k=0}^{n} \binom{n}{k} a^{n-k}(1-a^2)^{k/2} H_{n-k}(x) H_k(y). \tag{A.21}$$

In particular

$$H_n\left(\frac{x+y}{\sqrt{2}}\right) = 2^{-n/2} \sum_{k=0}^{n} \binom{n}{k} H_{n-k}(x) H_k(y)$$

$$H_m(x) H_n(x) = \sum_{k=0}^{m \wedge n} 2^k k! \binom{m}{k} \binom{n}{k} H_{m+n-2k}(x) \tag{A.22}$$

$$\int_{-\infty}^{\infty} H_n(ax + \sqrt{1-a^2}\, y) H_k(y) e^{-y^2}\, dy = \frac{\sqrt{\pi}\, 2^k n!}{(n-k)!} a^{n-k}(1-a^2)^{k/2} H_{n-k}(x),$$

$$k \leq n \tag{A.23}$$

$$\int_{-\infty}^{\infty} H_m(x) H_n(x) e^{-x^2}\, dx = \delta_{n,m}\, 2^n n!\, \sqrt{\pi} \tag{A.24}$$

$$\int_{-\infty}^{\infty} H_m\left(\frac{x}{\sqrt{2}}\right) H_n\left(\frac{x}{\sqrt{2}}\right) e^{-x^2/2}\, dx = \sqrt{2\pi}\, \delta_{n,m}\, 2^n n! \tag{A.24'}$$

If $\xi_n(x) = \dfrac{1}{\sqrt{2^n n!}\, \sqrt[4]{\pi}} H_n(x) e^{-x^2/2}$, then

$$\int_{-\infty}^{\infty} \xi_m(x) \xi_n(x)\, dx = \delta_{m,n}; \tag{A.25}$$

$\{\xi_n; n \geq 0\}$ is a complete orthonormal system in $L^2(\mathbf{R})$

$$\begin{cases} \xi_n'(x) = \sqrt{\dfrac{n}{2}}\,\xi_{n-1}(x) - \sqrt{\dfrac{n+1}{2}}\,\xi_{n+1}(x) \\ x\xi_n(x) = \sqrt{\dfrac{n}{2}}\,\xi_{n-1}(x) + \sqrt{\dfrac{n+1}{2}}\,\xi_{n+1}(x), \qquad n \geq 1 \end{cases} \quad (A.26)$$

$$\frac{1}{\sqrt{2\pi}} \int_{-\infty}^{\infty} e^{ixy} H_n(y) e^{-y^2/2}\, dy = i^n H_n(x) e^{-x^2/2} \tag{A.27}$$

$$\frac{1}{\sqrt{2\pi}} \int_{-\infty}^{\infty} e^{ixy} H_n(y) e^{-y^2}\, dy = \frac{1}{\sqrt{2}}(ix)^n e^{-x^2/4} \tag{A.28}$$

$$\frac{1}{\sqrt{2\pi}} \int_{-\infty}^{\infty} e^{ixy} y^n e^{-y^2}\, dy = \frac{1}{\sqrt{2}}\left(\frac{i}{2}\right)^n H_n\!\left(\frac{x}{2}\right) e^{-x^2/4} \tag{A.29}$$

$$\frac{1}{\sqrt{2\pi}} \int_{-\infty}^{\infty} (x+iy)^n e^{-y^2/2}\, dy = 2^{-n/2} H_n\!\left(\frac{x}{\sqrt{2}}\right) \tag{A.30}$$

$$\frac{1}{\sqrt{2\pi}} \int_{-\infty}^{\infty} H_n\!\left(\frac{i}{\sqrt{2}}x + y\right) e^{-y^2/2}\, dy = i^n H_n\!\left(\frac{x}{\sqrt{2}}\right). \tag{A.31}$$

(ii) Hermite Polynomials with Parameter

Definition.

$$H_n(x; \sigma^2) = \frac{(-\sigma^2)^n}{n!}\, e^{x^2/2\sigma^2}\, \frac{d^n}{dx^n}\, e^{-x^2/2\sigma^2}, \quad \sigma > 0,\ n \geq 0.$$

Generating function

$$\sum_{n=0}^{\infty} t^n H_n(x; \sigma^2) = e^{-\sigma^2 t^2/2 + tx}, \qquad t \in \mathbf{C},$$

$$H_n(x; \sigma^2) = \frac{\sigma^n}{n!\, 2^{n/2}}\, H_n\!\left(\frac{x}{\sqrt{2}\sigma}\right). \tag{A.32}$$

EXAMPLES.

$$H_0(x; \sigma^2) \equiv 1, \quad H_1(x; \sigma^2) = x,$$

$$H_2(x; \sigma^2) = \frac{1}{2}x^2 - \frac{1}{2}\sigma^2, \quad H_3(x; \sigma^2) = \frac{1}{6}x^3 - \frac{1}{2}\sigma^2 x,$$

$$H_4(x; \sigma^2) = \frac{1}{24}x^4 - \frac{1}{4}\sigma^2 x^2 + \frac{1}{8}\sigma^4.$$

A.5 Formulae for Hermite Polynomials

$$H_n''(x;\sigma^2) - \frac{x}{\sigma^2} H_n'(x;\sigma^2) + \frac{n}{\sigma^2} H_n(x;\sigma^2) = 0 \qquad (\text{' means } d/dx) \quad (A.33)$$

$$H_n'(x;\sigma^2) = H_{n-1}(x;\sigma^2) \quad (A.34)$$

$$H_{n+1}(x;\sigma^2) - \frac{x}{n+1} H_n(x;\sigma^2) + \frac{\sigma^2}{n+1} H_{n-1}(x;\sigma^2) = 0 \quad (A.35)$$

$$\sum_{k=0}^{n} H_{n-k}(x;\sigma^2) H_k(y;\tau^2) = H_n(x+y; \sigma^2 + \tau^2) \qquad \text{[S. Kakutani (1950)]}$$
$$(A.36)$$

$$H_m(x;\sigma^2) H_n(x;\sigma^2) = \sum_{k=0}^{m \wedge n} \frac{\sigma^{2k}(m+n-2k)!}{k!(m-k)!(n-k)!} H_{m+n-2k}(x;\sigma^2) \quad (A.37)$$

$$\frac{1}{\sqrt{2\pi}\sigma} \int H_m(x;\sigma^2) H_n(x;\sigma^2) e^{-x^2/2\sigma^2} dx = \delta_{m,n} \frac{\sigma^{2n}}{n!}. \quad (A.38)$$

If $\eta_n(x;\sigma^2) = \frac{\sqrt{n!}}{\sigma^n} H_n(x;\sigma^2)$, then $\{\eta_n; n \geq 0\}$

is a complete orthonormal system in $L^2\left(\mathbf{R}, \frac{1}{\sqrt{2\pi}\sigma} e^{-x^2/2\sigma^2} dx\right). \quad (A.39)$

(iii) Complex Hermite Polynomials [K. Itô (1953a)]

Definition.

$$H_{p,q}(z,\bar{z}) = (-1)^{p+q} e^{z\bar{z}} \frac{\partial^{p+q}}{\partial \bar{z}^p \partial z^q} e^{-z\bar{z}}, \qquad z \in \mathbf{C}, \, p, q \geq 0.$$

Generating function

$$\sum_{p,q=0}^{\infty} \frac{\bar{t}^p t^q}{p! q!} H_{p,q}(z,\bar{z}) = e^{-t\bar{t} + t\bar{z} + \bar{t}z}, \qquad t \in \mathbf{C}$$

$$\begin{cases} H_{p,q}(z,\bar{z}) = \sum_{k=0}^{p \wedge q} (-1)^k \frac{p! q!}{k!(p-k)!(q-k)!} z^{p-k} \bar{z}^{q-k} \\ H_{p,q}(z,\bar{z}) = \overline{H_{q,p}(z,\bar{z})}. \end{cases} \quad (A.40)$$

EXAMPLES.

$$H_{0,0}(z,z) \equiv 1, \quad H_{p,0}(z,\bar{z}) = z^p,$$
$$H_{1,1}(z,\bar{z}) = z\bar{z} - 1, \quad H_{2,1}(z,\bar{z}) = z^2\bar{z} - 2z,$$
$$H_{2,2}(z,\bar{z}) = z^2\bar{z}^2 - 4z\bar{z} + 2,$$
$$H_{3,1}(z,\bar{z}) = z^3\bar{z} - 3z^2, \quad H_{3,2}(z,\bar{z}) = z^3\bar{z}^2 - 6z^2\bar{z} + 6z,$$
$$H_{3,3}(z,\bar{z}) = z^3\bar{z}^3 - 9z^2\bar{z}^2 + 18z\bar{z} - 6.$$

$$\begin{cases} \dfrac{\partial^2}{\partial z\, \partial \bar{z}} H_{p,q}(z, \bar{z}) - \bar{z} \dfrac{\partial}{\partial \bar{z}} H_{p,q}(z, \bar{z}) + q H_{p,q}(z, \bar{z}) = 0 \\ \dfrac{\partial^2}{\partial \bar{z}\, \partial z} H_{p,q}(z, \bar{z}) - z \dfrac{\partial}{\partial z} H_{p,q}(z, \bar{z}) + p H_{p,q}(z, \bar{z}) = 0 \end{cases} \quad (A.41)$$

$$\begin{cases} \dfrac{\partial}{\partial z} H_{p,q}(z, \bar{z}) = p H_{p-1,q}(z, \bar{z}) \\ \dfrac{\partial}{\partial \bar{z}} H_{p,q}(z, \bar{z}) = q H_{p,q-1}(z, \bar{z}) \end{cases} \quad (A.42)$$

$$\begin{cases} H_{p+1,q}(z, \bar{z}) - z H_{p,q}(z, \bar{z}) + q H_{p,q-1}(z, \bar{z}) = 0 \\ H_{p,q+1}(z, \bar{z}) - \bar{z} H_{p,q}(z, \bar{z}) + p H_{p-1,q}(z, \bar{z}) = 0 \end{cases} \quad (A.43)$$

$\left\{ \dfrac{1}{\sqrt{p!\, q!}} H_{p,q}(z, \bar{z});\ p \geq 0,\ q \geq 0 \right\}$ is a complete orthonormal system in the Hilbert space $L^2\left(\mathbf{C},\, \dfrac{i}{2\pi} e^{-z\bar{z}}\, dz \wedge d\bar{z}\right)$ (A.44)

$$\sum_{p+q=n} \frac{n!}{p!\, q!} H_{p,q}(x, x) = H_n(x), \qquad x \text{ real.} \quad (A.45)$$

Remark. Complex Hermite polynomials with parameter.

Definition.

$$H_{p,q}(z, \bar{z}; \sigma^2) = \frac{(-\sigma^2)^{p+q}}{p!\, q!} e^{z\bar{z}/\sigma^2} \frac{\partial^{p+q}}{\partial \bar{z}^p\, \partial z^q} e^{-z\bar{z}/\sigma^2}, \qquad \sigma > 0,\ p, q \geq 0$$

Generating function

$$\sum_{p,q=0}^{\infty} \bar{t}^p t^q H_{p,q}(z, \bar{z}; \sigma^2) = e^{-\sigma^2 t\bar{t} + t\bar{z} + \bar{t}z}, \qquad t \in \mathbf{C}.$$

Bibliography

The following is a list of papers and books that are referred to in our discussion or that deal with related topics. Those originally written in Russian are listed here in English if English translations are available.

Aczél, J. (1966). Functional Equations and Their Applications. Academic, New York.

Araki, H. (1971), On representations of the canonical commutation relations. *Commun. Math. Phys.* **20**, 9–25.

Bachelier, L. (1941), Probabilités des oscillations maxima. *C.R. Académie Sci. Paris* **212**, 836–838 (Erratum: **213**, 220).

Balakrishnan, A. V. (1974), Stochastic optimization theory in Hilbert space 1. *Appl. Math. Optimization* **1**, 97–120.

Bochner, S. (1932), Vorlesungen über Fouriersche Integrale. Leipzig, Akademische Verlagsgesellschaft.

— (1955), Harmonic Analysis and the Theory of Probability. Univ. of Calif. Press, Berkeley, CA.

Brown, R. (1828), A brief account of microscopical observations made in the months of June, July, and August, 1827, on the particles contained in the pollen of plants; and on the general existence of active molecules in organic and inorganic bodies. *Philos. Mag. Ann. of Philos. New ser.* **4**, 161–178.

Cameron, R. and Martin, W. T. (1944), Transformations of Wiener integrals under translations. *Ann. Math. 2* **45**, 386–396.

— (1945), Fourier–Wiener transforms of analytic functionals. *Duke Math. J.* **12**, 489–507.

— (1947a), Fourier–Wiener transforms of functionals belonging to L_2 over the space C. *Duke Math. J.* **14**, 99–107.

— (1947b), The orthogonal development of non-linear functionals in series of Fourier–Hermite functionals. *Ann. Math. 2* **48**, 385–392.

Chung, K. L., Erdös, P., and Sirao, T. (1959), On the Lipschitz's condition for Brownian motion. *J. Math. Soc. Japan* **11**, 263–274.

Doob, J. L. (1953), Stochastic Processes. Wiley, New York.

Dynkin, E. B. and Yushkevich, A. A. (1969), Markov Processes. Theorems and Problems. Plenum, New York (Russian original: Издательство Наука, 1967).

Feller, W. (1950), An Introduction to Probability Theory and Its Applications, Vol. I. Wiley, New York (third edition, 1968).

— (1966), *ibid.*, vol. II (second edition, 1971).

Freedman, D. (1971), Brownian Motion and Diffusion. Holden-Day, San Francisco.

Fürth, R. (1956), Albert Einstein. Investigations on the Theory of the Brownian Movement. Dover, New York (Translated by Cowper, A. D.).

Gel'fand, I. M. and Yaglom, A. M. (1960), Integration in functional spaces and its applications in quantum physics. *J. Math. Phys.* **1**, 48–69.

Gel'fand, I. M. and Vilenkin, N. Ya. (1964), Generalized Functions, Vol. 4, Applications of Harmonic Analysis. Academic, New York (Russian original: Государственное Издательство, Физико-Математ. Литературы, 1961).

Helgason, S. (1962), Differential Geometry and Symmetric Spaces. Academic, New York and London.

Hida, T. (1960), Canonical representations of Gaussian processes and their applications. *Mem. Coll. Sci. Univ. Kyoto, Ser. A* **33**, 109–155.

— (1967), Finite dimensional approximations to white noise and Brownian motion. *J. Math. Mech.* **16**, 859–866.

— (1970a), Harmonic analysis on the space of generalized functions. *Theor. Probab. Appl.* **15**, 117–122.

— (1970b), Stationary Stochastic Processes. Princeton Univ. Press, Princeton, NJ.

— (1971a), Quadratic functionals of Brownian motion. *J. Multivar. Anal.* **1**, 58–69.

— (1971b), Complex white noise and infinite dimensional unitary group. Lecture note, Nagoya Univ., Nagoya, Japan.

— (1973a), A role of Fourier transform in the theory of infinite dimensional unitary group. *J. Math. Kyoto Univ.* **13**, 203–212.

— (1973b), Functionals of complex white noise. Proc. Symp. on continuum Mechanics and related problems of Analysis. Vol. 1, Tbilisi, USSR, 355–366.

— (1975), Analysis of Brownian Functionals. Carleton Math. Lecture Notes No. 13.

— (1978), Causal calculus in terms of white noise. Bielefeld Encounters in Physics and Mathematics II, held at Bielefeld, December 1978.

— (1978), Generalized Brownian functionals. Complex Analysis and its Applications (in Russian), Jubilee Vol. of Academician I. N. Vekua, Nauka, Moskow, 586–590.

— (1978), White noise and Lévy's functional analysis. Lecture Notes in Mathematics, Springer Verlag, No. 695, 155–163.

Hida, T. and Ikeda, N. (1967), Analysis on Hilbert space with reproducing kernel arising from multiple Wiener integral. Proc. Fifth Berkeley Symp. on Math. Statist. and Probability. Vol. 2, Part 1, 117–143.

Hida, T., Kubo, I., Nomoto, H., and Yoshizawa, H. (1968), On projective invariance of Brownian motion. *Publ. Res. Inst. Math. Sci. Ser. A* **4**, 595–609.

Hida, T. and Nomoto, H. (1964), Gaussian measure on the projective limit space of spheres. *Proc. Japan Acad.* **40**, 301–304.

Hida, T. and Streit, L. (1977), On quantum theory in terms of white noise. *Nagoya Math. J.* **68**, 21–34.

Hitsuda, M. (1968), Representation of Gaussian processes equivalent to Wiener process. *Osaka J. Math.* **5**, 299–312.

Ikeda, I., Hida, T., and Yoshizawa, H. (1962), Theory of flows I (in Japanese). Seminar on Probability, Vol. 12, Probability Seminar.

Itô, K. (1951a), On stochastic differential equations. Mem. Amer. Math. Soc. No. 4.

— (1951b), Multiple Wiener integral. *J. Math. Soc. Japan* **3**, 157–169.

— (1953a), Complex multiple Wiener integral. *Japan. J. Math.* **22**, 63–86.

— (1953b), Stationary random distributions. *Mem. Coll. Sci., Univ. of Kyoto, Ser. A* **28**, *Math.* 209–223.

— (1953c), Theory of Probability (in Japanese). Iwanami, Tokyo.

Itô, K. (1956), Spectral type of the shift transformation of differential processes with stationary increments. *Trans. Amer. Math. Soc.* **81**, 253–263.

— (1961), Lectures on stochastic processes. Tata Institute of Fundamental Research.

Itô, K. and McKean Jr., H. P. (1965), Diffusion processes and their sample paths. Springer-Verlag, Berlin, New York.

Iwahori, N. (1957), Theory of Lie groups I, II (in Japanese). Iwanami series of modern Applications of Mathematics A 3, I, II.

Kac, M. (1951), On some connections between probability theory and differential and integral equations. Proc. Second Berkeley Symp. on Math. Statist. and Probability, 189–215.

— (1959), Probability and Related Topics in Physical Sciences. Interscience, New York.

Kailath, T. (1971), The structure of Radon–Nikodym derivatives with respect to Wiener and related measures. *Ann. Math. Statist.* **42**, 1054–1067.

Kakutani, S. (1948), On equivalence of infinite product measures. *Ann. of Math.* **49**, 214–224.

— (1950), Determination of the spectrum of the flow of Brownian motion. *Proc. Nat. Acad. Sci. USA* **36**, 319–323.

Kallianpur, G. (1970), The role of reproducing kernel Hilbert spaces in the study of Gaussian processes. *Adv. Probabil. Related Topics* **2**, 49–83. Dekker, New York.

Kallianpur, G. (1970), The role of reproducing kernel Hilbert spaces in the study of Gaussian processes. In Adv. Probabil. Related Topics **2**, 49–83. Dekker, New York.

Колмогоров, А. Н. (1958), Новый метрический инвариант транзитивных динамических систем и автоморфизмов пространств дебега. *Док. Акад. НАУК, СССР* **119**, 861–864.

Lévy, P. (1937), Théorie de l'addition des variables aléatoires. Gauthier-Villars, Paris.

— (1940), Le mouvement brownien plan. *Amer. J. Math.* **62**, 487–550.

— (1948), Processus stochastiques et mouvement brownien. [2ème éd. revue et augmenté (1965).] Gauthier-Villars, Paris.

—(1951), Problèmes concrets d'analyse fonctionelle. Gauthier-Villars, Paris.

— (1953), La mesure de Hausdorff de la courbe du mouvement brownien. *G. Ist. Ital. Attuari* **16**, 1–37.

— (1954), Le mouvement brownien. Mém. des Sci. Math. CXXXVI. Gauthier-Villars, Paris.

— (1956), A special problem of Brownian motion, and a general theory of Gaussian random functions. Proc. Third Berkeley Symp. on Math. Statist. and Probability, Vol. 2, 133–175.

— (1957), Fonctions aléatoires à corrélation linéaire. *Illinois J. Math.* **1**, 217–258.

— (1963), Le mouvement brownien fonction d'un ou de plusieurs paramètres. *Rend. Mat.* **22** (1–2), 24–101.

Linnik, Yu. V. (1964), Decomposition of Probability Distributions. Oliver and Boy Ltd. (Russian original: Разложения Вероятностных Законов, Москва, 1960).

Lukacs, E. (1970), Characteristic Functions. 2nd ed. Griffin, London.

McKean Jr., H. P. (1963), Brownian motion with a several-dimensional time. *Theor. Probability Appl.* **8**, 335–354.

— (1969), Stochastic Integrals. Academic, New York.

Miller Jr., W. (1968), Lie Theory and Special Functions. Academic, New York.

— Symmetries of differential equations. I. The heat equation. Lecture notes, Univ. Minnesota, Minneapolis.

Minlos, R. A. (1959), Generalized random processes and their extension to a measure. Selected transl. in Math. Statist. and Probability. Vol. 3, 291–313, 1962 (Russian original: Труды Московского Мат. Общества. Том. 8).

Nelson, E. (1967), Dynamical Theories of Brownian Motion. Princeton

Univ. Press, Princeton, N.J.
Neveu, J. (1964), Bases mathématiques du calcul des probabilités. Masson et Cie, Paris.
Orihara, A. (1966), Hermite polynomials and infinite dimensional motion group. *J. Math. Kyoto Univ.* **6**, 1–12.
Paley, R. E. A. C. and Wiener, N. (1934), Fourier transforms in the complex domain. *Amer. Math. Soc. Colloq. Pub.* XIX.
Prokhorov, Yu. V. (1956), Convergence of random processes and limit theorems in probability theory. *Theor. Probability Appl.* **1**, 157–214.
Rozanov, Ju. A. (1971), Infinite-dimensional Gaussian distributions. Transl. Amer. Math. Soc. (Russian original: Труды мат. инст. Стеклова. CVIII, 1968).
Sato, H. (1971), One-parameter subgroups and a Lie subgroup of an infinite dimensional rotation group. *J. Math. Kyoto Univ.* **11**, 253–300.
Schwartz, L. (1950–1), Théorie des distributions. Hermann Cie (2ème éd. 1966).
Schweber, Silvan S. (1966), An Introduction to Relativistic Quantum Field Theory. 2nd ed. Harper International Ed. Harper & Row and John Weatherhill, Inc., Tokyo.
Shepp, L. A. (1966), Radon–Nikodym derivatives of Gaussian measures. *Ann. Math. Statist.* **37**, 321–354.
Sirao, T. (1960), On the continuity of Brownian motion with a multi-dimensional parameter. *Nagoya Math. J.* **16**, 135–156.
Sinai, Ja. G. (1961), Dynamical systems with countably multiple Lebesgue spectrum. *Izv. Mat. Nauk USSR* **25**, 899–924. *Amer. Math. Soc. Transl.* **39** (2), 83–110 (1961).
Smithies, F. (1958), Integral Equations. Cambridge Univ. Press.
Takenaka, S. (1977), On projective invariance of multi-parameter Brownian motion. *Nagoya Math. J.* **67**, 89–120.
Titchmarsh, E. G. (1946), Eigenfunction Expansions Associated with Second-Order Differential Equations. Clarendon Press, Oxford.
Totoki, H. (1971), Introduction to Ergodic Theory (in Japanese). Kyoritsu, Tokyo.
Umemura (Yamasaki), Y. (1965), Measures on infinite dimensional vector spaces. *Pub. Res. Inst. Math. Sci. Kyoto Univ., Ser. A* **1**, 1–47.
von Neumann, J. (1931), Die Eindeutigkeit der Schrödingerschen Operatoren. *Math. Ann.* **104**, 570–578.
Weyl, H. (1928), Gruppentheorie und Quantenmechanik (English ed.: Transl. by H. P. Robertson, Methuen, 1931).
Wiener, N. (1923), Differential space. *J. Math. and Phys.* **2**, 131–174.
— (1928–29), Hermitian polynomials and Fourier analysis. *J. Math. and Phys.* **8**, 70–73.
— (1938), The homogeneous chaos. *Amer. J. Math.* **60**, 897–936.
— (1948), Cybernetics, or Control and Communication in the Animal and the Machine. Wiley, New York (2nd ed. 1962).

— (1958), Nonlinear Problems in Random Theory. Technology Press of MIT, Cambridge.
Widder, D. V. (1946), Laplace Transform. Princeton Mathematical Series 6. Princeton University Press, Princeton, N.J.
Wiener, N. and Akutowicz, E. J. (1957), The definition and ergodic properties of the stochastic adjoint of a unitary transformation. *Rend. Circ Mat. Palermo* **6**, 205–217.
Xia Dao-Xing (1972), Measure and Integration Theory on Infinite-Dimensional Spaces. Academic, New York (Transl. by E. J. Brody).
Yoshizawa, H. (1970), Rotation group of Hilbert space and its application to Brownian motion. Proc. of the International Conference on Functional Analysis and related topics, 1969, Tokyo, 414–423.
Yosida, K. (1951), Functional Analysis I (in Japanese), Iwanami, Tokyo.
— (1965), Functional Analysis. Springer-Verlag, Berlin. (3rd ed., 1971, New York.)

Index

additive process 45, 157
anharmonic ratio 110
arc sine law 107
average power 190

basic commutation relations 265
Bochner-Minlos theorem 116, 227, 307
Bochner theorem 8
Borel-Cantelli lemma 24
Brownian bridge 18, 108, 197
Brownian functional viii, 280
  generalized 281, 287
Brownian motion 44
  complex 73, 236
  construction of 63, 69
  Einstein ix
  flow of 144, 196
  Lévy 299
  multi-, n-dimensional parameter 298
  n-dimensional 51
  with a reflecting barrier 99
$\dot{B}(t)$-derivative 290

Cameron-Martin 180
canonical commutation relations 223
canonical kernel 148
canonical representation 111, 148
causal 310

analysis 280
calculus 289
central limit theorem 25
Chapman-Kolmogorov equation 76, 87, 101
characteristic function 8
characteristic functional 114, 115, 122
characterizations of Gaussian
  distributions 36
Chung-Erdös-Sirao 63
compact-open topology 185, 254
completely monotonic 209
complex Fourier-Hermite
  polynomial 243, 244, 257. 262
complex Gaussian random variable 233
complex Gaussian system 233
complex Hermite polynomial 236, 242, 313
complex multiple Wiener integral 244
complex white noise 238
complex Wiener integral 240
conditional expectation 20
conditional probability 21
convergence of a sequence of random
  variables in law, in probability,
  almost surely, with probability one,
  in the sense of L^2, in mean
  square 27
countable (σ-) Lebesque 200, 263
counting model 4

321

covariance function 12
covariance functional 125
Cramèr, H. 38
cumulant (semi-invariant) 37, 166, 168, 170
cyclic subspace 199
cyclic vector 199
cylinder set 11, 46, 117

deterministic (process) 12
differential entropy 37
differential space (Wiener space) xii
diffusion process 103
dilation (tension) 193, 203, 263
discrete spectrum 200, 263
distribution, probability 7, 122
  infinite-dimensional 13
  of $\{X(t)\}$ 11
distribution function 8, 10
Doob, J. L. 54, 131, 152, 156, 293, 295
Doob's inequality 58
Dvoretsky-Erdös-Kakutani 52
Dynkin, E.B. 96
Dynkin formula 97

entropy 37
equally dense 283
equivalence of measures 14, 173, 216
expectation (mean) 3
exponential function 133, 170, 172, 180, 241
exponential (map) 92, 98

Feller, W. 4, 76, 105
flat spectrum 131
flow 142, 143
  Kolmogorov 145
  of Brownian motion 144
  of complex Brownian motion 257
  of Ornstein-Uhlenbeck process 196, 202–3
Fock space 141, 165, 246
Fourier-Hermite polynomial 135, 181, 220, 222, 262, 285, 310
Fourier-Mehler transform 260, 270
Fourier-Wiener transform 182, 183, 218, 292

free particle 275
functional of Brownian motion (Brownian functional) vii

gauge transform 256
Gaussian 31, 125
Gaussian distribution 31
  standard 33
Gaussian measure 125
Gaussian process 13, 36, 157
Gaussian random variable 31
  system 31
Gauss kernel 45, 94, 100, 268
Gauss transform 182
Gelfand, I.M. - N. Ya. Vilenkin 227
generalized random measure 288
generalized random process 13
generalized stochastic process (random distribution) 122
  Gaussian 125
  stationary 130
generating function 139, 181, 311, 312, 313, 314
generator 91
G-ergodic 207
G_f^*-ergodic 208
Glivenko, V. 9
G-quasi-invariant 214
Green measure 89
Green operator 89

Hahn-Hellinger theorem 199, 206
harmonic functional 291
harmonic oscillator 277
Heisenberg group 259, 266
Hermite polynomial
  with parameter, generating function of 163, 312
Hida, T. 142, 148, 192, 212
Hida, T. - N. Ikeda 142
Hille-Yosida theorem 86, 92, 99
Hitsuda, M. 174
holomorphic functional 251
holomorphic polynomial 250

ideal gas 5
independent 2, 3

independent value at every moment 126
infinite dimensional motion group 218
  Laplace Beltrami operator 221
  rotation group 186
  unitary group 282
  $U(\mathcal{S}_c)$ 254
initial distribution 76
innovation 121
integral representation 141, 246, 282
  kernel of 141, 246, 282
irreducible representation of G_n 226
Itô, K. 46, 73, 111, 152, 175, 313
Itô, K. - H.P. McKean 76, 80, 105, 113, 175

Kac, M. 6, 39, 103, 106, 112
Kac theorem, formula 103
Kakutani, S. 313
Kakutani theorem 15, 212, 216
Kallianpur, G. 142
Khintchine, A. 57
Kolmogorov extension theorem 11, 33, 46, 128
Kolmogorov flow 146, 202
Kolmogorov 0-1 law 29
Kolmogorov's inequality 24
Kubo, I. 192

Laguerre function 309
Langevin equation 176
Laplace-Beltrami operator 222
Laplacian operator, infinite dimensional 291
law of the iterated logarithm 61
Lee-Wiener electrical circuits 308
Lévy, P. x, xi, 9, 38, 42, 61, 81, 84, 86, 110, 111, 127, 148, 168, 188, 281, 283, 289, 293, 300
Lévy measure 126
Lévy process 126
Lévy's inversion formula 8
Lévy's method 63
Linnik, Yu. V. 42
Lukacs, E. 42

Markov process 13
Markov property(ies) 75, 77, 300

Markov time 78, 96
Martingale 54, 149, 293, 297
  super 157, 294
  sub 58, 294
McKean, H.P. Jr. 58, 111, 156, 175, 300
mean (expectation) 3
mean (valeur moyenne) x
Miller, W. Jr. 266, 268, 278
Minlos theorem 119
modified Fredholm determinant 166
moment 3
multiple Wiener integral 136
multiplication 257
multiplication by $\dot{B}(t)$ 291
multiplicity 200
μ-topology 186

Nomoto, H. 192, 212
non-deterministic (process) 12
non-differentiability 52
nonlinear circuit 205
normal functional 289
nuclear Lie group 226
  irreducible representation of 230
nuclear space (countably Hilbert) 114, 301
  examples of 301

Ornstein-Uhlenbeck process 177
Ottaviani's inequality 30
$O^*(E^*)$-ergodic 207, 211
$O^*(E^*)$-invariant 187, 209
$O^*(E^*)$-invariant measure 211

Paley-Wiener's method 67
particle in a constant external field 277
polynomial 133, 180, 240
probability space 2
  in the weak sense 7, 118
projective invariance of Brownian motion 46, 110, 197, 198, 299
Prokhorov, Yu. V. 46
pure point spectrum 200
purely non-deterministic 12, 146, 149

quadratic functional 165
  exponential of 170
quasi-invariant 214, 218

Rademacher functions 4
random distribution 13, 130
  stationary 130
random measure 131
random variable 3
  multi-dimensional 9
random walk 5
real-valued functional 247
reflection principle 82
regular functional 289
reproducing kernel 139, 281
reproducing kernel Hilbert space 139, 281
resolvent equation 89
rotation 185
  finite dimensional 188
  group 186

sample function, sample path 11, 51, 300
  iterated logarithm 57
  non-differentiability 52
  uniform continuity 61
  variation 53
Sato, H. 190
scale change 190
Schrödinger equation 274
self-reproducing property 38
semi-group 80
separable (stochastic process) 297
shift 191, 257
Sirao, T. 63, 300
space
  D_0 304
  $\mathcal{D}(K)$ 303
  $\hat{\mathcal{D}}(\pi)$ 302
  $\mathcal{S}$(Schwartz space) 255, 305
  $\mathcal{S}_c$ 254
  $\mathfrak{a}$ 307
space-time transform 45, 98
spectrally equivalent 200
spectral type 200
stable process 126
stationary process 12, 143, 147
  weakly 12, 143

strictly 12
stochastic area 169
stochastic differential equation 111, 175
stochastic integral 152
stochastic process 10
  continuous parameter 10
  discrete parameter 10
strong law of large numbers 25, 116, 127
strong Markov process 79
sum of independent random variables 29
symmetry group (of differential operator) 267, 275

Takenaka, S. 299, 301
Tchebychev's inequality 24
temporally homogeneous 76
T-quasi-invariant 214
transformation $\mathcal{T}$ 137, 292
transition probability 76
  density 76
translation 212

Umemura, Y. 207, 215, 221, 227
unitary equivalence 200, 230, 231
unitary invariant 200
unitary representation 215, 218
  cyclic 229
  irreducible 226
upcrossing number 295

variance 3
variation 53

weak convergence of distributions 9
weak law of large numbers 24
Weyl, H. 221
Weyl commutation relation 223, 259
white noise 124, 127, 131, 205, 229
  discrete parameter 19, 128
  input 310
  periodic 124
  transformation of 185, 212
  with variance σ^2 122
Wiener, N. xii, 127, 137, 151, 167, 205, 207, 261
Wiener integral 111

Wiener-Itô theorem
  (decomposition)   134, 137, 308
Wiener measure   ii, 50, 125, 173
  conditional   78, 109, 111
  n-dimensional   51
Wiener process   44, 298
  with multi-dimensional parameter   298

Wiener's nonlinear circuit (theory)   175, 205, 308
Wiener's probability space   4

Yoshizawa, H.   185, 192
Yosida-Hille theorem   92

Applications of Mathematics

Vol. 1
Deterministic and Stochastic Optimal Control
By **W.H. Fleming** and **R.W. Rishel**
1975. ix. 222p. 4 illus. cloth

Vol. 2
Methods of Numerical Mathematics
By **G.I. Marchuk**
1975. xii, 316p. 10 illus. cloth

Vol. 3
Applied Functional Analysis
By **A.V. Balakrishnan**
1976. x, 309p. cloth

Vol. 4
Stochastic Processes in Queueing Theory
By **A.A. Borovkov**
1976. xi, 280p. 14 illus. cloth

Vol. 5
Statistics of Random Processes I
General Theory
By **R.S. Liptser** and **A.N. Shiryayev**
1977. x, 394p. cloth

Vol. 6
Statistics of Random Processes II
Applications
By **R.S. Liptser** and **A.N. Shiryayev**
1978. x, 339p. cloth

Vol. 7
Game Theory
Lectures for Economists and Systems Scientists
By **N.N. Vorob'ev**
1977. xi, 178p. 60 illus. cloth

Vol. 8
Optimal Stopping Rules
By **A.N. Shiryayev**
1978. x, 217p. 7 illus. cloth

Vol. 9
Gaussian Random Processes
By **I.A. Ibragimov** and **Y.A. Rosanov**
1978. approx. 290p. cloth

Vol. 10
Linear Multivariable Control: a Geometric Approach
By **W.M. Wonham**
1979. ix, 326p. 27 illus. cloth

Vol. 11
Brownian Motion
By **T. Hida**
1980. xvi, 325p. 13 illus. cloth

Vol. 12
Conjugate Direction Methods in Optimization
By **M. Hestenes**
1980. x, 336p. 22 illus. cloth

Vol. 13
Stochastic Filtering Theory
By **G. Kallianpur**
1980. approx. 304p. cloth

Vol. 14
Controlled Diffusion Processes
By **N.V. Krylov**
1980. approx. 320p. cloth